Properties of Water from Numerical and Experimental Perspectives

Editor

Fausto Martelli

IBM Research

CRC Press

Taylor & Francis Group

Boca Raton   London   New York

CRC Press is an imprint of the
Taylor & Francis Group, an **informa** business

A SCIENCE PUBLISHERS BOOK

Cover Credit: Cover Illustration reproduced by the kind courtesy of the editor.

First edition published 2022
by CRC Press
6000 Broken Sound Parkway NW, Suite 300, Boca Raton, FL 33487-2742

and by CRC Press
4 Park Square, Milton Park, Abingdon, Oxon, OX14 4RN

© 2022 Taylor & Francis Group, LLC

CRC Press is an imprint of Taylor & Francis Group, LLC

Library of Congress Cataloging-in-Publication Data (applied for)

ISBN: 978-0-367-13802-8 (hbk)
ISBN: 978-1-032-32851-5 (pbk)
ISBN: 978-0-429-02866-3 (ebk)

DOI: 10.1201/9780429028663

Typeset in Times New Roman
by Radiant Productions

Preface

Water is essential to life as we know it. Because of this, it has been extensively studied, yet it remains incompletely understood. For example, water does not always behave in accordance with theories of thermodynamics, but deviates from them under particular conditions. Upon cooling, the density of a "normal" liquid increases monotonically and, eventually, the liquid freezes in a denser crystal that sinks. This is not the case for water: upon cooling, the density of liquid water increases but, at about 4 degrees Celsius, the density starts to decrease and the liquid freezes in a crystal that floats instead of sinking. As a result, the surface of water freezes during winter, while the depths maintain a comfortable constant temperature of approximately 4 degrees Celsius that allows life to advance. If water were a "normal" material, it would freeze from the bottom up, killing most aquatic life. These are just two of the many anomalies of water (scientists have now counted at least 72 of them), and scientists are working hard to understand their underlying sources.

Aimed at researchers entering the fascinating world of water, this book provides a broad overview of the properties of water under different conditions and probed with different techniques, both numerical and experimental.

Fausto Martelli

Contents

CHAPTER 1

Theoretical X-ray Absorption Spectroscopy of Liquid Water Using First-Principles Calculations

Fujie Tang* and Xifan Wu

Introduction

Liquid water is one of the most important liquids on the Earth. It seems to be simple, because of its chemical component, which contains two hydrogen atoms and one oxygen atom. However, the complete understanding of water at the molecular level is still a challenging task, due to its unique hydrogen-bond (HB) network. The basic nature and extent of the HB network is under debate now (Pettersson et al. 2016, Jungwirth and Tobias 2006, Bellissent-Funel et al. 2016, Ball 2008, Werber et al. 2016, Salucci et al. 2003, Eaves et al. 2005, Luzar and Chandler 1996). The structural origins of the well-known anomalies of liquid water are still not universally agreed upon within the scientific community. The HB network of liquid water was first proposed as a locally tetrahedral network by Bernal and Fowler in 1930s with the help of X-ray diffraction spectroscopy (Bernal and Fowler 1933). Since then, a number of experimental and theoretical studies have been carried out in this field and the results are in favor of the well-known tetrahedral structures of liquid water. Among these experimental methods, the spectroscopic methods, especially, the X-ray absorption/Raman spectroscopy has been proven to be a good indicator of the local chemical environment. In these processes, X-ray interacts with the core orbitals of the water molecules, the energy differences in the core orbital make the X-ray radiation highly sensitive to the element and local chemical environment (Wernet et al. 2004, Prendergast and Galli 2006, Tse et al. 2008, Nilsson et al. 2016, Fransson et al. 2016, Chen et al. 2010, Sun et al. 2017, 2018, Kraus et al. 2018, Pettersson et al. 2019).

In general, a X-ray absorption process could be understood as follows: within its near-edge region, a high-energy photon is absorbed and promotes a core

Department of Physics, Temple University, Philadelphia, PA 19122, USA.

* Corresponding author: xifanwu@temple.edu

electron to empty states, which should be allowed under the selection rules. X-ray Raman scattering (XRS) is a technique which can provide similar information to X-ray absorption spectroscopy (XAS). The core processes occur at a time scale of femtoseconds that is much faster than any molecular vibrational movements, that is, the vibration of HB network of water. As such, XAS could capture the instantaneous snapshots of the structure of liquid water during the X-ray measurement. Moreover, it is found that the pre-edge feature of the XAS spectra for liquid water comes from the bound exciton while the main edge and post-edge features originate from the contributions of the resonant excited states, the first one is more sensitive to the local structure compared to the latter two. Therefore, XAS spectroscopy could be good for probing the local structure of liquid water.

Recently, a large number of studies were published about the examination of the HB network in liquid water and ice with X-ray spectroscopies, especially XAS and XRS technics (Wernet et al. 2004, Prendergast and Galli 2006, Tse et al. 2008, Nilsson et al. 2016, Fransson et al. 2016, Chen et al. 2010). Together with theoretical calculations, debates about the nature of HB networks have also arisen. Wernet et al. (2004) claim the water molecules in liquid phase bind on average to two others, forming chains and rings, which is different from the near-tetrahedral structure of water. The argument was made based on the comparison of the intensity of the pre-edge peak of liquid water between the one obtained from experimental data and the one from theoretical calculation. The pre-edge peak is prominent in liquid water and is believed to be absent in bulk ice in this study. However, a systematic XRS study, reported by Tse and coworkers (2008), showed the presence of the pre-edge peak with different intensities, not only in the liquid water, but also in hexagonal (Ih), cubic (Ic), low-density amorphous (LDA), and high-density amorphous (HDA). These different views highlight the importance for obtaining the accurate theoretical XAS spectra to resolve the puzzle.

Based on the motivation to model the XAS spectra at the highest accurate level, tremendous efforts have been made from the theoretical side in two parallel ways. Simulating the XAS spectra of water is a highly challenge task in theory:spectral calculations are used, based on electronic structure theory and molecular modelling of the water structure by using molecular dynamics (MD) simulation to compute the XAS spectra. Also, the quality of the computed XAS spectra has improved from the qualitative to the semiquantitative level, and has finally reached the quantitative level. To describe the proper configurations present within the liquid, it is better to average over structural fluctuations in the phase space, which are taken from a MD simulation trajectory. Since the XAS lineshape is very sensitive to the local environment as well as the intermediate and long-range interactions, the proper characterization of molecular structures is important. The MD simulation snapshots of liquid water could be obtained from a classical water model-based MD simulation or *ab initio* MD (AIMD) simulation, the latter is a parameter-free calculation.

After obtaining a reliable molecular structure, a theoretical method, which can describe the X-ray absorption process, is required as well. The computational method which treats the electron-hole exciton at the core level is a challenging task since it will face the following complications: (1) The targeted core excited states are embedded in many valence electrons, which should be allowed to relax in the presence of the core-hole via some assumptions, (2) the excited core states are screened by the valence electrons and the screening effects are complicated, (3) the final states should be described by using the quasi-particle orbital instead of the Kohn-Sham orbital at the conduction band, (4) the computational cost in calculating XAS spectra for liquid water will be huge since the number of atoms are much larger than that of the crystalline materials with symmetry, because we need to sample enough numbers of oxygen atoms, which are present in the different chemical environments of the disordered system. These limitations strongly restrict the calculation of XAS for water. Therefore, the proper approaches or approximations to accurately assess the core excited states are highly desired.

In this chapter, we will first discuss the efforts to compute XAS spectra based the accurate characterization of the electronic structure theory, from density functional theory (DFT) to the many-body perturbative theory, the GW approximation plus the Bethe-Salpeter Equation (GW-BSE) approach (and its relevant approximations), then we discuss the molecular modelling of the water structure, which is another important issue for computing XAS spectra. Lastly, we will list some perspectives about the theoretical calculations for gauging the XAS spectra of water.

Electronic Structure Aspect I: Half Core Hole, Full Core Hole, and Excited-state Core Hole Approaches

Introduction to Theory for Computing XAS on the Level of DFT

We have discussed above the challenges inaccurately computing the XAS spectra from first principles. In this section, we will review the early efforts to compute the XAS spectra from the electronic structure aspect. Many groups, including the L.G.M. Pettersson group (Wernet et al. 2004), Roberto Car's group (Hetényi et al. 2004), and Giulia Galli's group (Prendergast and Galli 2006) tried to compute the XAS spectra of liquid waterbased on different approximations and compared it with the experimental spectra to understand the controversies and debates about the structures of water.

Firstly, in the X-ray absorption process, a core electron is excited to a near-edge conduction state. The spectroscopic shape is determined by the core hole state and the valence relaxation near the conduction state. The accurate XAS including the electron–hole interaction requires solving a Bethe-Salpeter equation. While these calculations should use a dielectric function to describe the screened electron–hole interaction, and they are very computationally demanding, as one needs to calculate each individual excitation. As discussed above, early efforts had been made by several groups to reduce the computation cost by using a core hole approximation approach instead of directly solving the Bethe-Salpeter equation (BSE). Apparently, these approaches based on DFT lead to computationally efficient schemes. Here we will briefly explain how they work. In principle,the X-ray absorption cross-section could be determined by using Fermi's golden rule:

$$\sigma(\omega) = 4\pi^2 \, \alpha_0 \hbar \omega \sum_f |M_{i \rightarrow f}|^2 \, \delta(E_f - E_i - \hbar\omega), \qquad (1)$$

where E_f and E_i associated with the final and initial states, respectively. In these approximations, the initial and final states are represented by Kohn-Sham orbitals, the initial state is fixed as the 1 s eigenstate of the oxygen atom, while the final state is various among three approaches, which will be explained later. $M_{i\rightarrow j}$ are the transition matrix elements between the initial state $|\phi\rangle_i$ and the final state $|\phi\rangle_f$, which could be evaluated within the electric-dipole approximation as $M_{i\rightarrow j} \sim \langle\phi_i|x|\phi_f\rangle$. This method describes the final XAS state in Eq. (1) as one long-lived oxygen core hole and retains a localized atomic character, instead of the real electron-hole quasi-particle excitation described by the BSE treatment.

Within the DFT framework, the relaxation of the valence electrons because of the presence of the core hole could be solved self-consistently. Slater had proposed the transition state method for calculating core excitation energies (Slater 1972, Slater and Johnson 1972). In this method, half an electronic occupation is promoted from a core state to an excited state, the energy of the exciton is the difference between the final and initial eigenstates. The accuracy of excitation obtained in this way could be up to the second order in the occupation changes of the transition orbitals. The transition state method is convenient and accurate enough in most cases, however, it won't be able to handle the many excited states' system because its self-consistent field calculation will collapse for the excited states, leading to the nonorthogonality of these states (Triguero et al. 1998).

Another approach to compute XAS spectra within the DFT framework comprises the time-dependent density functional theory (TDDFT) approach. In the TDDFT approach, one doesn't need to deal with each state separately. For TDDFT calculation, the choice of the exchange-correlation functional is rather important (Besley and Asmuruf 2010). However, the lack of the relaxation effects and the self-interaction error inherent in approximate functionals result in an underestimation of core-excitation energies, which needs to shift the computed spectra to align the experimental spectra or estimate the necessary energy shift using ΔSCF calculation. So far, only a few studies have used the TDDFT method to study the XAS spectra of liquid water (Fransson et al. 2016, Brancato et al. 2008). We will not discuss the application of XAS spectra for water using the TDDFT approximation here because it is beyond the bounds of this study.

In the transition potential method, the potential corresponds to the transition hole state with a half electron on the hole orbital (Triguero et al. 1998, Stener et al. 1995, Zhang et al. 2016). In fact, the transition potential method is widely used in the XAS calculation (Zhang et al. 2016). The half core hole approximation (HCH) is similar to the transition potential method, as it too operates without half an electron occupied in the first empty state. The full core hole approximation (FCH) is similar to the static exchange method (Ågren et al. 1997, 1994). This could be explained as follows, the occupied orbitals of an N-electron core-excited system are represented by the occupied orbitals of the (N-1)-electron ionic system with the corresponding core hole, which means a full core electron is removed and a restricted open-shell Hartree-Fock calculation is performed to obtain the occupied orbitals of the ionic system (Hunt and Goddard 1969). The excited-state core hole approximation (XCH) is similar to the FCH, as the final states are computed in the presence of a core hole with an excited electron in the lowest unoccupied orbital (Prendergast and Galli 2006). The HCH approach seems to work well in the case of small molecules and clusters while the FCH seems to be preferred for condensed phase systems. The

FCH should reproduce the higher-energy excitations in more delocalized states, and in this state of high energy, it should yield the same results as the XCH approach because of the limited impact of the localized excited electron presented at the lowest unoccupied state. In the following section, we will discuss the difference of XAS spectra of liquid water and ice based on HCH, FCH, and XCH approaches.

XAS Spectra Based on Half Core Hole, Full Core Hole, and Excited Core Hole Approximations

First, let us focus on the experimental XAS spectra of bulk ice and bulk water. We adopted the data from Wernet et al. (2004), and replotted in Fig. 1 (a) and (d). Figure 1 (a) shows the XAS spectrum of crystalline ice, and Fig. 1(d) shows the XAS and XRS spectra of water. The spectra can be divided into three main regions: the pre-edge (533–536 eV), the main edge (537–539 eV), and the post-edge (from 539 eV and beyond). The bulk ice spectrum is dominated by more prominent intensity in the post-edge region, while the main edge structure is relatively weak. Comparatively, the liquid water spectrum has a clearer feature in the pre-edge region, its dominant peak in the main edge region, and a less prominent peak comparing with bulk ice in the post-edge region.

The structure of molecular crystals with metal surface from the study by Wernet et al. (2004) is not either the [0001] surface of hexagonal ice [Ih] or the [111] surface of cubic ice [Ic] because it was prepared by molecular deposition on Pt [111] surface. In order to compare the experimental spectra of ice with theoretical spectra, it would be better to use the hexagonal ice and cubic ice at the same time. As such, we adopted the corresponding theoretical data (HCH, FCH, and XCH approaches) from Wernet et al. (2004) and Prendergast and Galli (2006) and plotted them in the Fig. 1(b)

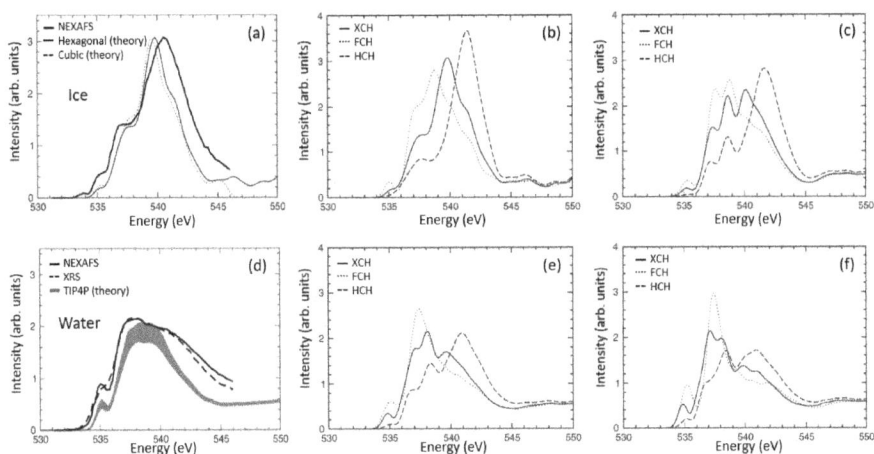

Fig. 1: (a) The XAS spectra of crystalline ice, (d) the XAS and XRS spectra of liquid water. Data was adopted from Wernet et al. (2004). The comparison of XAS with various approximations, XCH, FCH, and HCH; (b) hexagonal ice; (c) amorphous ice; (e) liquid water; (f) liquid water sampling only broken-hydrogen-bonded species. Reproduced with permission from Prendergast and Galli (2006). Copyright [2006] [American Physical Society].

hexagonal ice, Fig. 1(c) amorphous ice, Fig. 1(e) liquid water, and Fig. 1(f) liquid water sampling only broken-hydrogen-bonded species, respectively.

Now we could comment on the impact of different approaches (HCH, FCH, and XCH) on the XAS spectra shape. Firstly, the HCH approach tends to overestimate the main-edge peak and underestimate the pre-edge intensity owing to the reduced binding energy of the HCH in the excited oxygen atom. The FCH approach tends to overestimate the intensity at and near the onset, and consequently underestimate the main-edge peak height because of the oscillator strength sum rule. The XCH approach yields the XAS spectrum of water in between the spectra provided by the HCH and FCH approaches. Note that the XAS spectrum calculated by the XCH approach based on a classical potential water model, which simulating a standard, quasi-tetrahedral structure of liquid water, reaches a reasonable agreement with the experimental XAS spectrum. The XAS spectrum of liquid water calculated using the HCH approach based on a standard, quasi-tetrahedral structure of water appears to be ice-like XAS spectra compared to the experimental data (Wernet et al. 2004). Furthermore, we want to note that the agreement between the XAS spectra calculated based on the DFT method and quasi-tetrahedral structure of water and experimental XAS spectra is at the semiquantitative level. As we have discussed in the introduction, it could be improved by using the excitation theory, for example, the GW method, and advanced modelling methods of liquid water to capture the local structure of water which is closest to that of real water. In the next section, we will first discuss the progress of theoretical XAS calculation of liquid water using GW methods.

Electronic Structure Aspect II: Quasi-particle Approach

Green's Function, GW Approximation, and the Bethe-Salpeter Equation

In the section, we will review the impact of electronic structure on XAS calculation in the quasi-particle approach. Actually, at the DFT level, we can easily handle the wavefunctions which were involved in Eq. (1). The problem with using the DFT method to treat the quasi-particle interaction is that the corresponding effective single particle eigenvalues from the Kohn-Sham equation could not be generally justified to be explained as the quasi-particle electron–hole interaction energies and in fact, there were significant errors, such as the underestimation of the semiconductor band gaps (Kohn 1999). A much older idea compared to the DFT method, is that of the long-range, and relatively strong, Coulomb forces, which with a surrounding charge cloud of the other electrons could screen the individual electrons. As such, this electron plus its screening cloud can be considered as the quasi-particle (Hedin and Lundqvist 1970, Hedin 1965, Onida et al. 2002, Aryasetiawan and Gunnarsson 1998, Hybertsen and Louie 1986).

By using the quasi-particle terminology, one can describe the response of interacting particles. The mathematical description of quasi-particles is based on the single-particle Green's function G $(r, t; r', t')$, which describes the probability amplitude for the propagation of an electron from position r' at time t' to position r at time t. In the many-body perturbation theory approach, the nonlocal energy-dependent electron self-energy operator is the key quantity. This is because the exact determination of Green's function requires the understanding of the quasi-particle self-energy. The *GW* approximation for the electron self-energy proposed by Hedin has been widely used in the calculation of quasi-particle energies in real materials. The most accurate method of calculating XAS spectra can be carried out by solving the Bethe-Salpeter Equation (BSE) and using Hedin's *GW* approximation of the quasi-particle (Rohlfing and Louie 2000, Hybertsen and Louie 1986). In this section of the chapter, we will try to review the successfulness of calculating XAS by this GW-BSE approach and its similar approximations.

First of all, let's focus on the basic ideas how to compute the energy of the electron excitations, which also could be called as the quasi-particle. For the details of the Green's function formalism, there are a lot of comprehensive papers have well summarized in the literatures (Onida et al. 2002, Aryasetiawan and Gunnarsson 1998). In many-body theory, the quasi-particle energy could be computed by evaluating the amplitude of the particle via the single-particle Green's function:

$$iG\ (x,\ t;\ x',\ t') = \langle N|T\{\psi(x,\ t)\psi^{\dagger}(x',\ t')\}|N\rangle$$

$$= \begin{cases} \langle N|\psi(x,\ t)\psi^{\dagger}(x',\ t')|N\rangle,\ \text{for } t > t' \text{ (electron)} \\ -\langle N|\psi^{\dagger}(x',\ t')\psi(x,\ t)|N\rangle,\ \text{for } t < t' \text{ (hole)} \end{cases} \qquad (2)$$

where $|N\rangle$ is the N-electron ground state, $\psi(x,\ t)$ is a field operator in the Heisenberg representative where an electron annihilates at $(x,\ t)$, and T is the time-ordering operator. The physical meaning of the Green's function is that for $t' > t$ it is the probability amplitude that a hole created at x will propagate to x' and for $t > t'$, the probability amplitude is that an electron created at x' will propagate to x. As such, Green's function describes the photoemission and inverse photoemission processes. Based on Eq. (2), one could obtain the dispersion relation and lifetime of the quasi-particle excited state. Moreover, the quasi-particle energies E_{nk} and wavefunction ψ_{nk} could be calculated by solving a Schrodinger-like equation:

$$(T + V_{ext} + V_{H})\psi_{k}(r) + \int dr\ \Sigma(r,\ r';\ E_{nk})\psi_{nk}(r') = E_{nk}\psi_{nk}(r) \qquad (3)$$

where T is the kinetic energy operator, V_{ext} is the external ionic potential, V_{H} is the electron Hartree potential, and Σ is the self-energy operator with all the many-body exchange and correlation effects. The key to solve the quasi-particle equation is to find a proper method to evaluate the self-energy operator Σ. In 1960s, Hedin et al. (1965) had proposed a series of self-consistent set of Dyson-like equations to obtain the electron self-energy. Furthermore, Louie and coworkers (Hybertsen and Louie 1986) proposed that in the practical calculation, the start point should be from a non-interacting or mean-field scenario, like the Kohn-Sham DFT system including the exchange and correlation approximations, instead of the methods used in standard textbooks, where the start point is often taken to be the non-interacting system of electrons under the potential $V_{ext}(r) + V_{H}(r)$.

In fact, the key quantity and equation in the descriptions of the single-particle excitation of a many-body system are the self-energy and Dyson's equation, respectively. The self-energy operator Σ contains all the complexity of the many-body interactions, it could not be done in the practical calculation. The most useful and simplest approximation of the self-energy operator is the so-called *GW* approximation which is taken to be the first order term in a series expansion in terms of the screened Coulomb interaction W and the dressed Green's function G of the electron. This was first proposed by L. Hedin in the 1960s (Hedin 1965). In terms of the mathematical equation, it takes the 0th order expansion of the vertex function in terms of W. The approximation for the polarizability P used in GW level is known as the random-phase approximation (RPA).

Although one can use the GW approximation to compute the quasi-particle energy, it is still extremely computationally demanding since one need to the self-consistent process for the Dyson's equation. A popular method which avoids the self-consistent process is the so-called G_0W_0 approximation. The essence of this approximation is that the assumption based on an effective single-particle potential $V^{XC}(\mathbf{r})$, which contains some of the exchange-correlation effects of the many-body system. Most of *ab-initio GW* application will carry out the self-consistent calculation by taking the DFT results as the mean field and evaluating the quasi-particle energy while keeping fixed its wavefunction (which equals to the DFT wavefunction). In other words, the G_0W_0 scheme for calculation of the quasi-particle energy E_n^{QP} as a first-order perturbation to the Kohn-Sham energy ε_n:

$$E_n^{QP} = \varepsilon_n + \langle \psi_n | \Sigma(E_n) - V_{XC} | \psi_n \rangle \qquad (4)$$

where V_{xc} is the exchange-correlation potential within DFT and ψ_n is the corresponding DFT wavefunction. Hybersten and Louie first introduced G_0W_0 methods for the study of solid-state physic in 1985 (Hybertsen and Louie 1986). The G_0W_0 approximation reproduces to within 0.1 eV the experimental band gaps for many semiconductors and insulators, and fixed the well-known bandgap problems. Note that for some systems, the quasi-particle wavefunction could differs significantly from the DFT wavefunctions, such as, the liquid water; one should solve the quasi-particle equation, Eq. (3), directly (Jain et al. 2014, Sun et al. 2017).

Above, we have discussed the basic ideas how to evaluate the quasi-particle energy in the many-body theory, and how to do the G_0W_0 in a practical way to simplify the calculation. Moreover, the Bethe-Salpeter equation, which describes the electron–hole excitation interaction by using the two-particle Green's function, has been proved to be the most accurate methodology commonly used to compute the optical response within the same level of approximation of GW. Although it is extremely computationally heavy for carrying out the GW-BSE calculation compared with a typical DFT calculation for the same system, almost an order of magnitude worse, the pioneering work has been done by Louie et al. (2000). Moreover, it has been proved that within the first order self-energy $\Sigma^{(1)} = G_0W_0$, the Bethe-Salpeter approach to two-particle excited states as the natural extension of GW approach for calculating one-particle excited state, has helped elucidate the optical spectra within a wide range of systems from nanomaterials to bulk semiconductors, surfaces and several attempts to liquid water system. Shirley and coworkers have used GW-BSE method to study the optical spectra and XAS spectra of liquid and solid water (Vinson et al. 2012). The huge computation cost still limits the application of such a method to the water system since the cell size of this study is limited to 17 water molecules, which is not efficient to capture the molecular structure of water.

COHSEX and Enhanced COHSEX Approximation

For the most practical calculations, the self-consistent procedure is complicated. The huge computational cost prohibits the *GW* method from being applied to complex systems. Several alternative attempts have been made based on the physically motivated approximations. Actually, the *GW* approximation could also be interpreted as a combination of the Hartree-Fock approximation and a dynamically screened Coulomb interaction (Onida et al. 2002). In this approach, the *GW* self-energy could be evaluated via the Coulomb-hole plus screened exchange (COHSEX) approximation:

$$\Sigma_{COHSEX}(r_1, r_2) = \Sigma_{COH}(r_1, r_2) + \Sigma_{SEX}(r_1, r_2) \tag{5}$$

With the conversion from time domain to the energy domain, the self-energy (Eq. (13)) could be wrote as:

$$\Sigma(r_1, r_2; E) = \frac{i}{2\pi} \int dE' \, e^{-i\eta E'} G(r_1, r_2; E - E') W(r_1, r_2; E') \tag{6}$$

Then we could rewrite the self-energy operator in terms of the Coulomb-hole (COH) contribution and the dynamically screened-exchange (SEX) contribution, which are:

$$\Sigma_{COH}(r_1, r_2; E) = \sum_{n,k} \psi_{n,k}(r_1)\psi^*_{n,k}(r_2) P \int_0^\infty dE' \frac{B(r_1, r_2; E')}{E - E_{n,k} - E'} \tag{7}$$

where $B(r_1, r_2; E')$ is the spectral function, defined as $B = \pi^{-1}|\text{Im } W|$, of the screened Coulomb interaction W. The P is the Cauchy principal value of the integration. And the SEX term is:

$$\Sigma_{SEX}(r_1, r_2; E) = -\sum_{n,k}^{OCC} \psi_{n,k}(r_1)\psi^*_{n,k}(r_2) W(r_1, r_2; E - E_{n,k}) \tag{8}$$

Furthermore, the static COHSEX approximation could be obtained by assuming the $E - E_{n,k} \to 0$ in Eqs. (7) and (8).

$$\Sigma_{COH}^{static}(r_1, r_2; E) = \frac{1}{2} \delta(r_1 - r_2) W_p(r_1, r_2; E = 0) \tag{9}$$

$$\Sigma_{SEX}^{static}(r_1, r_2; E) = -\sum_{n,k}^{OCC} \psi_{n,k}(r_1)\psi^*_{n,k}(r_2) W(r_1, r_2; E = 0) \tag{10}$$

where the $W_p = W - v$ and v is the bare Coulomb interaction. And here,

$$W(r_1, r_2) = \int \epsilon^{-1}(r_1, r) v(r, r_2) dr \tag{11}$$

ϵ is the dielectric constant. It has been proved by the numerical calculation that most of the error in the static COHSEX approximation comes from the COH contribution, while the SEX contribution yields relatively close results comparing with the full *GW* calculations (Kang and Hybertsen 2010). Moreover, the COHSEX approximation have several practical advantages compared with the full *GW* self-energy approach. The COHSEX self-energy operator is Hermitian, static (which does not depend on ω), and requires only the summation up to the number of occupied electronic states; instead the approach used in GW, a summation of all the states (occupied and unoccupied electronic states, which means convergence check for the number of unoccupied states is needed for the practical calculation). However, it is known that COHSEX approach overestimates the bandgap of several materials.

Kang and Hybertsen had proposed one strategy (Kang and Hybertsen 2010), called the enhanced COHSEX approach, to improve the accuracy of COH contribution: include a correction factor to the adiabatic $W_p(E = 0)$ term of Eq. (9). The correction factor is the wave-vector-resolved and energy-dependent ratio $f(q, E_{nk})$. As such, the dynamic screening is included in the original static COH term, the new COH will change to:

$$\Sigma_{COH}^{new}(r_1, r_2; E) = \frac{1}{2}\delta(r_1 - r_2)\int W_p(q; E = 0)f(\frac{q}{k_f})e^{-iq\cdot r}d\mathbf{q} \qquad (12)$$

where q is a plane wave and k_f is the Fermi vector. The correction factor is

$$f(x) = \frac{1+a_1x+a_2x^2+a_3x^3+a_4x^4+a_5x^5+a_6x^6}{1+bx+b_2x^2+bx^3+b_4x^4+b_5x^5+b_6x^6} \qquad (13)$$

where $a_1 = 1.9085$, $a_2 = -0.542572$, $a_3 = -2.45811$, $a_4 = 3.08067$, $a_5 = -1.806$, $a_6 = 0.410031$, $b_1 = 2.01317$, $b_2 = -1.55088$, $b_3 = 1.58466$, $b_4 = 0.368325$, $b_5 = -1.68927$, and $b_6 = 0.599225$.

XAS Spectra Based on COHSEX/Enhanced-COHSEX Approximations

In this section, we will discuss how to compute the XAS spectra of water within the COHSEX approach. In fact, as we have discussed, the most accurate method to compute XAS spectra is the GW-BSE approach, however, it requires a full GW calculation with an extension to a BSE treatment of the particle-hole interaction, energy-dependent quasi-particle self-energy corrections, and a screening of the core-hole with the RPA approximation. The whole process is extremely demanding in computation resources, so far only one work of XAS using GW-BSE approach has been reported (Vinson et al. 2012). The reported spectra of liquid water were in good agreement with experiment in the pre-edge region, while it overestimated the main edge feature and lacked a clear post-edge feature.

However, the most-used calculation approach for computing the XAS spectra of water is still the so-called frozen-core method. Instead of the DFT wavefunction treatment discussed in last section, one should use the quasi-particle wavefunction for the excited electron in the presence of a frozen core hole and its surrounding electrons sea. As such, the quasi-particle obeys the equation Eq. (3). The first XAS spectra of water calculated by using such an approach was first reported Chen et al. 2010. As we have explained before, the COHSEX approximation used in this work reduced the computation cost greatly compared to the full GW method. They had adopted the frozen-core approach plus the self-consistent calculation updated quasi-particle wavefunction to compute the XAS of water. Actually, when one takes a look at Eq. (11), it is clear that the COHSEX quasi-particle equation will become the Hartree-Fock equation when the screening effects are neglected, that is, when $\epsilon = 1$ as in the monomer. Furthermore, for the large band-gap insulators, like water and ice, the screening effect of electrons is small since the experimental dielectric constant from electronic contribution alone is quite small, which is $\epsilon_0 = 1.8$ and $\epsilon_0 = 1.7$ for water and ice. As such, they have proposed a rather crude homogeneous screening model, neglecting the local field effect, that is, in which $\epsilon (r_1, r_2) = \epsilon (|r_1 - r_2|)$. Based on this approach, they have successfully computed the XAS spectra of water and ice, as is shown in Fig. 2(a).

Here, we would like to make a comment on the calculated XAS spectra of liquid water and ice. Firstly, all the spectra were aligned at the onset of XAS spectra, and the calculated spectra had been multiplied by the scaling factor to adjust the near-edge peak intensity (at $T = 363$ K) to the corresponding experimental value (at the standard

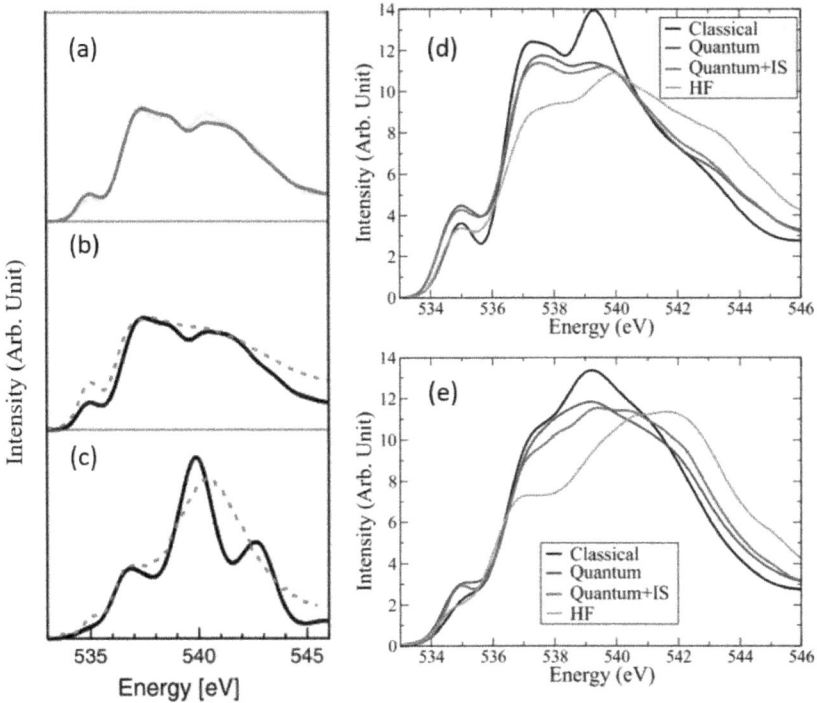

Fig. 2: (a) Calculated XAS spectra of liquid water at 330 K (black) and 360 K (gray). (b) The XAS spectra comparison between computed (solid line) at 363 K and experimental data (dashed line) at 290 K. (c) The different XAS spectra of ice, experimental data (dashed line) at 95 K, and computed data (solid line) based on a perfect ice structure. Reproduced with permission from Chen, Wu, and Car (2010). Copyright [2010] [American Physical Society]. Computed XAS spectra for water at 300 K (d) and ice Ih at 269 K (e) based on four schemes: (1) COHSEX approximation with classical nuclei and homogeneous screening model (black); (2) COHSEX approximation with quantum nuclei and homogeneous screening model (light black); (3) COHSEX approximation with quantum nuclei and inhomogeneous screening (IS) model (gray); (4) Hartree-Fock approximation with quantum nuclei (light gray). Reproduced with permission from Kong, Wu, and Car (2012). Copyright [2012] [American Physical Society].

temperature and pressure). As the figure shows, the calculated spectra of liquid water are in good agreement with the experimental spectra in terms of position, intensity, and spectra width among the pre-edge (533 ~ 536 eV), main edge (537 ~ 539 eV), and post-edge (from 539 eV and beyond) regions. This is a significant improvement compared the spectra based on XCH approach in Fig. 1, since the XCH approach is similar to current approach without taking into account the quasi-particle self-energy. A non-local self-energy operator causes higher energy states compared with experience-reduced exchange effects (i.e., a reduced attractive potential), leading to an overall increase of the spectral width. The temperature effects on the XAS spectra could also be observed from Fig. 2. As shown, by increasing the temperature, the pre-edge and main-edge peak intensities are enhanced while the post-edge peak intensity is reduced. Meanwhile, one can see that the calculated pre-edge peak is weaker than experimental data, while the intensity ratio is well reproduced. This underestimation of the pre-edge peak may come from the certain approximations used in the calculations, such as the zero-kelvin ice structure (since the vibrational effects of ice should enhance the pre-edge peak), and the assumption of a fully screened core hole. Furthermore, the HB network distortion and fraction of the broken bonds in the simulation with classical nuclei and the exchange-correlation functional at the level of PBE could be underestimated compared with experimental data, therefore, the temperature used in simulation is ~ 60 K higher than experimental temperature to obtain a softer water structure.

Here, we will discuss the efforts about including the inhomogeneous screening model to improve the quality of calculated XAS spectra. We would like to insist that for the liquid water, the inhomogeneous screening effect is modest while it could be neglected for ice, since ice has a less homogeneous microscopic structure. Before going into the details, we first go through the theoretical aspects of the inhomogeneous screening model used by Kong et al. 2012 for computing the XAS spectra. In general, the screening effects in the real space should not be considered as universal in the medium, it should be dependent on the local charge density. As such, the screened interaction within the Hybertsen-Louie (HL) ansatz (Hybertsen and Louie 1988)should be written as

$$W(r_1, r_2) = \frac{1}{2}\left(W(r_1 - r_2; \rho(r_2)) + W(r_2 - r_1; \rho(r_1))\right). \tag{14}$$

Here the screened term could be analytically defined as:

$$W(r_2 - r_1; \rho(r_1)) = \frac{1}{(2\pi)^3}\int \epsilon^{-1}[q; \rho(r)]v(q)e^{iq\cdot(r_1-r_2)}dq \tag{15}$$

For the dielectric function, it could be written based on the Bechstedt model (Bechstedt et al. 1992):

$$\epsilon[q; \rho(r)] = 1 + \left[(\epsilon_0 - 1)^{-1} + \alpha\left(\frac{q}{q_{TF}}\right)^2 + \frac{q^4}{4\omega_p^4}\right], \tag{16}$$

where q_{TF} and ω_p are the Thomas-Fermi wave vector and plasmon frequency. Then the screened interaction term could be rewritten as

$$W(\boldsymbol{r}_2 - \boldsymbol{r}_1; \rho(\boldsymbol{r}_1)) = \frac{v(\boldsymbol{r}_2 - \boldsymbol{r}_1)}{\epsilon_0} - \frac{1}{a(x_1 - x_2)|\boldsymbol{r}_2 - \boldsymbol{r}_1|} \times \left(\frac{e^{i(x_1)^{1/2}|r_2 - r_1|}}{x_1} - \frac{e^{i(x_2)^{1/2}|r_2 - r_1|}}{x_2} \right) \quad (17)$$

here $x_{1,2} = (-b \pm \sqrt{b^2 - 4bc})/2a$ and $a = 1/4\omega_p^2$, $b = \alpha/q_{TF}^2$, and $c = \epsilon_0/(\epsilon_0 - 1)$. Now the screened interaction W has been separated into two terms: one is the bare interaction divided by the macroscopic dielectric constant, the other is a local-density-dependent screened interaction, which is nonlocal but short-ranged in space.

Based on the inhomogeneous screening model, Kong et al. (2012) computed the XAS spectra of liquid water and ice, which are shown in Fig. 2. As we know, the screening effect is rather important to get the correct XAS spectra of water, and therefore,we would like to see how the inhomogeneous screening effects will change the shape of XAS spectra and comment on how the latter was improved by them. Let's first start with the Hartree-Fock, in this limit, the screening is absent. The quasi-particle excitation energies are overestimated, leading to too wide spectra comparing with the experimental spectra. Not only is the wide feature introduced by Hartree-Fock, but the post-edge is also more prominent than the near-edge in both water and ice. With the help of the COHSEX calculation, we could reduce the overall width and lower the post-edge below the near-edge in the liquid. The inhomogeneous screening model used here is a measure of the inhomogeneity of matter at the molecular scale, which renders screening less effective than in the corresponding uniform medium. However, the inhomogeneous screening effects are underestimated in the HL approach, and will be further improved upon using a more accurate theory. We could also expect that the effect will have a small impact on water, where the HL-induced change is minor, but for ice, the induced change is quite large and could be not negligible. One more thing we would like to comment on is that for ice XAS spectra, the flat feature around 537 eV could not be reproduced by using any screening model or quantum nuclei, which suggests that the advanced electronic approach and molecular modeling for ice XAS spectra are crucially needed.

Finally, we would like to emphasize the importance of self-consistently diagonalized QWs for computing the XAS spectra of liquid water. Actually, as we discussed in Section Green's Function, GW Approximation, and the Bethe-Salpeter Equation, the quasi-particle wavefunction is quite different from the Kohn-Sham wavefunction, which is essential for computing the XAS spectra. Here, we plotted the XAS spectra based on three different excited treatments: (1) the full core-hole (FCH) approximation, where both the energies and wavefunctions are generated from the unoccupied Kohn-Sham eigenstates, (2) the G_0W_0 approximation, where the energies are generated by using excited theory while the wavefunctions are kept the same as the Kohn-Sham wavefunctions, (3) the enhanced COHSEX approach, where both the energies and wavefunctions are generated by using the self-consistently diagonalized self-energy operator. These three plots together with the experimental spectra are shown in Fig. 3. It is well known that G_0W_0 approximation succeeds in

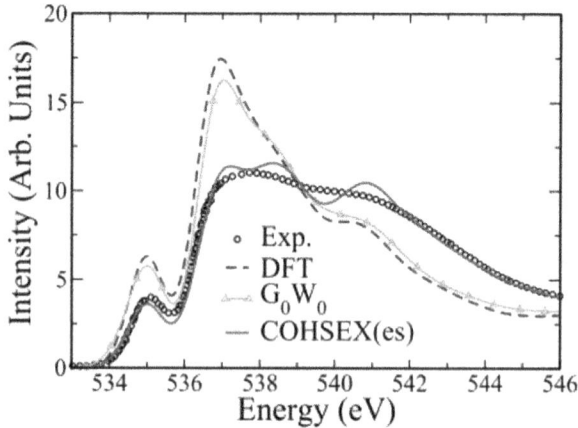

Fig. 3: The computed XAS spectra based on different excited states theory: (1) DFT (dashed line) method with FCH approximation without updating the wavefunction; (2) the G0W0 method (light gray) without the wavefunction; (3) the enhanced-COHSEX (gray) method with a self-consistent calculation updated wavefunction. The experimental XAS spectrum (black) is shown here for comparison. Reproduced with permission from Sun et al. (2017). Copyright [2017] [American Physical Society].

many aspects, such as in computing the band structure of water, aqueous system, nano-materials, organic materials, etc. (Cohen and Louie 2016). However, the computed XAS spectra are very sensitive to the QWs, because of the transition matrix element M_{ij} is based on the QWs. Actually, the calculated XAS spectra based on the FCH and G_0W_0 approaches sharing the similar use of the wavefunction result in very similar spectral shape. These two spectra significantly deviate from the experimental spectrum, while the computed XAS from a self-consistently diagonalized self-energy operator within the enhanced-COHSEX approach shows a much better agreement with the experimental spectra. The agreement emphasizes the importance of the self-consistent diagonalized self-energy operator for computing the XAS spectra of water.

Molecular Structure Aspect: AIMD Simulation

We have discussed above the theoretical calculation progress from the electronic structure aspect, that the self-consistently obtained quasi-particle approach within the GW-BSE framework is important to obtain the correct XAS spectra, and that many theoretical efforts have been made to implement the algorithm and adopt the approximation to reduce the computation cost. So far, the best result of liquid water XAS spectra could be obtained by using the so-called enhanced-COHSEX approximation (Sun et al. 2018, 2017). However, as we have discussed before, the accurate molecular modeling for water is as important as the electronic structure theory. In this section, we will discuss how the molecular modeling changes the shape of XAS spectra.

Let's go through the brief history of the molecular modeling of liquid water within the DFT framework. It is well known that within the generalized gradient approximation (GGA) (Perdew et al. 1996) level of theory, the liquid water structure modeling has several drawbacks (Wang et al. 2011, Møgelhøj et al. 2011, Yoo and Xantheas 2011, VandeVondele et al. 2005, Schwegler et al. 2004, Grossman et al. 2004, Asthagiri et al. 2003, Zhang et al. 2011, Distasio et al. 2014, Zhang et al. 2011, Yoo et al. 2009). For example, GGA will predict the over structured liquid water (Zhang et al. 2011a, Distasio et al. 2014, Zhang et al. 2011b, Yoo et al. 2009), post which one needs to increase the simulation temperature to get a softer HB structure which is closer to the experimental measurement (Distasio et al. 2014, Zhang et al. 2011a, Zhang et al. 2011b). The reason why GGA predicts the overstructured HB network compared to the experimental data is due to the lack of the intermediate van der Waals (vdW) interactions and the self-interaction error (Perdew and Zunger 1981) in the GGA functional. By including the vdW interactions, the number of water molecules in the interstitial region between the first and second coordination shells of water molecules is increased to better match with the experimental radial distribution function (RDF). Moreover, by introducing the hybrid functional, one could alleviate the self-interaction error (Distasio et al. 2014). Therefore, when the directional HB strength between the water molecules is weakened, it will be closer to the experimental data, which means the O-H covalent bond will be shorter compared to its length in the GGA case. Consequently, the protons will have a smaller chance of being donated to the neighbor water molecules. Furthermore, the nuclear quantum effects (NQEs) are rather important for the light mass of molecules, for example, the more delocalized protons via NQEs will introduce the unexpected effects in the HB network in addition to soften or strengthen the liquid water structures. As such, the accurate modeling of liquid water should include the NQEs to simulate the XAS spectra. One should keep in mind that compared to the bound exciton described by the pre-edge, the exciton states in the main edge and post-edge of the XAS spectra are the exciton resonant states, which are sensitive to the intermediate- and long-range of the HB network, therefore, a larger simulation box is important to capture these delocalized excited states. In the following section, we will overview the XAS spectra calculation from these aspects.

Hydrogen-bond Network Probed by the Different XAS Regions from Pre-edge, Main edge, and Post-edge

In order to discuss the impact of different molecular modeling approaches on the XAS shape, it is better to understand how the HB structure of liquid water is probed by the different XAS features, the pre-edge, the main edge, and the post-edge. Actually, the quasi-particle wavefunction (QWs) can be strongly perturbed by the local chemical environment. The three features, pre-edge, main edge, post-edge of XAS have strong molecular signatures related to different spatial regions of the HB network: the pre-edge has a $4a_1$ character while the main edge and post-edge features share the same b_2 character. Both of these characters originate from the molecular excitations in the gas phase. To quantitatively study the spatial regions with respect to the different XAS edges, Sun et al. (2017) have presented a study of the density distributions of QWs as a function of oxygen-oxygen distance. The plot can be found in Fig. 4. In order to get a better view of how QWs delocalizes in the spatial space, they also presented the oxygen-oxygen radial distribution function (RDF) as well as a schematic which shows an excited oxygen with QWs distributed within the HB network. As one can

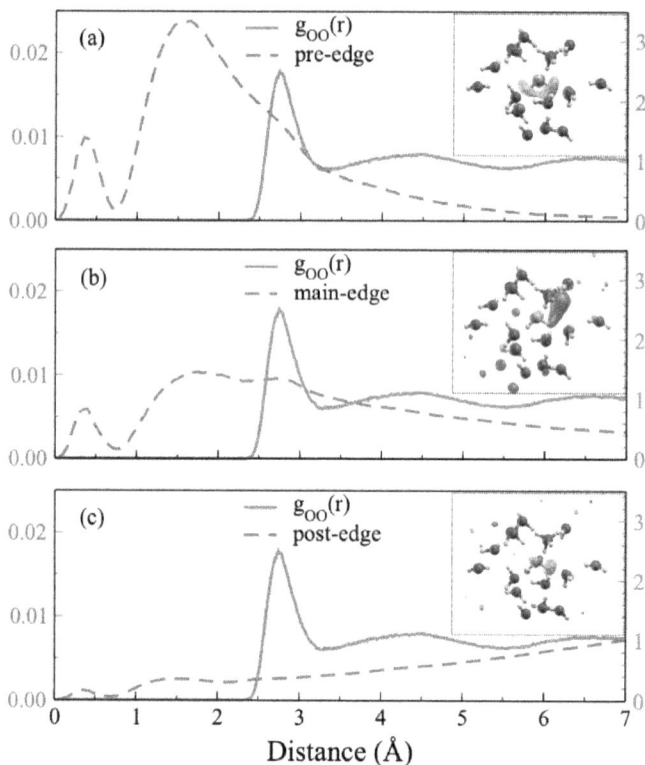

Fig. 4: The density distribution (dashed line) of quasi-particle wavefunctions from the regions: (a) pre- edge, (b) main edge, and (c) post-edge changes with oxygen-oxygen distance calculated from the PBE0+vdW AIMD trajectory by using the enhanced COHSEX method. The oxygen-oxygen radial distribution function (solid line) from the PBE0+vdW AIMD trajectory. The representative quasi-particle wavefunctions from the three edges around the excited water molecule. The background water molecules come from within the second coordination shell of the excited oxygen atom. Red, white, and yellow atoms are the oxygen, hydrogen, and oxygen with a core hole. Quasi-particle wavefunction with signs are shown in blue and green. Reproduced with permission from Sun et al. (2017). Copyright [2017] [American Physical Society].

see, the three distribution plots indicate that the density distribution of QWs become more delocalized from the pre-edge to the main edge and the post-edge.

Let's first take a look at Fig. 4 (a), which shows the QWs density distribution of the pre-edge. The QWs density distribution has the highest peak at 1.7 Å, and is mostly localized in the range of 2.75 Å (the first peak position of $g_{oo}(r)$). The inserted figure indicates that the QWs of the pre-edge contains elements from the first excited state of a water molecule in the gas phase with $4a_1$ symmetry. This result is consistent with the study reported by Chen et al. 2010, which assigns the pre-edge to a bound exciton state, where the electron orbitals are mostly localized with the first coordinate shell. As such, the pre-edge feature is strongly affected by the short-range structural ordering, such as the covalent bond strength and the broken HBs around the excited oxygen atom. One can expect that with the help of the hybrid DFT functional and vdW interaction to describe the HB network, the simulated water structure will produce an improved computed pre-edge feature of XAS for water in both energies and intensities.

Now we will talk about the density distribution of the QWs of the main edge, as one can see in Fig. 4(b), where the plot is more delocalized than that of the pre-edge. As evidenced by the inserted schematic, the QW could not only be found on the excited molecule itself but also on its first and second shell neighbors. Furthermore, the QW of the main edge clearly shows the b_2 character, which is consistent with the fact that the main edge originates from the second excited state of a water molecule in the gas phase. By comparing the QWs' density distributions in pre-edge and main edge, one can find that the main edge is more localized between the first and second coordination shells of the liquid water structure. This fact indicates that the main edge feature of XAS will be sensitive to the intermediate-range order of the liquid water structure, for example, the water molecules in the interstitial region.

Finally, we turn our focus on the post-edge feature, where one can clearly see that the density distribution of the QWs of the post-edge is much more delocalized compared to the QWs of the pre-edge and the main edge. However, QWs in the post-edge still holds the b_2 character. This delocalization nature is further evidenced by the increased density of QW as a function of the distance away from the excited water molecule. The post-edge feature is strongly correlated with the water molecules in the long-range order and essentially a resonant exciton state.

Calculated XAS Spectra Based on Different AIMD Trajectories

In this section, we will discuss the calculated XAS spectra based on three levels of XC functional approximation AIMD trajectories, from PBE to PBE+vdW, to PBE0+vdW. At the same time, we will compare the molecular structures from these different AIMD trajectories in terms of g_{oo}[®]. The results are shown in Fig. 5. We will show how the computed XAS spectra improve with increasing molecular model accuracy.

First of all, the computed XAS spectra based on the PBE AIMD trajectory shows the largest discrepancies when compared with the experimental data of the three XC functionals. One can find several differences between the calculated spectra and the experimental spectra. First, the computed XAS shows a lower intensity pre-edge peak compared with experimental data. Second, the computed spectrum has a blue-shifted main-edge peak (538.5 eV) compared with the experimental main-edge peak at around 537.5 eV. Third, the computed spectrum shows almost similar intensities of main edge and post-edge peak, which contradicts the experimental finding that the main edge is more prominent than the post-edge in liquid water. These discrepancies could be understood as follows: the pre-edge peak of XAS spectra for Ice Ih shows a more prominent peak of the post-edge than that of the main edge, indicating that the HB network from the PBE AIMD trajectory is overstructured. As the $g_{oo}(r)$ computed from the PBE AIMD trajectory significantly deviates from experimental data, the first and second peaks of $g_{oo}(r)$ are overestimated, and the first minimum is largely underestimated. As such, the average number of HBs per water molecule are found to be 3.76 in the PBE AIMD trajectory based on Chandler's criterion (Luzar and Chandler 1996), which is the highest number among the functionals reported here.

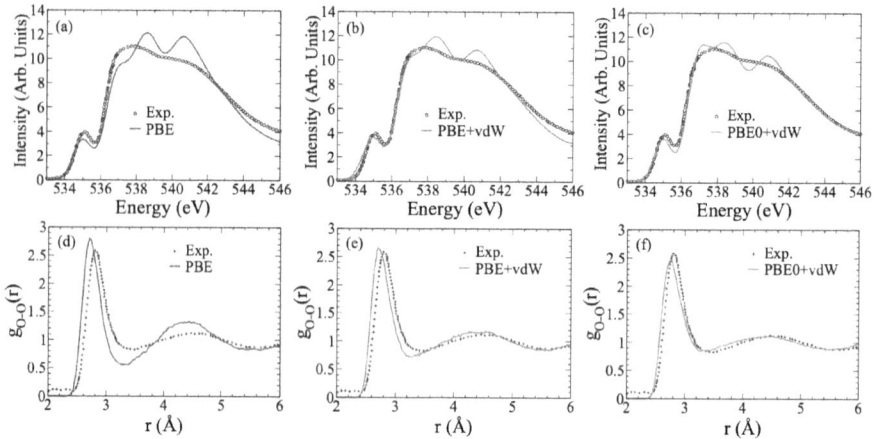

Fig. 5: The computed XAS and oxygen-oxygen radial distribution function $g_{OO}(r)$ from three different XC functionals: PBE, PBE+vdW, and PBE0+vdW. The experimental XAS and $g_{OO}(r)$ are shown as well for comparison. Reproduced with permission from Sun et al. (2017). Copyright [2017] [American Physical Society].

Therefore, both the calculated XAS spectra and oxygen-oxygen radial distribution function $g_{OO}(r)$ support the overstructured HB network in the PBE AIMD simulation.

Next, as the data shows in Fig. 5(b), the calculated XAS spectra from the PBE+vdW AIMD trajectory are largely improved upon with respect to the experimental data compared those obtained from the PBE trajectory. The improvements include a higher pre-edge peak, a shift of the main-edge peak to the lower energy side, and a lower postage intensity. The greater agreement between the calculated XAS spectra and experimental data could be understood by the fact that the better description of the HB network, including the vdW interactions in the AIMD simulation. The added vdW forces strengthens the non-directional attractive interactions among the water molecules, which greatly increase the number of the water molecules in the interstitial region. Therefore, the PBE+vdW $g_{OO}(r)$ is in closer agreement with the experimental data vis-a-vis within the first and second coordinate shells. The increased number of water molecules in the interstitial region weakens the HB among the water molecules in the first coordinate shell, leading to a lower first peak in $g_{OO}(r)$. The average number of the HB per water molecule is 3.56, almost 5% smaller than the one of PBE. In other words, an excited oxygen atom is surrounded by a more disordered water environment and therefore, one can find that the pre-edge intensity is increased. In order to verify this finding, Sun and coworkers (2017) selected two excited water molecules, one with a broken HB, and the other one with four intact HBs, calculated the density distribution of their QWs. The QW of the pre-edge from the excited molecule with the broken HBs is more localized with an enhanced p character because of the more disordered short-range molecular environment. As such, the transition amplitude of the pre-edge become larger. As a result, the pre-edge intensity is increased with more broken HBs. Furthermore, the main-edge peak is also enhanced by the increased number of water molecules in the interstitial region because of the added vdW interaction in the simulation. In order to verify how the water molecules in the interstitial region affect the main edge features, they also chose two representative excited water molecules and calculated their density distribution of QWs, one of the excited water molecules has four intact HBs, while the other one has an additional water molecule partner in the interstitial region. They found that the density of the main edge QW of the excited molecule with one additional water molecule partner in the interstitial region is more localized compared with the one with four intact HBs. Based on the discussion regarding the pre-edge region, it is seen that the increased number of water molecules in the interstitial region leads to larger amplitudes of the transition matrix elements in the main edge. As such, this discussion clarifies the reason why the improved main edge features happen, and the post-edge feature whose QW is orthogonal to the main edge, is predicted to have a lower spectral intensity.

Furthermore, we would like to comment on the XAS spectra of water based on molecular configurations generated by using the highest level of XC functional, PBE0+vdW. The plots are shown in Fig. 5(c). With the mix of a fraction of the exact exchange in the hybrid functional, PBE0 overcomes the self-interaction error and lowers the possibility of a hydrogen atom being donated to a neighboring water molecule. As such, the covalent bonds of the water molecule are strengthened, therefore, the length of the OH bond becomes shorter and the directional HB is weakened. Overall, the average HB number per water molecule is 3.48, which is the smallest among three XC functionals. The water structure is softened with a larger faction of broken HBs, which is further supported by the lower first peak of $g_{OO}(r)$. Moreover, the width of the first peak of $g_{OO}(r)$ is broadened, and the position is closer to the experimental data, which agrees with the weakened directional HB strength. The improved HB network of water will improve the quality of the calculated XAS spectra as well. This is because, first of all, the improved HB network, or the softened water structure, indicates the more disordered environment with which an excited oxygen is surrounded. Based on the same argument in the previous paragraph, the QW of the pre-edge with broken HBs is more localized with an enhanced p character, leading to the more prominent pre-edge intensity with more broken HBs. Therefore, the pre-edge peak of the PBE0+vdW XAS spectra is closer to the experimental spectra. Second, since the directional HB strength is weakened in PBE0+vdW simulation, more nonbonded water molecules appear in the interstitial region, evidenced by the fact that the first minimum of $g_{OO}(r)$ is increased to get closer to the experimental data. As such, the intermediate range HB network is improved as well, leads to the increased intensities from calculations to match the experimental data from 537 to 538 eV. This is because that the main edge features of XAS, which are sensitive to the intermediate range HB network, based on the same argument in the last paragraph. Third, the post-edge features have lower intensity and shift to higher energies. The PBE0+vdW AIMD trajectory captures more prominent features of the main edge than that of the post-edge. The width of the XAS is broadened in the high-energy region from 544 to 546 eV, getting close to the experimental spectra.

As we have discussed, the precise picture of NQEs on the XAS is rather important for understanding the details of the molecular structure of liquid water. The predicted XAS spectrum based on the PBE0+vdW AIMD trajectory is indeed in agreement with the experimental spectrum, however, the overestimated spectral intensity in the main edge and post-edge are still presented in the predicted spectrum. Therefore, the disagreement reflects the nature of NQEs. The delocalized protons via NQEs could either weaken or strengthen the HB structures, which is strongly dependent on the an harmonicity of the potential energy. Furthermore, the main edge and post-edge contributions are different from the ones of the pre-edge contribution, the latter is bound exciton. The excited resonant states from the main edge and post-edge are sensitive to the intermediate and long-range HB network, and this sensitivity requires a much larger simulation supercell to compute the XAS spectra. Sun and coworkers (2018) calculated the XAS spectra of liquid water, which takes NQEs in account via the path-integral molecular dynamics (PIMD) with MB-pol many-body potential. The computed XAS spectra based on MD and PIMD are plotted in Fig. 6. The theoretical spectrum obtained from PIMD simulation is in excellent agreement with the experimental spectrum. The broadened pre-edge reflects the proton fluctuations in the covalence of the water molecule. Protons approach the acceptor oxygen atoms with a much shorter HB distance due to the NQEs, as such, the "ice-like" spectral feature is enhanced with respect to that of main edge. In the spectrum from the MD simulation, the energy of the post-edge is underestimated, and both the main edge and post-edge show overestimated intensity compared with the experimental data. In the main edge region, the two subpeaks are present and separated by a valley, which is absent in the experimental spectrum. The spectrum from the PIMD simulation is in almost quantitative agreement with the experimental spectrum.

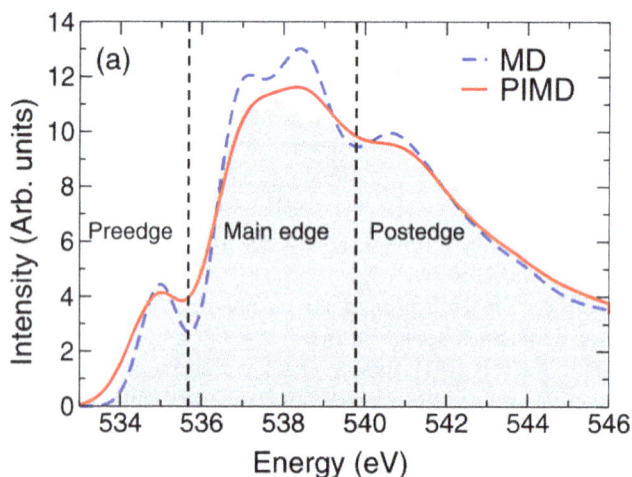

Fig. 6: The computed XAS spectra based on MD (blue) and PIMD (red) simulation at 298 K. The experimental XAS data is shaded. Reproduced with permission from Sun et al. (2018). Copyright [2018] [American Physical Society].

Perspectives

In this chapter, we have summarized the recent progress about the theoretical calculation of XAS spectra of water. Two parallel aspects have been made for obtaining the accurate XAS spectra: first, the electronic structure theory has been used for characterizing the properties of the excited core states, and the mean field theory was used at the beginning for computing the initial and final states. Later, the real excited state theories, that is, the *GW* approach and its simple approximations, COHSEX and enhanced COHSEX approximations, have been used for utilizing the final state, together with the frozen-core approximation to compute the quasi-particle wavefunctions and energies. Second, since the molecular structure snapshots are taken from the MD simulation trajectory, different levels of the MD simulations have been discussed, varying from the classical force field model to the AIMD simulation with the XC level from PBE, PBE+vdW, and PBE0+vdW. NQEs' effects on the XAS spectra have been checked via a MB-pol-based PIMD simulation.

Although computed XAS spectra based on these approximations are mostly in agreement with the experimental spectra, some limitations still exist. For example, the accurate quasi-particle characterization should be obtained using the standard GW-BSE approach, while the huge computation cost limits its application. Only one GW-BSE calculation has been made in water for XAS spectra calculation, however, its simulation box containing only 17 water molecules strongly limits its accuracy and interpretation (Vinson et al. 2012). Therefore, one benchmark calculation based on GW-BSE approximation conducting in a larger simulation cell will be highly desired, which could be a good justification for the currently used COHSEX and enhanced COHSEX approximations. Moreover, since the GW-BSE approximation will not use any empirical screening model, it will be a good chance to compute the XAS spectra for ice to resolve the flat feature at the main edge region. Furthermore, we expect that the real GW-BSE approach with AIMD simulation for computing XAS spectra could be used for other materials, which could not be possible otherwise, that is, based on current empirical screening models.

References

Ågren, Hans, Vincenzo Carravetta, Olav Vahtras and Lars G. M. Pettersson. 1994. Direct, atomic orbital, static exchange calculations of photoabsorption spectra of large molecules and clusters. Chemical Physics Letters 222: 75–81.

Ågren, Hans, Vincenzo Carravetta, Olav Vahtras and Lars G. M. Pettersson. 1997. Direct SCF direct static-exchange calculations of electronic spectra. Theoretical Chemistry Accounts 97: 14–40.

Aryasetiawan, F. and O. Gunnarsson. 1998. The GW Method. Reports on Progress in Physics 61: 237–312.

Asthagiri, D., Lawrence R. Pratt and J. D. Kress. 2003. Free energy of liquid water on the basis of quasichemical theory and *ab initio* molecular dynamics. Physical Review E 68: 41505.

Ball, Philip. 2008. Water as an active constituent in cell biology. Chemical Reviews 108: 74–108.

Bechstedt, F., R. Del Sole, G. Cappellini and Lucia Reining. 1992. An efficient method for calculating quasiparticle energies in semiconductors. Solid State Communications 84: 765–70.

Bellissent-Funel, Marie Claire, Ali Hassanali, Martina Havenith, Richard Henchman, Peter Pohl, Fabio Sterpone, David Van Der Spoel, Yao Xu and Angel E. Garcia. 2016. Water determines the structure and dynamics of proteins. Chemical Reviews 116: 7673–97.

Bernal, J. D. and R. H. Fowler. 1933. A theory of water and ionic solution, with particular reference to hydrogen and hydroxyl ions. The Journal of Chemical Physics 1: 515–48.

Besley, Nicholas A. and Frans A. Asmuruf. 2010. Time-dependent density functional theory calculations of the spectroscopy of core electrons. Physical Chemistry Chemical Physics 12: 12024–39.

Brancato, Giuseppe, Nadia Rega and Vincenzo Barone. 2008. Accurate density functional calculations of near-edge x-ray and optical absorption spectra of liquid water using nonperiodic boundary conditions: The role of self-interaction and long-range effects. Physical Review Letters 100: 107401.

Chen, Wei, Xifan Wu and Roberto Car. 2010. X-ray absorption signatures of the molecular environment in water and ice. Physical Review Letters 105: 017802.

Cohen, Marvin L. and Steven G. Louie. 2016. Fundamentals of Condensed Matter Physics. Fundamentals of Condensed Matter Physics. Cambridge: Cambridge University Press.

Distasio, Robert A., Biswajit Santra, Zhaofeng Li, Xifan Wu and Roberto Car. 2014. The individual and collective effects of exact exchange and dispersion interactions on the *ab initio* structure of liquid water. Journal of Chemical Physics 141: 84502.

Eaves, J. D., J. J. Loparo, C. J. Fecko, S. T. Roberts, A. Tokmakoff and P. L. Geissler. 2005. Hydrogen bonds in liquid water are broken only fleetingly. Proceedings of the National Academy of Sciences of the United States of America 102: 13019–22.

Fransson, Thomas, Yoshihisa Harada, Nobuhiro Kosugi, Nicholas A. Besley, Bernd Winter, John J. Rehr, Lars G. M. Pettersson and Anders Nilsson. 2016a. X-ray and electron spectroscopy of water. Chemical Reviews 116: 7551–69.

Fransson, Thomas, Iurii Zhovtobriukh, Sonia Coriani, Kjartan T. Wikfeldt, Patrick Norman and Lars G. M. Pettersson. 2016b. Requirements of first-principles calculations of X-ray absorption spectra of liquid water. Physical Chemistry Chemical Physics 18: 566–83.

Grossman, Jeffrey C., Eric Schwegler, Erik W. Draeger, François Gygi and Giulia Galli. 2004. Towards an assessment of the accuracy of density functional theory for first principles simulations of water. Journal of Chemical Physics 120: 300–311.

Hedin, Lars. 1965. New method for calculating the one-particle green's function with application to the electron-gas problem. Physical Review 139: A796–823.

Hedin, Lars and Stig Lundqvist. 1970. Effects of electron-electron and electron-phonon interactions on the one-electron states of solids. *In*: Frederick Seitz, David Turnbull and B. T. Henry (eds.). Solid State Physics. Solid State Physics Ehrenreich, 23: 1–181. Academic Press.

Hetényi, Balázs, Filippo De Angelis, Paolo Giannozzi and Roberto Car. 2004. Calculation of near-edge x-ray-absorption fine structure at finite temperatures: Spectral signatures of hydrogen bond breaking in liquid water. Journal of Chemical Physics 120: 8632–37.

Hunt, William J. and William A. Goddard. 1969. Excited states of H2O using improved virtual orbitals. Chemical Physics Letters 3: 414–18.

Hybertsen, Mark S. and Steven G. Louie. 1986. Electron correlation in semiconductors and insulators: Band gaps and quasiparticle energies. Physical Review B 34: 5390–5413.

Hybertsen, Mark S. and Steven G. Louie. 1988. Model dielectric matrices for quasiparticle self-energy calculations. Physical Review B 37: 2733–36.

Jain, Manish, Jack Deslippe, Georgy Samsonidze, Marvin L. Cohen, James R. Chelikowsky and Steven G. Louie. 2014. Improved quasiparticle wave functions and mean field for G0W0 calculations: Initialization with the COHSEX operator. Physical Review B 90: 115148.

Jungwirth, Pavel and Douglas J. Tobias. 2006. Specific ion effects at the air/water interface. Chemical Reviews 106: 1259–81.

Kang, Wei and Mark S. Hybertsen. 2010. Enhanced static approximation to the electron self-energy operator for efficient calculation of quasiparticle energies. Physical Review B 82: 195108.

Kohn, W. 1999. Nobel Lecture: Electronic structure of matter—wave functions and density functional. Reviews of Modern Physics 71: 1253–66.

Kong, Lingzhu, Xifan Wu and Roberto Car. 2012. Roles of quantum nuclei and inhomogeneous screening in the x-ray absorption spectra of water and ice. Physical Review B 86: 134203.

Kraus, Peter M., Michael Zürch, Scott K. Cushing, Daniel M. Neumark, Stephen R. Leone, Peter M. Kraus, Michael Zürch, Stephen R. Leone and Daniel M. Neumark. 2018. The ultrafast x-ray spectroscopic revolution in chemical dynamics. Nature Reviews Chemistry 2: 82–94.

Luzar, Alenka and David Chandler. 1996. Hydrogen-bond kinetics in liquid water. Nature 379: 55–57.

Møgelhøj, Andreas, André K. Kelkkanen, K. Thor Wikfeldt, Jakob Schiøtz, Jens Jørgen Mortensen, Lars G.M. Pettersson, Bengt I. Lundqvist, Karsten W. Jacobsen, Anders Nilsson and Jens K. Nørskov. 2011. *Ab initio* van der waals interactions in simulations of water alter structure from mainly tetrahedral to high-density-like. Journal of Physical Chemistry B 115: 14149–60.

Nilsson, A., S. Schreck, F. Perakis and L. G. M. Pettersson. 2016. Probing water with x-ray lasers. Advances in Physics: X 1: 226–45.

Onida, Giovanni, Istituto Nazionale, Roma Tor Vergata, Ricerca Scientifica, and I-Roma. 2002. Electronic excitations : Density-functional versus many-body green's-function approaches. Reviews of Modern Physics 74: 601.

Perdew, J. P. and Alex Zunger. 1981. Self-interaction correction to density-functional approximations for many-electron systems. Physical Review B 23: 5048–79.

Perdew, John P., Kieron Burke and Matthias Ernzerhof. 1996. Generalized gradient approximation made simple. Physical Review Letters 77: 3865–68.

Pettersson, Lars G. M., Yoshihisa Harada and Anders Nilsson. 2019. Do x-ray spectroscopies provide evidence for continuous distribution models of water at ambient conditions? Proceedings of the National Academy of Sciences 116: 201905756.

Pettersson, Lars Gunnar Moody, Richard Humfry Henchman and Anders Nilsson. 2016. Water—The most anomalous liquid. Chemical Reviews 116: 7459–62.

Prendergast, David and Giulia Galli. 2006. X-ray absorption spectra of water from first principles calculations. Physical Review Letters 96: 215502.

Rohlfing, Michael and Steven G. Louie. 2000. Electron-hole excitations and optical spectra from first principles michael. Physical Review B 62: 4927–44.

Salucci, P., F. Stel, M. I. Wilkinson, N. W. Evans, G. Gilmore, R. P. Saglia, O. Gerhard et al. 2003. Ultrafast hydrogen-bond dynamics in the infrared. Science 301: 1698–1702.

Schwegler, Eric, Jeffrey C. Grossman, François Gygi and Giulia Galli. 2004. Towards an assessment of the accuracy of density functional theory for first principles simulations of water. II. Journal of Chemical Physics 121: 5400–5409.

Slater, J. C. and K. H. Johnson. 1972. Self-consistent-field Xα cluster method for polyatomic molecules and solids. Physical Review B 5: 844–53.

Slater, John C. 1972. Statistical exchange-correlation in the self-consistent field. *In*: Per-Olov, B. T. (ed.). Advances in Quantum Chemistry. Advances in Quantum Chemistry Löwdin, 6: 1–92. Academic Press.

Stener, M., A. Lisini and P. Decleva. 1995. Density functional calculations of excitation energies and oscillator strengths for C1s → π* and O1s → π* excitations and ionization potentials in carbonyl containing molecules. Chemical Physics 191: 141–54.

Sun, Zhaoru, Mohan Chen, Lixin Zheng, Jianping Wang, Biswajit Santra, Huaze Shen, Limei Xu, Wei Kang, Michael L. Klein and Xifan Wu. 2017. X-ray absorption of liquid water by advanced *ab initio* methods. Physical Review B 96: 104202.

Sun, Zhaoru, Lixin Zheng, Mohan Chen, Michael L. Klein, Francesco Paesani and Xifan Wu. 2018. Electron-hole theory of the effect of quantum nuclei on the x-ray absorption spectra of liquid water. Physical Review Letters 121: 137401.

Triguero, L., L. Pettersson and H. Ågren. 1998. Calculations of near-edge x-ray-absorption spectra of gas-phase and chemisorbed molecules by means of density-functional and transition-potential theory. Physical Review B 58: 8097–8110.

Tse, John S., Dawn M. Shaw, Dennis D. Klug, Serguei Patchkovskii, György Vankó, Giulio Monaco and Michael Krisch. 2008. X-ray raman spectroscopic study of water in the condensed phases. Physical Review Letters 100: 095502.

VandeVondele, Joost, Fawzi Mohamed, Matthias Krack, Jürg Hutter, Michiel Sprik and Michele Parrinello. 2005. The influence of temperature and density functional models in *ab initio* molecular dynamics simulation of liquid water. Journal of Chemical Physics 122: 14515.

Vinson, J., J. J. Kas, F. D. Vila, J. J. Rehr and E. L. Shirley. 2012. Theoretical optical and x-ray spectra of liquid and solid H2O. Physical Review B 85: 045101.

Wang, Jue, G. Román-Pérez, Jose M. Soler, Emilio Artacho and M. V. Fernández-Serra. 2011. Density, structure, and dynamics of water: the effect of van der waals interactions. Journal of Chemical Physics 134: 024516.

Werber, Jay R., Chinedum O. Osuji and Menachem Elimelech. 2016. Materials for next-generation desalination and water purification membranes. Nature Reviews Materials 1: 16018.

Wernet, Ph, D. Nordlund, U. Bergmann, M. Cavalleri, N. Odelius, H. Ogasawara, L. Å Näslund et al. 2004. The structure of the first coordination shell in liquid water. Science 304: 995–99.

Yoo, Soohaeng, Xiao Cheng Zeng and Sotiris S. Xantheas. 2009. On the phase diagram of water with density functional theory potentials: The melting temperature of ice Ih with the perdew-burke-ernzerhof and becke-lee-yang-parr functionals. Journal of Chemical Physics 130: 221102.

Yoo, Soohaeng and Sotiris S. Xantheas. 2011. Communication: The effect of dispersion corrections on the melting temperature of liquid water. Journal of Chemical Physics 134: 121105.

Zhang, Cui, Davide Donadio, François Gygi and Giulia Galli. 2011a. First principles simulations of the infrared spectrum of liquid water using hybrid density functionals. Journal of Chemical Theory and Computation 7: 1443–49.

Zhang, Cui, Jun Wu, Giulia Galli and François Gygi. 2011b. Structural and vibrational properties of liquid water from van der waals density functionals. Journal of Chemical Theory and Computation 7: 3054–61.

Zhang, Yu, Weijie Hua, Kochise Bennett, Shaul Mukamel, Jianjun Cheng and Timothy J. Deming. 2016. Nonlinear spectroscopy of core and valence excitations using short x-ray pulses: Simulation challenges. *In*: Nicolas Ferré, Michael Filatov and Miquel Huix-Rotllant (eds.). Density-Functional Methods for Excited States, 310: 273–345. Cham: Springer International Publishing.

CHAPTER 2

Dynamic Crossovers in Water under Extreme Conditions

Paola Gallo,* Gaia Camisasca and *Mauro Rovere*

Introduction

The anomalous behaviour of water in different portions on its phase diagram has been a challenge for experimental and theoretical research for a long time. Since water plays a key role in our life, it is important to understand its properties under the different thermodynamic conditions. The subject of supercooled liquid in particular has attracted researchers' attention for anomalies not found in other liquids (Debenedetti 1996, Gallo et al. 2016).

Among others, just above the melting line it is well known that there is a change in the signs of thermal expansion, and this change is such that the transition to the ice phase occurs with a lowering of the density. In the P-T plane, for a given pressure, the density increases at decreasing temperature, it reaches a maximum, and then it starts to decrease, so it is possible to define a line of maximum density (TMD line) where the thermal expansion from positive becomes negative.

Also, just before melting, anomalous increases of specific heat and isothermal compressibility start and they continue in the supercooled region (Speedy and Angell 1976). This kind of increase of the thermodynamic response functions usually indicates the presence of a second-order phase transition. The occurrence of this type of phase transition in a metastable region is an interesting and unique issue for research in the field of critical phenomena. Anomalies are also present in the dynamical properties of supercooled water like the diffusion coefficient and relaxation times extracted from the correlation function of density fluctuations.

The difficulties in understanding the properties of water in approaching the glassy state are due to the tendency of water to crystallize below the homogenous

Dipartimento di Matematica e Fisica, Università degli Studi Roma Tre, Via della Vasca Navale 84, 00146 Roma, Italy.
* Corresponding author: paola.gallo@uniroma3.it

nucleation line (232 K at ambient pressure). However, there is no reason that prevents water from being experimentally supercooled down to the glass transition temperature (Speedy et al. 1996, Debenedetti 2003, Smith and Kay 1999). In fact, the previously known nucleation line has been recently trespassed with water kept as liquid (Kim et al. 2017, Gallo and Stanley 2017).

Starting from the low-temperature side of the phase diagram, glassy states of water exist and amorphous ice is present in two main forms: high density amorphous (HDA), discovered in 1984 (Mishima et al. 1984) and low density amorphous (LDA), discovered in 1982 (Mayer and Brüggeller 1982). For this reason, it is common to say that water presents polyamorphism.

By increasing the temperature starting from the glasses, water is found to be an ultraviscous liquid but then the system crystallizes (at around 160 K at ambient pressure). So, it exists in a region called *no man's land*, delimited by two curves of temperature versus pressure, the nucleation line TH and the line of crystallization from below Tx, where it is difficult to perform experiments (Gallo and Stanley 2017). The definition of this region and the access paths are subject of continuous investigation (Stern et al. 2019, Gallo and Sciortino 2019).

Different scenarios have been proposed for explaining the anomalies of water, among them we have the retracing spinodal scenario, the critical point free scenario, and the singularity free scenario (Debenedetti 2003, Gallo et al. 2016).

Recent experiments (Kim et al. 2017, Woutersen et al. 2018) and simulations (Palmer et al. 2018, Palmer et al. 2014) especially support the hypothesis formulated for the first time by Poole et al. (1992). According to this scenario, liquid supercooled water in approaching the glassy state can exist in two coexisting forms: the low density liquid (LDL) and the high density liquid (HDL). LDL and HDL are supposed to be the extrapolation of the two coexisting states of glassy water, the LDA and the HDA mentioned above.

A first-order transition taking place between LDA and HAD has been observed in experiments (Mishima et al. 1985, Mishima and Stanley 1998a, Mishima and Stanley 1998b, Winkel et al. 2008, Kim et al. 2009, Loerting et al. 2011, Winkel et al. 2011).

At increasing temperatures, LDA and HDA glasses transforms into LDL and HDL, respectively. The LDL/HDL coexistence curve terminates in a critical point, called the liquid-liquid critical point (LLCP). The presence of the LLCP could explain the anomalous increase of the thermodynamic response functions. LDL is characterized by a local ordered hydrogen bond network with a tetrahedral order extending to the second coordination shell. HDL, instead, has a more disordered hydrogen bond structure with a tetrahedral order limited to the first coordination shell. The LDL-HDL coexistence and the LLCP would be located in the *no man's land*. By extrapolating experimental data, Mishima and Stanley (1998a) approximately located the LLCP at $T \approx 220$ K and $P \approx 100$ MPa. In spite of the progress done in the last years, the research in this field is still in need of new contributions (Gallo et al. 2019), in particular concerning the idea that liquid water would be a mixture of the high entropy component HDL and the low entropy component LDL. Each component prevails on the other depending on the thermodynamic conditions determining the behaviour of the liquid (Anisimov et al. 2018, Russo and Tanaka 2014, Nilsson and Pettersson 2011). It has become usual to say that water is a polymorphic liquid (Gallo et al. 2019).

In this framework of critical phenomena,the dynamics of water plays a pivotal role, also due to its recently discovered connection with thermodynamics that we will describe in this chapter.

In the first section, we review the main results on the slow dynamics of supercooled bulk water; its behaviour is fragile and follows the MCT of glassy dynamics in the region of mild supercooling (Gallo et al. 1996, Sciortino et al. 1996). In the subsequent section, we show that upon further cooling, the dynamics change from the fragile behavior to a strong behavior and we will show how this crossing is characterized by hopping processes (DeMarzio et al. 2017a). We also discuss results that connect the relaxation of the density fluctuations to the presence of a Widom line emanating from the LLCP (Xu et al. 2005, Gallo and Rovere 2012, De Marzio et al. 2016).

In the third section, we consider the other side of the phase diagram, the supercritical region, where water shows peculiar properties that are of interest for extraction of coal, waste disposal, and other industrial processes. Recent studies of experimental data and computer simulation found analogies in the behaviour of water in approaching the liquid-gas critical point from the supercritical region and in approaching the LLCP upon cooling from a high temperature (Gallo et al. 2014, Corradini et al. 2015). In both cases the dynamic crossover corresponds to the crossing of a Widom line.

In a number of applications in biology, chemistry, geology, water is either in the solution or in contact with different substrates. Experiments and computer simulations tested how the dynamical properties of confined water are modified with respect to bulk. The rest of the chapter focuses on studies that were done in particular for water confined in silica pores, water in solutions with ions and water in contact with biomolecules. In these environments supercooling in experiments is often easier, and we will see in this chapter how and to what extent the thermodynamic and dynamics scenarios of bulk supercooled water can be fitted to the cases of confinement and solutions.

Slow Dynamics of Supercooled Water and MCT

Fluids are characterized by a homogeneous constant density. At the microscopic level, however, there are spontaneous fluctuations of their density. These fluctuations decay with a process where the particles diffuse from a region of higher density to a region of lower density. So, the relaxation to equilibrium of the density fluctuations is connected to diffusion.

In the early stage of the diffusion process, a particle moves without interacting with the others. This is called ballistic regime, which is characterized by a free motion. Then the particle starts to collide with the other particles, so the diffusion process becomes a random walk. This is called the Brownian regime. The diffusion is described by the mean square displacement (MSD) $<\Delta r^2(t)>$ where $\Delta r(t) = r(t) - r(0)$ is the distance traveled by the single particle. In normal conditions in a liquid, after the initial increase in the ballistic regime, the MSD follows a linear behaviour as function of time

$$< \Delta r^2 (t) > = 6Dt \tag{1}$$

where D is the diffusion coefficient. Equation (1) derives from the Einstein theory of Brownian diffusion.

In connection to diffusion, it is possible to observe the relaxation of the density fluctuations.

These fluctuations are described by the density correlation functions or the Van hove functions. In the following section, we will consider in particular the van Hove self correlation function (VHSCF) defined as

$$G_s(r, t) = \frac{1}{N} < \sum_i \delta(r + r_i (0) - r_i(t)) > \tag{2}$$

and its Fourier transform, the selfintermediate scattering function (SISF) that can be measured in neutron scattering experiments defined as

$$F_s(Q, t) = \frac{1}{N} < \sum_i e^{iQ[r_i(t) - r_i(0)]} > \tag{3}$$

In liquids, in normal conditions, the SISF is in the ballistic regime in a very short period of time. Post this, the SISF shows a rapid exponential decay, called à la Debye, that corresponds to the Brownian diffusion.

Liquids upon supercooling below the melting temperature and in approaching the glassy state show a slowdown of dynamical behaviour. This phenomenon was the topic of many theoretical and experimental research studies, see for example (Angell 1995, Ediger et al. 1996). Great progress has been made in understanding the phenomenology due to the Mode Coupling Theory (MCT) of glassy dynamics introduced by Goetze (Goetze 2008).

The concept of "cage effect" was introduced by the MCT in order to explain the slowing down of the dynamics in the supercooled region. Upon supercooling, the particle after the ballistic regime does not start to diffuse with random collision but for a certain period, it is trapped in a transient cage formed by its nearest neighbors. The MSD becomes almost constant with the development of a plateau that increases its length at decreasing temperature. After the plateau, in correspondence with the breaking of the cage, the MSD starts to increase linearly following Eq. (1). The corresponding effect in the SISF is the development of a plateau here as well. MCT predicts that the signature of the cage effect is a double relaxation for the SISF. After the ballistic motion the SISF enters in the β regime, the region of the plateau, that persists until the cage relaxes. Then the function decays with a stretched exponential characterized by a relaxation time. As the temperature decreases, the length of the plateau and the relaxation time increases. The MCT in its so-called ideal formulation predicts a crossover at temperature T_c for which the relaxation time diverges with a power law

$$\tau \sim (T - T_c)^{-\gamma} \tag{4}$$

where the exponent $\gamma > 1.766$.

The Stokes-Einstein equation relates the diffusion coefficient and the relaxation time $D \sim 1/\tau$ Consequently, a power law behaviour it is also expected for the diffusion coefficient, in particular:

$$D \sim (T - T_c)^\gamma \qquad (5)$$

This asymptotic divergence of τ corresponds to a crossover from an ergodic to a non-ergodic regime where the liquid undergoes a structural arrest. MCT give precise predictions of the behaviour of the density correlators in approaching T_c. The crossover to the non-ergodic regime implies singularities with critical exponents. Experiments and computer simulations found that the MCT is able to describe very well the liquid dynamics in the mild supercooled region. We will discuss later how, in approaching the glass transition, the ergodicity is restored by hopping processes treated in the extended formulation of the theory.

Also, for supercooled liquid water, a close agreement with the predictions of the MCT in Molecular Dynamics (MD) simulations performed with the SPC/E potential was found (Gallo et al. 1996, Sciortino et al. 1996). Further computational studies performed with different potential models and experiments on the dynamics of supercooled water confirmed the agreement with MCT (Xu et al. 2005, Gallo and Rovere 2012, De Marzio et al. 2016, Dehaoui et al. 2015, Torre et al. 2004). As mentioned above, the MCT behaviour is reflected in the SISF defined earlier in Eq. (3). Now it is assumed that the coordinates in Eq. (3) are the positions of the oxygen atoms, so the SISF represents the translational dynamics of the center of mass of the molecules. We show in the upper left panel of Fig. 1, as an example, the SISF obtained for water simulated with the TIP4P/2005 (Abascal and Vega 2005) potential at a density of 1 g/cm³ calculated at $Q_0 = 0.225$ nm⁻¹ corresponding to the position of the maximum of the static structure factor (De Marzio et al. 2016). At Q_0, it is expected that all the effects are enhanced, since the associated wavelength corresponds to the maximum short range order.

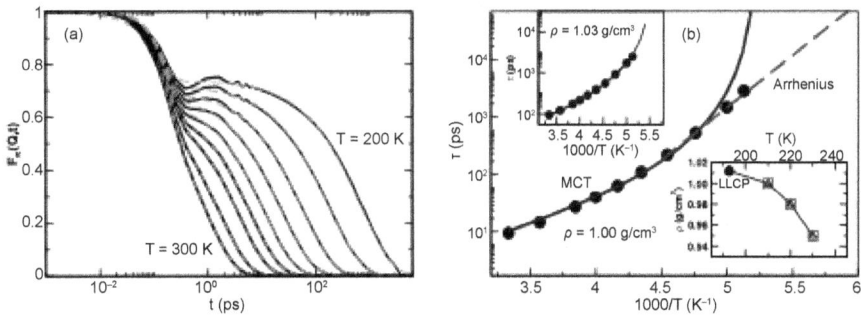

Fig. 1: (a) Self intermediate scattering function for $Q = 0.225$ nm (corresponding to the maximum of the oxygen structure factor) as function of time for decreasing temperatures from below, at density $\rho = 1.00$ g/cm³. Continuous lines are results from simulations. Long-dashed lines are fit to Eq. (6). (b) in the main frame the relaxation time as a function of 1000/T at $\rho = 1.00$ g/cm³ (filled circles) together with the fit to the fragile behaviour, Eq. (4) (continuous line), and to the strong behaviour, Eq. (8) (long-dashed line). In the upper inset the same quantity at $\rho = 1.03$ g/cm³. Symbols are extracted from simulation data and the continuous line is the fit to Eq. (4). In the lower inset the Widom line (triangles), the FSC points (squares) and the position of the LLCP (filled circle) (De Marzio et al. 2016).

The initial ballistic regime is evident at all temperatures. At ambient temperature, the SISF after the ballistic period decays with a simple exponential function, the decay à la Debye. When the temperature decreases, a plateau starts to appear and its length in time increases upon cooling. This is the β regime, the period when the molecule is trapped inside the cage. Finally, the SISF enters in the α decay characterized by a stretched exponential. The model used to fit the curves in the figure takes into account the predictions of MCT and it is realized with the formula (Gallo et al. 1996)

$$f_s(Q, t) = (1 - A(Q)) \exp\left(-(t/\tau_s)^2\right) + A(Q)\exp\left(-(t/\tau)^\beta\right) \tag{6}$$

where τ_s is the fast relaxation short time, τ is the α relaxation time and β is the Kohlrausch exponent. A(Q) is the Lamb-Mossbauer factor given by

$$A(Q) = A_0 \exp\left(-(a^2Q^2/3)\right) \tag{7}$$

where a is the radius of the cage.

From the fit of the curves the behaviour of τ(T) can be obtained and it shows a singular behaviour in approaching the temperature T_C with the predicted power law, Eq. (4). The fit is shown in the right panel of Fig. 1 and the resulting values for the TIP4P/2005 potential at density of 1 g/cm³ are T_C = 190.8 K and the exponent is γ = 2.94 in agreement with the scaling law of MCT (De Marzio et al. 2016). This increase of the relaxation time as a power law in the mild supercooled regime of water has also been found experimentally (Torre et al. 2004).

In the next section, we will discuss how the hopping modifies the MCT behaviour in the deep supercooled region.

Dynamic Crossovers in Supercooled Water and the Widom Line

As said above, to explain the anomalous behaviour of supercooled water, recent experimental and computer simulations supported the scenario of the presence of a LLCP (Gallo et al. 2016, Kim et al. 2017, Gallo and Stanley 2017, Woutersen et al. 2018). In critical phenomena, upon starting from the critical point and in entering in the one phase region, the coexistence curve extrapolates in a line formed by the maxima of the correlation length. This line is called the Widom Line (WL). The WL is expected therefore to emanate also from the LLCP (Franzese and Stanley 2007, Xu et al. 2005, Corradini et al. 2010) with important consequences for the supercooled water's thermodynamics and dynamical properties. The WL in the case of LLCP is expected to mark the border between the region of higher temperature where the liquid is HDL-like and the region of lower temperature where the liquid is LDL-like. Because when it is close to the critical point, the correlation length determines the behaviour of the thermodynamic response functions, the maxima of the specific heat or the isothermal compressibility can be used to locate the WL (Xu et al. 2005, Corradini et al. 2010, Abascal and Vega 2010).

The dynamics of supercooled water can be interpreted in terms of the MCT in its ideal version down to temperatures in which activated processes intervene. This happens for many glass formers in the deep supercooled region and it gives rise to hopping phenomena. Hopping affects the long-time behaviour of the time correlation function. In particular, the relaxation time as a function of temperature, extracted from the SISF, deviates from the MCT behaviour, and below a certain temperature it displays Arrhenius behaviour

$$\tau = \tau_0 \exp\left(-\frac{E_A}{k_B T}\right) \tag{8}$$

where E_A is the activation energy.

In analogy with the Angell classification, the liquid in the MCT regime is called fragile, while when its relaxation time follows the Arrhenius formula is called strong. So the hopping phenomena induce a fragile to strong crossover (FSC).

The FSC has been first hypothesized for water as an interpretation/extrapolation of experimental results (Angell 1993) and it was later found to take place for the first time in simulations of SPC/E water (Starr et al. 1999). Experimentally, it was found in confined water (Faraone et al. 2004, Liu et al. 2005) and recently in the bulk phase (Xu et al. 2016). It is of very relevant interest that this crossover happens at the crossing of the WL (Xu et al. 2005). The connection between the change of the relaxation time behaviour and the presence of the WL was determined in a number of computer simulations on supercooled water performed with different potential models like ST2 and TIP5P (Xu et al. 2005), TIP4P (Gallo and Rovere 2012), TIP4P/2005 (DeMarzio et al. 2016, De Marzio et al. 2017a), and in recent experiments (Taschin et al. 2020).

The important consequences of this finding are that we can identify HDL and LDL as the fragile and the strong components, respectively, and that hopping is more favoured where water is more ordered and less dense. In Fig. 1, in the right panel, the relaxation time as a function of the inverse temperature as extracted from the fit of the SISF calculated with the TIP4P/2005 for the density $\rho = 1.00$ g/cm^3 has been reported. We have already said that in the mild region of supercooling it can be fitted with Eq. (2). The deviation from the MCT behaviour with a crossover to the Arrhenius behaviour is evident. The activation energy in Eq. (8) for the 1 gr/cm^3 isochore of TIP4P/2005 of the Figure 1b is $E_A = 45.29$ kJ/mol and the FSC occurs at $T_S = 210$ K. In the low inset, it is shown how the line of the FSC strictly follows the WL estimated from the maxima of the specific heat. The LLCP of TIP4P/2005 has been determined to be at $T = 200$ K and $\rho = 0.99$ g/cm^3 (Vega 2010). For the density $\rho = 1.03$ g/cm^3 there is no sign of the FSC, as shown in the upper inset of Fig. 1. This is expected for isochores that do not cross the WL, as it happens for the $\rho = 1.03$ g/cm^3 isochore.

The presence of hopping processes as being the mechanism that induces the FSC has been explicitly demonstrated in successive computer simulations of the same system with the calculations of the Van Hove self correlation functions (VHSCF) (De Marzio et al. 2017a). In Fig. 2, examples of the VHSCF are shown for supercooled water at density $\rho = 1$ g/cm^3 simulated with the TIP4P/2005 in the same range of the results presented in Fig. 1. Here and in the following sections, the calculations discussed are for the motions of oxygen. The VHSCF contains the information on the microscopic dynamic of the single particle as a function of time. It gives the probability that a particle that started at t = 0 from the origin is at the position r after time t. So, from this function, it is possible to obtain a detailed description

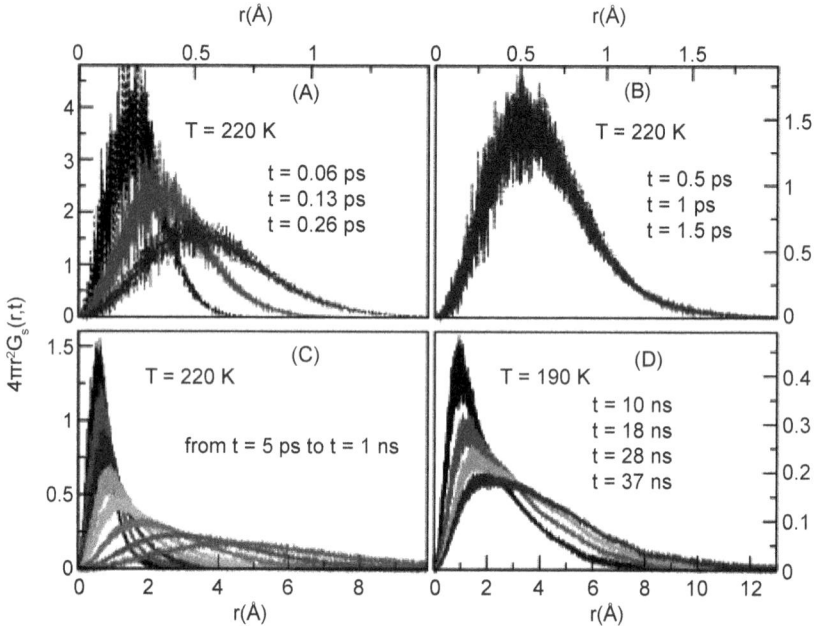

Fig. 2: Van Hove self correlation functions for $\rho = 1$ g/cm³. For T = 220 K the three time windows predicted by the MCT (ballistic regime (A), caging regime (B) and cage relaxation regime (C)) are shown. For T = 190 K cages are frozen and hopping peaks appear at long times (D) (De Marzio et al. 2017a).

of its time evolution. The panels A, B and C show the VHSCF for T = 200 K. The system is in the fragile zone. In A, the particle moves in the ballistic regime, and the VHSCF at increasing time becomes broader and its peak shifts. Then the molecules enter in the zone of the plateau of the SISF. The molecule is trapped in the cage. As a consequence, the VHSCF does not show any change in position and the curves calculated at different times become indistinguishable as shown in panel B. After some time, the cage relaxes, and the molecules enter in the zone of the α decay. Now, as shown in panel C, the molecule moves, the peak of the VHSCF shifts and decays since the system reaches equilibrium and for t → ∞ VHSCF goes to zero. The plot in the panel C must be compared with the one in panel D. In D the temperature is well below the FSC, $T_s = 210$ K, and just below T_c, and according to the predictions of MCT in its ideal version, the cage would be frozen. We observe that now the molecule moves slowly compared with the α decay in C. In D, after more than 30 ns, the peak shifted 0.2 nm compared with the shift of more than 0.9 nm after 1 ns in C. It is important to notice the different form of the peaks in panel C and D. In C, the molecule is in a regime of diffusion and the VHSCFs keep a regular form while in the panel D shoulders appear to deform the functions. The shoulders indicate the presence of hopping. The cage is still present and the only way in which the system can relax to equilibrium is by means of activated processes. The temperature of appearance of the hopping processes is consistent with the FSC crossover found in the relaxation time.

The period when the particle is trapped in the cage can be interpreted with the hypothesis that it is under the effect of a harmonic potential such that

$$G_S(r, t) = \left(\frac{m\omega^2}{4k_B}\right)^{(3/2)} e^{-m\omega^2 r^2/(4k_B T)} \tag{9}$$

where ω is the frequency of oscillation inside the cage.

Another way to explain this phenomenology is to think in terms of the so-called potential energy landscape (PEL). The PEL was introduced by Stillinger (1995) and it is, in the multidimensional phase space of the coordinates, the surface generated by the potential energy of each particle. It is independent of temperature and it is usually formed by a succession of barrier and valleys. At high temperatures, the particles explore the region of the PEL where they can move without much constrains so that the ergodicity is ensured. At decreasing temperature, the particles start to feel the presence of barriers and valleys. In particular, the MCT regime takes place in a range of temperature where the particles are trapped in some minima for a short period of time. By lowering more the temperature, the only mechanism for the particle to escape and avoid a transition to a non-ergodic regime is to hop above the barriers between the minima. In the framework of this interpretation, the real non-ergodicity is reached in approaching the glass transition (Stillinger and Debenedetti 2001).

The connection between dynamical and structural properties and thermodynamics was confirmed by a simulation study of the two body entropy in supercooled water (Gallo and Rovere 2015). The two body entropy (TBE) calculated from the radial distribution function of the oxygen, was connected to the relaxation time of density fluctuations. Evidence was found that the behaviour of the TBE changes when the dynamics crosses from the fragile MCT regime to the Arrhenius regime. We will discuss the TBE in the paragraph of the ionic solutions.

Dynamic Crossover in Supercritical Water and Widom Line

There is large interest in water in the supercritical region of its phase space due to applications in waste disposal and geochemical processes. Recently studies also observed that in supercritical water, the WL emanating from the liquid-gas critical point separates a liquid-like regime from a gas-like regime in the single phase portion of the phase diagram above the critical point (Gallo et al. 2014). In particular, a crossover in the behaviour of the experimental viscosity and in the diffusion, coefficient obtained from simulation has been found (Gallo et al. 2014, Corradini et al. 2015). The WL has been traced by looking at the maxima of the constant pressure specific heat and the isothermal compressibility. As said before, the line of those maxima can be considered to follow closely the line of the maxima of the correlation length and they collapse into a single line close to the critical point.

As shown in Fig. 3, the viscosity of water along a given isobar changes its curvature at the WL temperature. In the supercritical region above the WL, the viscosity decreases with temperature, a behaviour typical of the liquid phase, while after crossing of the WL, the slope is the one typical of a gas phase. With the constraint of very low pressure, the viscosity collapses at the dilute gas limit. The

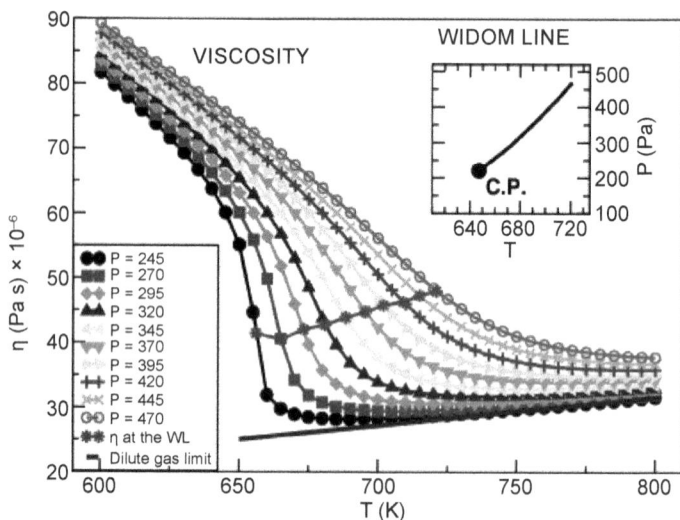

Fig. 3: Viscosity of supercritical water as function of the temperature for different values of the pressure in experiments (NIST Chemistry WebBook http://webbook.nist.gov/chemistry/fluid/, 2011). The line connecting the stars represents the value of the viscosity across the WL and it happens at the flexus of the curves. The WL represented in the inset starts from the critical point (C.P.) and divides the liquid-like zone at a low pressure from the gas-like zone at high pressure (Gallo et al. 2014).

inset in the figure shows the values of pressure and temperatures of the WL points (Gallo et al. 2014).

In a computer simulation, by using different potential models, similar results are found with a change in the slope of the diffusion coefficient from liquid-like to gas-like at the crossing of the calculated WL (Gallo et al. 2014, Corradini et al. 2015). So, in the supercritical portion of the phase space by crossing the WL emanating from the liquid-gas critical point crossover from a liquid-like to a gas-like of dynamical quantities was observed both in the experiment and in the computer simulation a. This is in accordance with the results of the simulations in the supercooled region of water, where the WL separates the LDL-like from the HDL-like behaviour. These findings, confirming the picture of the WL as a line where also dynamics shows a crossover, support the hypothesis of the LLCP in supercooled water where the WL has been recently measured (Kim et al. 2017, Gallo and Stanley 2017).

We conclude saying that another dynamic crossover, the Frenkel line, that separates rigid and non-rigid fluids, has been found in the phase diagram of supercritical water (Cockrell et al. 2020, Yang et al. 2015).

Supercooled Confined Water

As said above the LLCP and all the phenomena connected to it are supposed to happen in a region difficult to accede for experiments due to the nucleation processes that drive the supercooled liquid into the ice phase. From this point of view, confined water preserves the liquid state more easily upon supercooling.

How the static and dynamical properties of water change under the effect of confinement is still an open field of research that has close connections with technological problems, chemistry, biology. It has to be taken into account that there are a number of different features that can induce modifications in the behaviour of water. They are, among others, the geometry of the environment and the type of atoms of the substrate. The surface interacting with water could be, for instance, hydrophilic or hydrophobic, or, as in the case of some biomolecules, it can be both in its interaction with water. The case of biomolecules will be considered later.

Here we restrict our analysis to water inside hydrophilic pores of silica for which a number of computer simulations and experiments were performed to study the dynamics of supercooled water (Gallo et al. 2012, Gallo et al. 2010a, Gallo et al. 2000a, Gallo et al. 2000b, Faraone et al. 2004, Liu et al. 2005, Mallamace et al. 2006, Zhang et al. 2009, Liu et al. 2006, Wang et al. 2015, Xu et al. 2016, Zanotti et al. 1999).

An insightful simulation study of water confined *in silica* was done on a model of water confined in a porous material called Vycor (Gallo et al. 2000a, b). It is a porous silica glass obtained with a special procedure that consists in performing a spinodal decomposition of melted SiO_2 and B_2O_3. The final silica glass is characterized by a well-defined structure of the pores. From the point of view of the simulation, the important features are that Vycor has a strong capability to absorb water and the pores of diameter of 0.4 nm do not change size when filled with water.

To build a realistic model in simulation a special procedure was used (Spohr et al. 1999). A crystal of β cristobalite was melted at 6000 K and supercooled to obtain a cubic cell of silica glass at 300 K. The size of the cube was around 60 nm. Inside it was carved a cylindrical pore of 2 nm of radius. The surface of the pore was prepared first by excluding all the silicon atoms that have less than four oxygens as neighbours. After this, two types of oxygens were found. The bridging oxygens (BO) bonded to two silicons and not bridging oxygens (NBO) bonded to only one silicon; the not bridging were saturated with hydrogens (acidic hydrogens AH) to imitate the preparation of the surface done in experiments. The final concentration of the different atoms on the surface was in agreement with the experimental numbers.

A similar procedure was also done to build a model for the MCM41. This is a silica porous material characterized by a periodic arrangement of pores and for this reason frequently used as a substrate for confining fluids. In particular, a series of experiments have been performed on supercooled water confined in this material. In these experiments a FSC line pointing at the LLCP was detected (Liu et al. 2005). In the simulations, the MCM41 pore was assumed to have a radius of 0.10 nm. The density profile of water confined in this hydrophilic pore is shown at full hydration in the upper inset of Fig. 4 (Gallo et al. 2010a). A similar profile (not shown) was found for the Vycor pore (Spohr et al. 1999). The water molecules form hydrogen bonds with the silica surface in such way that a 6 Å layer of water close to the substrate is extremely slowed down. This layer (bound water) screens the hydrophilic surface so that the more internal layer is formed by a liquid state (free water) whose behaviour was found to be similar to bulk water. This behaviour was evidenced in Vycor by the study of both the structural and dynamical properties of water (Spohr et al. 1999, Gallo et al. 2000a,b, Gallo et al. 2002a,b, Gallo and Rovere 2003a,b). Later, it was also observed in MCM41. The difference between bound and free water are evidenced in the different trend of their dynamical properties. The behaviour of the

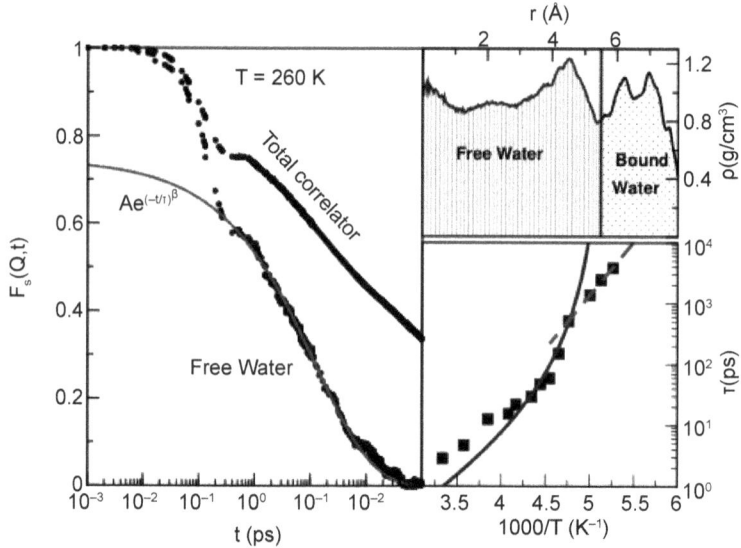

Fig. 4: In the left panel both the total SISF of all the supercooled water molecules in the pore (total correlator) and the SISF of the free water from simulations are shown (symbols), along with the fit to the stretched exponential function (continuous line) for T = 260 K. In the upper panel on the right the density profile along the pore radius is shown together with the division of water in free water and bound water. The lower panel on the right shows the relaxation time as function of the inverse temperature together with the fit of both the fragile behaviour, Eq. (4) (continuous line on the left) and the strong behaviour, Eq. (8) (dashed lines on the right) (Gallo et al. 2012a, Gallo et al. 2010a).

SISF of free water can be interpreted in terms of the MCT in analogy with bulk water (Gallo et al. 2000a,b, Gallo et al. 2010a, Gallo et al. 2012), while the bound water appears to be already very close to a glassy state (Gallo et al. 2002b, Gallo and Rovere 2003a,b, Rovere and Gallo 2003, Gallo et al. 2010b).

In Fig. 4, the case of MCM41 is shown. Along the pore of MCM41, we can distinguish free and bound water. The SISF for the supercooled confined water for MCM41 is shown in the main panel of Fig. 4. The SISF of the total system does not decay, but if we separate the contribution of the free water layer, we find a behaviour similar to supercooled bulk water in it. The SISF of the free water follows the MCT behaviour in the mild supercooled regime. The relaxation time is shown in the inset. In a range of temperatures, it can be fitted with Eq. (4), the asymptotic divergence is at T_c = 195 K with γ = 3.2. At T = 210 K, hopping effects induce an FSC crossover and the relaxation time can be fitted with the Arrhenius formula Eq. (8). The values found in the simulation of MCM41 are very similar to the ones found in the experiment (Farone et al. 2004, Liu et al. 2005).

An analysis of the VHSCF of water in Vycor (Gallo and Rovere 2003) clearly shows that along the pore there are two distinct subset of water molecules. The total VHSCF as function of the distance shows a double peak structure. By separating the contributions from the bound and the free water it is evidenced that the peak at shorter distance derives from the bound water, while the VHSCF of free water has a peak at longer distance. During the same period of time, the molecules in the bound water traveled much less than the molecules in the free water. The bound water is characterized by a very localized VHSCF that rapidly decays to zero and that can clearly be distinguished from that of the free water. The latter has in fact a more distributed VHSCF that decays slowly.

For a better understanding of bound water, it is possible to study the system at low hydration. At this level of hydration, in fact, almost all the water molecules are in the layer closer to the substrate. Studies at low hydration on water confined in Vycor show details about the properties of bound water and the low hydration regime is interesting for a better understanding of the effects of the substrate on the confined water (Gallo et al. 2002b).

With a layer analysis it was also possible to study the local arrangement of the molecules. By calculating the distribution of the angle between two vectors joining the oxygen atom of a water molecule with the oxygen atoms of two nearest neighbour molecules, it was found that the distribution changes with respect to bulk water (Gallo et al. 2002b). In bulk water, at ambient conditions, due to the tetrahedral short range order, this angle results to be around 103° with a secondary peak at 53° due to the presence of interstitial molecules. At the lowest level of hydration investigated the peak was found at 90°, indicating a strong distortion of the tetrahedral arrangement of the water molecules. This is a consequence of the formation of hydrogen bonds between some of the water molecules and the substrate. Another interesting quantity explored is the residence time (Gallo et al. 2002b). It is evaluated as the average time that a molecule spends in a zone along the pore. We recall that the diameter of the pore is 0.4 nm. The longest residence time is found at the contact with the substrate, increasing as the hydration decreasing. The distribution function of the residence time close to the Vycor surface shows a power law behaviour given by $P(t_{res}) = at^{-\mu}$ in a layer from 0.14 nm. This is connected to a not-Brownian diffusion as found in experiments where water is in contact with globular proteins (Rocchi et al. 1998). In the layer from the origin to 0.16 nm, the distribution of the residence time becomes exponentially $P(t_{res}) = be^{-Bt}$ and the long-time diffusion results are Brownian. The calculation of the MSD for low hydration in the pore after the initial ballistic regime found the presence of a plateau connected to the cage effect. At the end of the plateau, however (Gallo et al. 2002b, Gallo and Rovere 2003a,b):

$$< \Delta r^2(t) > \sim t^\alpha \quad \alpha < 1 \tag{10}$$

This anomalous diffusion is also found in water in contact with MCM41 (Gallo et al. 2010b) and with proteins and biomolecules (Camisasca et al. 2018b, Iorio et al. 2019a, Rocchi et al. 1998).

The diffusion is considered anomalous when the MSD for along time is proportional to t^α with $\alpha \neq 1$. If $\alpha < 1$ it is termed sub diffusion, while $\alpha > 1$ is termed super diffusion. The diffusion mechanism is related to the distribution of the residence time and for sub-diffusion it is expected that $\alpha = \mu - 1$, this relation is approximately satisfied at low hydration.

Supercooled Aqueous Solutions of Electrolytes

There is a great interest in solutions of water with ions. Water is a very good solvent for ionic substances. Ions usually dissolve in water and become strong electrolytes, that is, electrically conductive solutions. Aqueous solutions of ions are present in different types of materials, and they are the subject of many research studies in chemistry, biology, and chemical engineering. It is important to understand how the behaviour of water can be modified by its interaction with ions since this is somehow the basic problem that needs to be solved in order to understand a number of phenomena related to the presence of ionic aqueous solutions.

Hofmeister classified ions according to whether an ion enhances or weakens the hydrogen bond network of the surrounding water. Therefore, these ions are termed, respectively, as structure makers or structure breakers. The structure makers are strongly hydrated; from one side they are able to break the HB in the surrounding molecules, but the rest of the water remains with a good short range order. With the structure breakers, however, since they interact weakly with the water molecules, they induce a sort of disorder in the tetrahedral arrangement of the solvent.

The Hofmeister classification (or series) has been used since long a time in the interpretation of the properties of ionic solutions. Although, more recently, however, this scheme has been challenged by a number of experimental studies (Waluyo et al. 2014). The Hofmeister classification can be used for a first approach but it does not take into account appropriately the effects of the thermodynamic conditions and the ionic concentration. Results of experiments and computer simulations show that ions are able to perturb water even beyond the first shell. Ions' effect on water is similar to the application of pressure, see for example (Gallo et al. 2011a, Corradini et al. 2010a).

Ionic aqueous solutions are relevant also in the study of the water anomalies in supercooled conditions. MD studies of aqueous solutions of NaCl upon supercooling found that the effect of ions is to shift the phase diagram of water, making it possible to perform experimental observations in the *no man's land*, in particular, to explore the LDL-HDL coexistence (Corradini et al. 2010a). These results are in agreement with simulations of water in a solution with hydrophobic solutes where the LLCP was determined with a shift with respect to pure water for the same potential model and for solvophobic solutions for a short range water potential (Corradini et al. 2012, Corradini et al. 2010c). An important issue obtained in computer simulation is that the increase of the NaCl concentration reduces the thermodynamic region of the LDL water (Corradini and Gallo 2011). In a more recent interpretation of water as a two-state liquid (Nilsson and Pettersson 2015), it would be interesting to better understand the interplay between the dissolved ions and water as a mixture of the LDL and HDL components.

Recent experiments with X-ray absorption spectroscopy on aqueous solutions of alkali halides (Walujo et al. 2014) confirmed the result of computer simulation concerning the effect of ions on the LDL form of water. The experiments carried on the different ionic solutions show that alkali cations have the effect of *structure breaking*.

An FSC was found in experiments on aqueous solution of LiClat an eutectic concentration (Turton et al. 2012). In a solution of NaCl at a concentration of 0.67 mol/kg, the presence of a LLCPat T = 200 K and ρ = 0.99 g/cm^3 was determined by MD with a TIP4P model for water and also the WL was evaluated (Corradini et al.

2010b). A study of the dynamics upon supercooling has been performed on this solution in the range of density from 0.97 to 1.05 and a range of temperature from 300 K to 190 K (Gallo et al. 2013). The behaviour of the relaxation time extracted from the SISFs shows a good agreement with MCT for the densities explored $\rho = 0.92, 0.95, 0.97, 0.98, 1.05$ g/cm³. In the region of deep supercooling, the SISFs deviate from the MCT and it appears to transition to the Arrhenius behaviour. The FSC temperatures are reported in the upper-left panel of Fig. 5 for the first four densities. For $\rho = 1.05$ g/cm³, the relaxation time follows the MCT in all the ranges. This is in agreement with the position of the LLCP. In the upper-left panel of Fig. 5, an interesting comparison between the points of the FSC and the WL was reported. The FSC takes place at the crossing of the WL as it was found in bulk water.

The strict connection between dynamical and thermodynamics properties is further demonstrated by looking at the entropy behaviour. The excess entropy of the liquid is defined as

$$S_{exc} = S - S_{id} \tag{11}$$

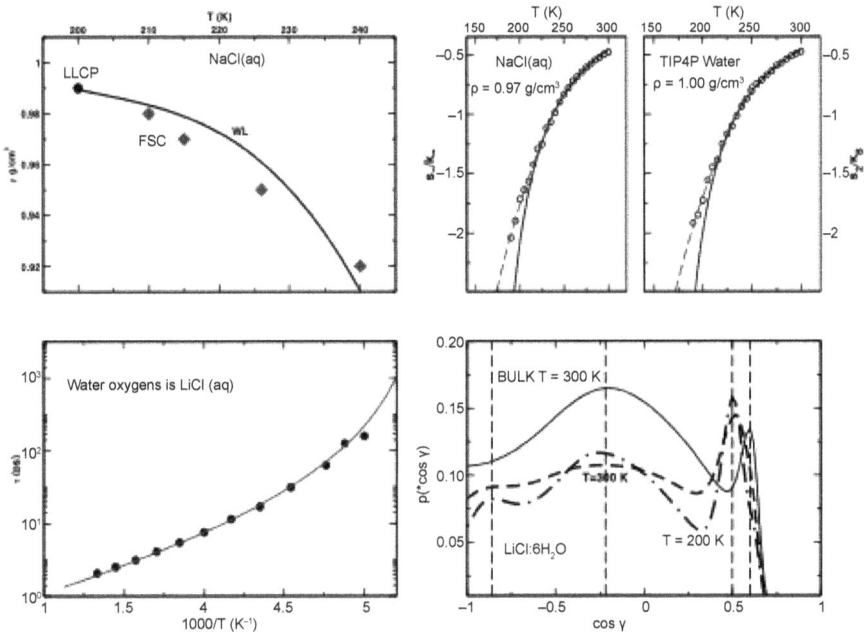

Fig. 5: The left and central upper panels refer to NaCl(aq) at concentration 0.67 mol/kg and density $\rho = 0.97$ g/cm³: on the left is the Widom line starting from the position of the LLCP (black point), the diamonds are the points of the FSC (Gallo et al. 2013); in the middle is the two body entropy of NaCl(aq) upon supercooling compared with the same quantity for TIP4P bulk water (right upper panel). The open circle is the s_2, where the continuous line is the fit to the fragile VFT formula, Eq. (17), while the broken line is the fit to the strong Arrhenius behaviour, Eq. (18) (Gallo et al. 2011b). The lower panels refer to the LiCl:6H$_2$O system: on the left the relaxation time as a function of 1000/T together with the fit to a fragile behaviour, Eq. (5); on the right the distribution function of the angle between three nearest neighbour water oxygens of both the solutions at two different temperatures (dashed lines) and the bulk water at ambient temperature (continuous line) (Camisasca et al. 2018a).

The excess entropy per particle $S_{exc} = S_{exc}/N$ can be obtained with an expansion in *n-body* terms. The formula for the two body entropy (TBE) is related to the radial distribution function $g_{\alpha\beta}(r)$ of the liquid and it is given by

$$S_{exc} = -2\pi\varrho k_B \sum_{\alpha\beta} x_\alpha x_\beta \int \{g_{\alpha\beta}(r)ln[g_{\alpha\beta}(r)] - [g_{\alpha\beta}(r) - 1]\}\ r^2\ dr \qquad (12)$$

where x_α is the concentration of the α component. It was found that the two body excess entropy of water is connected to the structural anomaly of the liquid (Johnson and Head-Gordon 2009, Errington and Debenedetti 2001, Errington et al. 2006, Saija et al. 2003). For the water component, oxygen's contribution is taken into account by assuming that the oxygens are located approximately in the centre of mass of the molecules. Also, for the ionic solution due to the low ion concentration the main contribution to s_2 comes from oxygen-oxygen radial distribution function. In pure water, simulated with the TIP4P model, looking at the $g_{OO}(r)$ at low temperatures, like T = 190 K, at increasing density from ρ = 0:86 g/cm³ up to ρ = 1.09 g/cm³, the structure moves from the LDL-like to the HDL-like (Gallo et al. 2011, Gallo and Rovere 2015). The main changes are evidenced in the region of the second peak, for HDL water it shifts to lower distances and its shape broadens.

The anomalies of water can be connected to the entropy through the equation

$$\left(\frac{\partial\rho}{\partial T}\right)_P = \rho^2 \left(\frac{\partial\rho}{\partial P}\right)_T \left(\frac{\partial s}{\partial\rho}\right)_T \qquad (13)$$

Now the stability condition requires that in the region of the anomalies where the thermal expansion is negative $(\partial s/\partial\rho)_T > 0$, this is equivalent to the condition for the excess entropy

$$\left(\frac{\partial s_{exc}}{\partial \ln\rho}\right)_T > 0 \qquad (14)$$

that gives a possibility to individuate the anomalous range from the RDF calculations by assuming $s_{exc} \approx s_2$. For TIP4P water and the NaCl(aq) considered here, this criterion was found good enough to give indications of the region of anomalous behaviour of TIP4P water and the NaCl(aq) (Gallo et al. 2011).

A relation between the excess entropy and the diffusion was introduced by Rosenfeld (Rosenfeld 1977, Rosenfeld 1999, Rosenfeld 2000, Dzugutov 1996) and it is based on the idea that the cage relaxation must be determined by the number of configurations available to the system through the formula

$$D \propto exp\left(^{as_{exc}}/_{k_B}\right) \qquad (15)$$

It can be assumed that a similar relation is valid for the relaxation time

$$\tau^{-1} \propto exp\left(^{bs_{exc}}/_{k_B}\right) \qquad (16)$$

A recent review of applications is given by Dyre (2018).

From these formulas it is clear that the FSC can be determined also by considering the behaviour of s_{exc} upon supercooling. From MD both in TIP4P bulk water and in the NaCl(aq) it was found that the $s_{exc} \approx s_2$ can be fitted with the Vogel-Fulcher-Tamman (VFT) formula (Angell 1995) valid for fragile liquid and equivalent to the MCT

$$\frac{s_2(T)}{k_B} = A - \frac{BT_0}{T - T_0} \tag{17}$$

where B is related to the fragility parameter and T_0 is the point of singularity derived in the theoretical approach to glass transition of Adam and Gibbs (Adam and Gibbs 1965).

At decreasing temperature, the behaviour of s_2 crosses to the Arrhenius strong formula

$$\frac{s_2(T)}{k_B} = C - \frac{E_A}{k_B T} \tag{18}$$

where E_A is the activation energy. The results of the calculation of the TBE for the NaCl(aq) at density $\rho = 0.97$ g/cm^3 and the bulk water for $\rho = 1$ g/cm^3 are shown in the upper-right panel of Fig. 5. The FSCs obtained with the equations (17–18) are in agreement with the results given by the analysis of the relaxation time obtained from the SISF (Gallo et al. 2011, Gallo and Rovere 2015).

Also, for confined water the calculation of the TBE reveals the connection of this structural quantity with the FSC as in bulk water (De Marzio et al. 2017b, Camisasca et al. 2017).

In the experimental studies of supercooled ionic aqueous solutions, the system of LiCl dissolved in water has attracted the attention of experimentalists since the discovery that LiCl-H$_2$O can be maintained in the liquid state in a large range of supercooled conditions (Moran 1956, Elarby-Aouizerat et al. 1988, Kobayashi and Tanaka 2011a,b). In particular, the LiCl:6H2O solution can be undercooled to the glass transition, easily avoiding crystallization;thus, LiCl:6H2O is considered a glass-forming liquid. Its glass transition temperature is $T_g = 135$ K. A large number of experiments were performed on this solution. Features of structural α relaxation of glass formers were recently revealed by the combinations of different experimental techniques like Brillouin scattering, photon correlation spectroscopy, inelastic UV scattering, and inelastic X-rays scattering (Santucci et al. 2009, Comez et al. 2012). From these experiments, there is evidence of a relaxation with a fragile character upon approaching the glass transition.

A recent MD simulation on LiCl (aq), motivated by the experiments done on this ionic solution, was performed on a concentration of 9.25 mol/kg, corresponding to the LiCL:6H2O system, at constant pressure of 1 bar (Camisasca et al. 2018a). As shown in the lower-left panel in Fig. 5, the SISF follows the MCT in all the ranges explored. By assuming that water is composed of a mixture of HDL and LDL forms at this concentration, the solute seems to enhance the HDL component of water similar to what found for NaCl (Corradini and Gallo 2011). The analysis of the thermodynamic and structural properties of this solution supports this conclusion. In the same figure, in the lower-right panel,the distribution of the angle gbetween three oxygens that reside on three nearest neighbor molecules is shown. In bulk water, the distribution shows a main peak at $\gamma = 103°$ representing the tetrahedral order and a secondary peak at 53° due to the interstitial molecules. In the LiCl aqueous solution, the main peak is at 60° close to the interstitial peak of bulk water and there is a secondary peak at 150°. The hydrogen bond network appears strongly perturbed by the interaction with the ions. Results on the RDF show changes in the structure compatible with the prevalence of the HDL component.

Supercooled Biosolutions

Water in contact with a biological substrate is a topic of great interest scientifically since it is well known that water plays an important role in many biological processes (Franks 2000). In particular, the studies on supercooled water in contact with biomolecules are of great relevance for cryoprotection.

Unlike the scientific interest in water being confined to inorganic mesoporous materials, the main interest in biology is devoted to the first layers of water molecules surrounding macromolecules. Very important roles are in fact played by these water molecules, called hydration water,which directly mediate interactions among biomolecules; see for example (dos Santos et al. 2019, Martelli et al. 2018).

It has been found that the structure of the hydration water could be strongly modified by its interaction with a substrate due to the formation of hydrogen bonds with the hydrophilic part of the macromolecule. Also, the water's dynamical properties are perturbed by the substrate as shown in a series of experimental studies (Paolantoni et al. 2009, Comez et al. 2013) and MD simulations (Magno and Gallo 2011).

In these studies, the translational dynamics of the whole water contained in solution with different biomolecules was studied. The main result is that water in bio-systems has two relaxation mechanisms, happening on two different timescales. The slower process, called long-relaxation, is missing in bulk water and has consequently been ascribed to hydration water only. The long process is sensitive to the molecules' complexity and to the temperature, becoming slower and slower as the molecules' complexity increases and the temperature decreases. The translational dynamics of this type of water was studied in detail by calculating the SISFs. When the second long-relaxation is taken into account, the model of Eq. (6) must be modified, and it becomes for the water contained in biosolutions (Magno and Gallo 2011):

$$f_s(Q,t) = (1 - A(Q) - A_l(Q))\exp(-(t/\tau_s)^2) + A(Q)\exp(-(t/\tau)^\beta) + A_l(Q)\exp(-(t/\tau_l)^{\beta_l}) \quad (19)$$

In Eq. (19), a second stretched exponential function is included for the long-relaxation. The main result is that the relaxation of the density fluctuations can be interpreted in terms of the MCT with evidence of the cage effect and the α-relaxation process, but for the strong interaction with the biomolecules a correction must be introduced to deal with the long-time relaxation of the density fluctuations.

In the cited studies, the water contained in the biosolution was studied as a whole. Therefore, the aprocess could not be assigned unambiguously only to bulk-like water. This left open the question if the hydration water relaxes with a single, longer mechanism (long-relaxation) or if the latter mechanism was observed in hydration water together with an α process.

Successive works of MD simulations (Corradini et al. 2013, Camisasca et al. 2016) gave the answer to this question through a selective calculation of SISFs of only hydration water, by excluding therefore bulk-like water from the calculation of the SISFs, and naturally studying the translational dynamics of hydration water.

Camisasca et al. (2016) performed a MD simulation study on water in a solution with the lysozyme protein. Hydration water is defined as the group of water molecules with the oxygen that resides in a layer of 0.6 nm around the biomolecule. The system is shown in the inset of Fig. 6. Water is simulated with the SPC/E potential while lysozyme with the CHARMM forcefield.

The SISFs of the lysozyme hydration water were calculated upon supercooling and are shown in the left panel Fig. 6 for temperatures ranging from 300 K down to 200 K. Upon decreasing temperature, the curves relax slower and slower, but in a different manner with respect to bulk water: they also show long-time stretched tails. These hydration water SISFs could be analyzed only by means of the model of Eq. (19). The main results here is that hydration water itself has two relaxation processes.

The time constants of two relaxations are shown in the Arrhenius plots of Fig. 6. In the middle panel of the figure, the α-relaxation behaviour has been shown. The timescale of this relaxation is similar to the timescale of the α-relaxation in bulk water, confined water, and water in electrolyte solutions. Besides, it also shows the analogous temperature behaviour of the α-relaxation of the water in these systems: it can be fitted with the power law of Eq. (4) at a higher temperature and with the Arrhenius law of Eq. (8) at a low temperature. An FSC was observed at 215 K in the hydration water, 10 degrees higher than the FSC of bulk SPC/E water at similar thermodynamic conditions.

The long-relaxation process is shown in the right panel of the same figure. This relaxation happens on a much slower timescale with respect to the α-relaxation and shows a completely different temperature behaviour with respect to the latter. In particular, it can be described with the Arrhenius law at high and low temperatures.

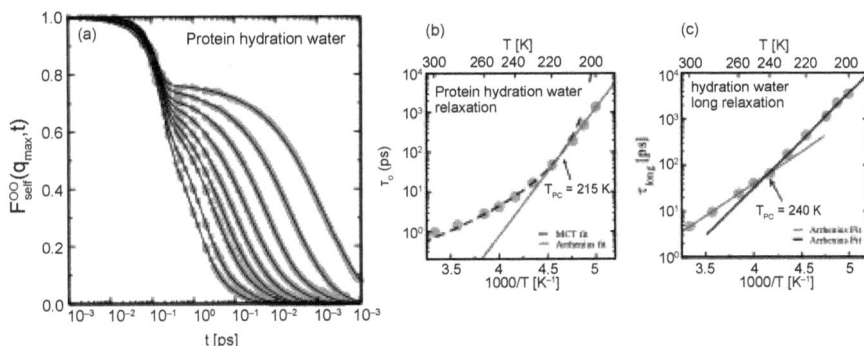

Fig. 6: Panel (a): SISFs of the oxygen atoms of lysozyme hydration water from T = 300 K (bottom curve) to T = 200 K (upper curve). The calculation is done at q_{max} = 0.25 nm^{-1}. Panel (b): α-relaxation time of hydration water, extracted from the fits of the curves shown in panel (a) with the model of Eq. (19). The dashed line is the high-temperature fit done via the power-law of Eq. (4), fragile behaviour, the continuous line is the low-temperature fit done via Eq. (8), strong behaviour. Panel (c): long-relaxation time of hydration water extracted from the fits of the curves shown in panel (a) with the model of Eq. (19). Continuous lines are Arrhenius fits at high temperatures (left) and low temperatures (right).
Reprinted from (Camisasca et al. 2016), with the permission of AIP publishing.

Nonetheless, the activation energies at high and low temperature are different, therefore a strong to strong crossover (SSC) is observed. In the case of the lysozyme hydration water this SSC is observed at 240 K (Camisasca et al. 2016).

Due to the different timescales of the α and long-relaxation, it was proposed that hydration water molecules relax with the α-process as if they were in bulk phase, but they become sensitive on longer timescales to the motion of the protein. The long relaxation was therefore inferred to be the outcome of the coupled dynamics of the protein with his hydration layer. The protein internal dynamics were characterized by quantifying the fluctuations of the protein structure during the simulations. These fluctuations are found to be small at low temperature and to increase linearly upon increasing temperature. Suddenly, around 240 K, a steep increase in the magnitude of these fluctuations was observed. This behaviour is also observed in actual experiments and is named protein-dynamical transition (PDT) (Doster et al. 1990). At temperature lower than the PDT, proteins are stiff and do not explicate their biological functions, which is possible only above the PDT when they become more flexible. Besides, only if proteins are hydrated do they show the PDT.

The activation of the protein happens in coincidence with the change of the activation energy of the long-relaxation time at the SSC. The dynamic coupling of the protein with its hydration water has its signature in the coincidence of the PDT (purely protein dynamics) with the SSC crossover of the long relaxation times of hydration water. Here it was fundamental to prove the dynamic coupling by observing a dynamic crossover in coincidence of the SSC temperature in some dynamical quantity calculated only upon protein atoms positions.

Successive studies show that the long relaxation is actually quite a general feature of hydration water. It is a long-time mechanism of relaxation also for the hydration water of smaller molecules than proteins, when they do form extended clusters (Iorio et al. 2019a,b). In particular, in these MD simulations, concentrated solutions of water with the sugar trehalose were studied. A cluster of trehalose molecules was observed in the simulation box and the structural fluctuations of this cluster considered as a whole show a similar behaviour of the protein fluctuations upon cooling, upon which a trehalose-dynamic transition was detected. Moreover, the translational dynamics of trehalose hydration water molecules shows the α and the long relaxation, and the latter shows an SSC in coincidence of the trehalose-dynamic. This striking result points to the fact that complex and big biomolecules like proteins, but also aggregate of smaller biomolecules,have their internal dynamics which causes the onset of the long-relaxation mechanism on the water molecules closer to their surface. Recently similar results were found also for aqueous solutions of flexible non-biological macromolecules (Zanatta et al. 2018).

Translational dynamics of protein hydration water was also studied by means of the VHSCF with reference to a study by Camisasca et al. (2018b). Here, from the time-behaviour of the position of the peaks of VHSCFs, it was shown that the dynamics of hydration water is subdiffusive and that hopping phenomena intervene in hydration water at a high temperature, in the time scale of the long relaxation. It must be said that this hopping mechanism is different from the hopping phenomena of the glassy dynamics because it does not happen on the nearest-neighbor's length scale but on longer distances allegedly related to the protein surface.

The presence of the long translational relaxation also has a great impact on the dynamics of the hydrogen bond networks. In particular, the network was studied for the trehalose-water system with reference to a study by Iorio et al. (2019b). Here, a new proof of the coupling of water-macromolecule dynamics via long relaxation is given by the results of the analysis of the hydrogen bond correlation functions. The relaxation time of the function describing the hydration water-hydration water hydrogen bonds shows an FSC at the same temperature of the FSC of the translational a process. The relaxation time of the function describing the hydration water-protein hydrogen bonds shows an SSC at the same temperature of the SSC of the translational long process.

These findings signal that the α process does not involve the biomolecules, but it is a characteristic relaxation of water and other glass-forming systems. The onset of the long process in hydration water signals the strong mutual influence between biomolecules and the first layers of water surrounding them.

Conclusions

In this chapter, we focused mainly on the studies on the dynamical properties of supercooled water. The anomalies of liquid water below the melting point in approaching the glass transition are of particular interest since the phenomenology is completely different with respect to other supercooled liquids. In recent times, new sophisticated methodologies (Stern et al. 2019, Kim et al. 2017, Gallo et al. 2017) opened up the possibilities of performing experiments in the no man's land driven by results of computer simulations, which have played a highly important role in understanding the complex phenomenology by suggesting different scenarios and relative interpretations.

This chapter contains a review of some important results obtained by studying the dynamical properties of supercooled water with computer simulations. Water has been studied in different frameworks of bulk water and water in contact with surfaces, with ions, or with biomolecules. One of the main issues is that in all the cases examined, the MCT is able to describe the relaxation of the density fluctuations in a wide region upon supercooling. In bulk water, there is evidence of a crossover from a fragile to strong behaviour of the relaxation time. In the thermodynamics space, the points where this crossover takes place coincide with the Widom line, that is, the line starting from a critical point and moving to the single phase region which collects the maxima of the correlation length. These findings support the hypothesis of a liquid-liquid coexistence with a liquid-liquid critical point (LLCP) in supercooled water. As shown in the paragraph on supercritical water, changes of water's dynamical quantities are found at the crossing of the Widom line relative to the liquid-gas critical point in accordance with the results obtained for supercooled water. So the strict connection between dynamics and thermodynamics of water is widely demonstrated in computer simulation. In the case of biomacromolecules, we also find a further strong to strong dynamic crossover connected to the PDT and coming from the coupling between water and protein surface movements.

Even if we are still far from a complete understanding of the behaviour of supercooled water, great progress has been done, starting from the hypothesis of the presence of a second critical point. Relaxation dynamics and its crossover in behaviour upon varying thermodynamic conditions is an important topic in this respect since, as we have reported, it is strictly related to thermodynamics. It is also very relevant that these studies have been extended beyond bulk water to the situations where water is in contact with substrates or in solutions, since this is an important topic of research connected to different applications.

References

Abascal, J. L. F. and C. Vega. 2005. A general purpose model for the condensed phases of water: TIP4P/2005. J. Chem. Phys. 123: 234505.

Abascal, J. L. F. and C. Vega. 2010. Widom line and the liquid-liquid critical point for the TIP4P/2005 water model. J. Chem. Phys. 133: 234502.

Angell, C. A. 1993. Water II is a strong liquid. J. Phys. Chem. 97: 6339–6341.

Angell, C. A. 1995. Formation of glasses from liquids and biopolymers. Science 267: 1924–1935.

Anisimov, M. A., M. Duška, F. Caupin, L. E. Amrhein, A. Rosenbaum and R. J. Sadus. 2018. Thermodynamics of fluid polyamorphism. Phys. Rev. X 8: 011004.

Camisasca, G., M. De Marzio, D. Corradini and P. Gallo. 2016. Two structural relaxations in protein hydration water and their dynamic crossovers. J. Chem. Phys. 145: 044503.

Camisasca, G., M. De Marzio, M. Rovere and P. Gallo. 2017. Slow dynamics and structure of supercooled water in confinement. Entropy 19: 185.

Camisasca, G., M. De Marzio, M. Rovere and P. Gallo. 2018a. High density liquid structure enhancement in glass forming aqueous solution of LiCl. J. Chem. Phys. 148: 222829.

Camisasca, G., A. Iorio, M. De Marzio and P. Gallo. 2018b. Structure and slow dynamics of protein hydration water. J. Mol. Liq. 268: 903–910.

Cockrell, C., O. Dicks, V. V. Brazhkin and K. Trachenko. 2020. Pronounced structural crossover in water at supercritical pressures. J. Phys.: Condens. Matt. In press.

Comez, L., C. Masciovecchio, G. Monaco and D. Fioretto. 2012. Progress in liquid and glass physics by brillouin scattering spectroscopy. pp. 1–77. In: Camley, R. E. and R. L. Stamps (eds.). Solid State Physics (Academic Press). Vol. 63.

Comez, L., L. Lupi, A. Morresi, M. Paolantoni, P. Sassi and D. Fioretto. 2013. More is different: Experimental results on the effect of biomolecules on the dynamics of hydration water. J. Phys. Chem. Lett. 4: 1188–1192.

Corradini, D., M. Rovere and P. Gallo. 2010. A route to explain water anomalies from results on an aqueous solution of salt. J. Chem. Phys. 132: 134508.

Corradini, D., P. Gallo and M. Rovere. 2010a. Molecular dynamics simulations of an aqueous solution of salts for different concentrations. J. Phys.: Condens. Matter 22: 284104.

Corradini, D., M. Rovere and P. Gallo. 2010b. A route to explain water anomalies from results on an aqueous solution of salt. J. Chem. Phys. 132: 134508.

Corradini, D., S. V. Buldyrev, P. Gallo and H. E. Stanley. 2010c. Effects of hydrophobic solutes on the liquid-liquid critical point. Phys. Rev. E 81: 061504.

Corradini, D. and P. Gallo. 2011. Liquid-Liquid critical point in NaCl aqueous solutions: Concentration effects J. Phys. Chem. B 115: 14161.

Corradini, D., P. Gallo, S. V. Buldyrev and H. E. Stanley. 2012. Fragile to strong crossover coupled to liquid-liquid transition in hydrophobic solutions. Phys. Rev. E 85: 051503.

Corradini, D., E. G. Strekalova, H. E. Stanley and P. Gallo. 2013. Microscopic mechanism of protein cryopreservation in an aqueous solution with trehalose. Sci. Rep. 3: 1218.

Corradini, D., M. Rovere and P. Gallo. 2015. The widom line and dynamical crossover in supercritical water: popular water models versus experiments. J. Chem. Phys. 143: 114502.

De Marzio, M., G. Camisasca, M. Rovere and P. Gallo. 2016. Mode coupling theory and fragile to strong transition in supercooled TIP4P/2005 water. J. Chem. Phys 144: 074503.

De Marzio, M., G. Camisasca, M. Rovere and P. Gallo. 2017a. Microscopic origin of the fragile to strong crossover in supercooled water: The role of activated processes. J. Chem. Phys. 146: 084502.

De Marzio, M., G. Camisasca, M. Martin Conde, M. Rovere and P. Gallo. 2017b. Structural properties and fragile to strong transition in confined water. J. Chem. Phys. 146: 084505.

Debenedetti, P. G. 1996. Metastable Liquids. Princeton University Press, Princeton, USA.

Debenedetti, P. G. 2003. Supercooled and glassy water. J. Phys. Condens. Matter 15: R1669–R1726.

Dehaoui, A., B. Issenmann and F. Caupin. 2015. Viscosity of deeply supercooled water and its coupling to molecular diffusion. Proc. Natl. Acad. Sci. U. S. A. 112: 12020–12025.

dos Santos, M. A. F., M. A. Habitzreuter, M. H. Schwade, R. Borrasca, M. Antonacci, G. K. Gonzatti et al. 2019. Dynamical aspects of supercooled TIP3P–water in the grooves of DNA. J. Chem. Phys. 150: 235101.

Doster, W., S. Cusack and W. Petry. 1990. Dynamic instability of liquid like motions in a globular protein observed by inelastic neutron scattering. Phys. Rev. Lett. 65: 1080–1083.

Dyre, J. C. 2018. Perspective: Excess-entropy scaling. J. Chem. Phys. 149: 210901.

Dzugutov, M. 1996. A universal scaling law for atomic diffusion in condensed matter. Nature 381: 137.

Ediger, M., C. A. Angell and S. R. Nagel. 1996. Supercooled liquids and glasses. J. Chem. Phys. 100: 13200–13212.

Elarby-Aouizerat, A., J. F. Jal, P. Chieux, J. M. Letoff, P. Claudy and J. Dupuy. 1988. Metastable crystallization products and metastable phase diagram of the glassy and supercooled aqueous ionic solutions of LiCl. J. Non-Cryst. Solids 104: 203–210.

Errington, J. R. and P. G. Debenedetti. 2001. Relationship between structural order and the anomalies of liquid water. Nature 409: 318–321.

Errington, J. R., T. M. Truskett and J. Mittal. 2006. Excess-entropy-based anomalies for a water like fluid. J. Chem. Phys. 125: 244502.

Faraone, A., L. Liu, C. Y. Mou, C. W. Yen and S. H. Chen. 2004. Fragile-to-strong liquid transition in deeply supercooled confined water. J. Chem. Phys. 121: 10843.

Franks, F. 2000. Water : A Matrix of Life. The Royal Society of Chemistry.

Franzese, G. and H. E. Stanley. 2007. The widom line of supercooled water. J. Phys.: Condens. Matter 19: 205126.

Gallo, P., F. Sciortino, P. Tartaglia and S.-H. Chen. 1996. Slow dynamics of water molecules in supercooled states. Phys. Rev. Lett. 76: 2730–2733.

Gallo, P., M. Rovere and E. Spohr. 2000a. Supercooled confined water and the mode coupling crossover temperature. Phys. Rev. Lett. 85: 4317–4320.

Gallo, P., M. Rovere and E. Spohr. 2000b. Glass transition and layering effect in confined water: A computer simulation study. J. Chem. Phys. 113: 11324.

Gallo, P., M. A. Ricci and M. Rovere. 2002a. Layer analysis of the structure of water confined in Vycor Glass. J. Chem. Phys. 116: 342.

Gallo, P., M. Rapinesi and M. Rovere. 2002b. Confined water in the low hydration regime. J. Chem. Phys. 117: 369.

Gallo, P. and M. Rovere. 2003a. Double dynamical regime of confined water. J. Phys.: Condens. Matter 15: 1521.

Gallo, P. and M. Rovere. 2003b. Anomalous dynamics of confined water at low hydration. J. Phys.: Condens. Matter 15: 7625.

Gallo, P., M. Rovere and S.-H. Chen. 2010a. Dynamic crossover in supercooled confined water: Understanding bulk properties through confinement J. Phys. Chem. Lett. 1: 729–733.

Gallo, P., M. Rovere and S.-H. Chen. 2010b. Anomalous dynamics of water confined in MCM-41 at different hydrations. J. Phys.: Condens. Matt. 22: 284102.

Gallo, P., D. Corradini and M. Rovere. 2011. Excess entropy of water in a supercooled solution of salt. Mol. Phys. 109: 2069.

Gallo, P., D. Corradini and M. Rovere. 2011a. Ion hydration and structural properties of water in aqueous solutions at normal and supercooled conditions: a test of the structure making and breaking concept. Phys. Chem. Chem. Phys. 13: 19814.

Gallo, P. and M. Rovere. 2012. Mode coupling and fragile to strong transition in supercooled TIP4P water. J. Chem. Phys. 137: 164503.

Gallo, P., M. Rovere and S.-H. Chen. 2012. Water confined in MCM-41: A mode coupling theory analysis. J. Phys. Condens. Matter 24: 064109.

Gallo, P., D. Corradini and M. Rovere. 2013. Fragile to strong crossover at the Widom line in supercooled aqueous solutions of NaCl. J. Chem. Phys. 139: 204503.

Gallo, P., D. Corradini and M. Rovere. 2014. Widom Line and Dynamical Crossovers: Routes to Understand Supercritical Water. Nat. Commun. 5: 5806.

Gallo, P. and M. Rovere. 2015. Relation between the two body entropy and the relaxation time in supercooled water. Phys. Rev. E 91: 012107.

Gallo, P., K. Amann-Winkel, C. A. Angell C. A., M. A. Anisimov, F. Caupin, C. Chakravarty et al. 2016. Water: A Tale of Two Liquids. Chem. Rev. 13: 7463–7500.

Gallo, P. and H. E. Stanley. 2017. Supercooled Water reveals its secrets. Science 358: 6370.

Gallo, P. and F. Sciortino. 2019. Several glasses of water but one dense liquid. Proceedings of the National Academy of Sciences 116: 9149–9151.

Gallo, P., T. Loerting and F. Sciortino. 2019. Supercooled water: A polymorphic liquid with a cornucopia of behaviours. J. Chem. Phys. 151: 210401.

Goetze, W. 2008. Complex Dynamics of Glass-Forming Liquids: A Mode-Coupling Theory. Oxford University Press.

Halalay, I. and K. A. Nelson. 1992. Time-resolved spectroscopy and scaling behavior in LiCl/H_{2} O near the liquid-glass transition. Phys. Rev. Lett. 69: 636–639.

Hofmeister, F. 1888. Zur lehre von der wirkung der salze. Archiv für experimentelle Pathologie und Pharmakologie 24(4-5): 247–260.

Iorio, A., G. Camisasca and P. Gallo. 2019a. Slow dynamics of hydration water and the trehalose dynamical transition. J. Mol. Liq. 282: 617–625.

Iorio, A., G. Camisasca, M. Rovere and P. Gallo. 2019b. Characterization of hydration water in supercooled water-trehalose solutions: The role of the hydrogen bonds network. J. Chem. Phys. 151: 044507.

Johnson, M. E. and T. Head-Gordon. 2009. Assessing thermodynamic-dynamic relationships for water like liquids. J. Chem. Phys. 130: 214510.

Kim, C. U., B. Barstow, M. W. Tate and S. M. Gruner. 2009. Evidence for liquid water during the high-density to low-density amorphous ice transition. Proceedings of the National Academy of Sciences. 106: 4596–4600.

Kim, K. H., A. Späh, H. Pathak, F. Perakis, D. Mariedahl, K. Amann-Winkel et al. 2017. Maxima in the thermodynamic response and correlation functions of deeply supercooled water. Science 358: 1589–1593.

Kobayashi, M. and H. Tanaka. 2011a. Possible link of the v-shaped phase diagram to the glass-forming ability and fragility in a water-salt mixture. Phys. Rev. Lett. 106: 125703.

Kobayashi, M. and H. Tanaka. 2011b. Relationship Between the Phase Diagram, the Glass-Forming Ability, and the Fragility of a Water/SaltMixture. J. Phys. Chem. B 115: 14077–14090.

Liu, L., S.-H. Chen, A. Faraone, C.-W. Yen and C.-Y. Mou. 2005. Pressure dependence of fragile-to-strong transition and a possible second critical point in supercooled confined water. Phys. Rev. Lett. 95: 117802.

Liu, Li, S.-H. Chen, A. Faraone, C.-W. Yen, C.-Y. Mou, A. I. Kolesnikov et al. 2006. Quasielastic and inelastic neutron scattering investigation of fragile-to-strong crossover in deeply supercooled water confined in nanoporous silica matrices. J. Phys.: Cond. Matter 18: S2261.

Loerting, T., K. Winkel, M. Seidl, M. Bauer, C. Mitterdorfer, P. H. Handle et al. 2011. How many amorphous ices are there? Phys. Chem. Chem. Phys. 13: 8783–8794.

Magno, A. and P. Gallo. 2011. Understanding the mechanisms of bioprotection: A comparative study of aqueous solutions of trehalose and maltose upon supercooling. J. Phys. Chem. Lett. 2: 977–982.

Mallamace, F., M. Broccio, C. Corsaro, A. Faraone, U. Wanderlingh, L. Liu et al. 2006. The fragile-to-strong dynamic crossover transition in confined water: nuclear magnetic resonance results. J. Chem. Phys. 124: 161102.

Mamontov, E. 2009. Diffusion dynamics of water molecules in a LiCl Solution: A Low-Temperature Crossover. J. Phys. Chem. B. 113: 14073–8.

Martelli, F., H. Y. Ko, C. C. Borallo and G. Franzese. 2018. Structural properties of water confined by phospholipid membranes. Frontiers of Physics 13: 136801.

Mayer, E. and P. Brüggeller. 1982. Vitrification of pure liquid water by high pressure jet freezing. Nature 298: 715–718.

Mishima, O., L. D. Calvert and E. Whalley. 1984. 'Melting ice' I at 77 K and 10 kbar: A new method of making amorphous solids. Nature 310: 393–395.

Mishima, O., L. D. Calvert and E. Whalley. 1985. An apparently first-order transition between two amorphous phases of ice induced by pressure. Nature 314: 76–78.

Mishima, O. and H. E. Stanley. 1998a. The relationship between liquid, supercooled and glassy water. Nature 396: 329–335.

Mishima, O. and H. E. Stanley. 1998b. Decompression-induced melting of ice IV and the liquid–liquid transition in water. Nature 392: 164–168.

Moran, H. E. 1956. System Lithium Chloride–Water. J. Phys. Chem. 60: 1666–1667.

Nilsson, A. and L. G. M. Pettersson. 2011. Perspective on the structure of liquid water. Chem. Phys. 389: 1–34.

Nilsson, A. and L. G. Pettersson. 2015. The structural origin of anomalous properties of liquid water. Nature Communications 6(1): 1–11.

Palmer, J. C., F. Martelli, Y. Liu, R. Car, A. Z. Panagiotopoulos and P. G. Debenedetti. 2014. Metastable liquid-liquid transition in a molecular model of water. Nature 510: 385–388.

Palmer, J. C., P. H. Poole, F. Sciortino and P. G. Debenedetti. 2018. Advances in computational studies of the liquid-liquid transition in water and water-like models. Chem. Rev. 118: 9129–9151.

Paolantoni, M., L. Comez, M. E. Gallina, P. Sassi, F. Scarponi, D. Fioretto et al. 2009. Light scattering spectra of water in trehalose aqueous solutions: Evidence for two different solvent relaxation processes. J. Phys. Chem. B 113: 7874–7878.

Poole, P. H., F. Sciortino, U. Essmann and H. E. Stanley. 1992. Phase behavior of metastable water. Nature 360: 324–328.

Rocchi, C., A. R. Bizzarri and S. Cannistraro. 1998. Water dynamical anomalies evidenced by molecular-dynamics simulations at the solvent-protein interface. Phys. Rev. E 57: 3315–3325.

Rosenfeld, Y. 1977. Relation between the transport coefficients and the internal entropy of simple systems. Phys. Rev. A 15: 2545.

Rosenfeld, Y. 1999. A quasi-universal scaling law for atomic transport in simple fluids. J. Phys: Condens. Matter 11: 5415.

Rosenfeld, Y. 2000. Excess-entropy and freezing-temperature scalings for transport coefficients: Self-diffusion in Yukawa systems. Phys. Rev. E 62: 7524.

Rovere, M. and P. Gallo. 2003. Strong layering effects and anomalous dynamical behaviour in confined water at low hydration. J. Phys.: Condens. Matter 15: S145.

Russo, J. and H. Tanaka. 2014. Understanding water's anomalies with locally favoured structures. Nat. Comm. 5: 3556.

Saija, F., A. M. Saitta and P. V. Giaquinta. 2003. Statistical entropy and density maximum anomaly in liquid water. J. Chem. Phys. 119: 3587.

Santucci, S., L. Comez, F. Scarponi, G. Monaco, R. Verbeni, J. F. Legrand et al. 2009. Onset of the alpha-relaxation in the glass-forming solution LiCl-6H2O revealed by Brillouin scattering techniques. J. Chem. Phys. 131: 154507.

Santucci, S., L. Comez, F. Scarponi, G. Monaco, R. Verbeni, J. F. Legrand, C. Masciovecchio, A. Gessini and D. Fioretto. 2009. Onset of the alpha-relaxation in the glass-forming solution LiCl-6H2O revealed by Brillouin scattering techniques. J. Chem. Phys. 131: 154507.

Sciortino, F., P. Gallo, P. Tartaglia and S.-H. Chen. 1996. Supercooled water and the kinetic glass transition. Phys. Rev. E 54: 6331–6343.

Smith, R. S. and B. D. Kay. 1999. The Existence of Supercooled Liquid Water at 150 K. Nature 398: 788–791.

Speedy, R. J. and C. A. Angell. 1976. Isothermal compressibility of supercooled water and evidence for a thermodynamic singularity at −45°C. J. Chem. Phys. 65: 851–858.

Speedy, R. J., P. G. Debenedetti, R. S. Smith, C. Huang and B. D. Kay. 1996. The evaporation rate, free energy, and entropy of amorphous water at 150 K. J. Chem. Phys. 105: 240–244.

Spohr, E., C. Hartnig, P. Gallo and M. Rovere. 1999. Water in porous glasses. A computer simulation study. J. Mol. Liq. 80: 165

Starr, F. W., F. Sciortino and H. E. Stanley. 1999. Dynamics of simulated water under pressure. Phys. Rev. E 60: 6757–6768.

Stern, J. N., M. Seidl-Nigsch and T. Loerting. 2019. Evidence for high-density liquid water between 0.1 and 0.3 GPa near 150 K. Proceedings of the National Academy of Sciences 116: 9191–9196.

Stillinger, F. H. 1995. A topographic view of supercooled liquids and glass formation. Science 267: 1935.

Stillinger, F. H. and P. G. Debenedetti. 2001. Supercooled liquids and the glass transition. Nature 410: 259–257.

Taschin, A., P. Bartolini, S. Fanetti, A. Lapini, M. Citroni, R. Righini, R. Bini and R. Torre. 2020. Pressure effects on water dynamics by time-resolved optical kerr effect J. Phys. Chem. Lett. 11: 3063–3068.

Torre, R., P. Bartolini and R. Righini. 2004. Structural relaxation in supercooled water by time-resolved spectroscopy. Nature 428: 296–299.

Turton, D. A., C. Corsaro, D. F. Martin, F. Mallamace and K. Wynnea. 2012. The dynamic crossover in water does not require bulk water. Phys. Chem. Chem. Phys. 14: 8067–8073.

Waluyo, I., D. Nordlund, U. Bergmann, D. Schlesinger, L. G. M. Pettersson and A. Nilsson. 2014. A different view of structure-making and structure-breaking in alkali halide aqueous solutions through x-ray absorption spectroscopy. J. Chem. Phys. 140: 244506.

Wang, Z., P. Le, K. Ito, J. B. Leão, M. Tyagi and S. H. Chen. 2015. Dynamic crossover in deeply cooled water confined in MCM-41 at 4 kbar and its relation to the liquid transition hypothesis. J. Chem. Phys. 143: 114508.

Winkel, K., E. Mayer and T. Loerting. 2011. Equilibrated high-density amorphous ice and its first-order transition to the low-density form. J. Phys. Chem. B 115: 14141–14148.

Winkel, K., M. S. Elsaesser, E. Mayer and T. Loerting. 2008. Water polyamorphism: Reversibility and (Dis)continuity. J. Chem. Phys. 128: 044510.

Woutersen, S., B. Ensing, M. Hilbers, Z. Zhao and C. A. Angell. 2018. A liquid-liquid transition in supercooled aqueous solution related to the HDA-LDA transition. Science 359: 1127–1131.

Xu, L., P. Kumar, S. V. Buldyrev, S.-H. Chen, P. H. Poole, F. Sciortino et al. 2005. Relation between the widom line and the dynamic crossover in systems with a liquid-liquid phase transition. Proc. Natl. Acad. Sci. U. S. A. 102: 16558–16562.

Xu, Y., N. G. Petrik, R. S. Smith, B. D. Kay and G. A. Kimmel. 2016. Growth rate of crystalline ice and the diffusivity of supercooled water from 126 to 262 K. Proc. Natl. Acad. Sci. USA 113: 14921.

Yang, C., V. V. Brazhkin, M. T. Dove and K. Trachenko. 2015. Frenkel line and solubility maximum in supercritical fluids. Phys. Rev. E 91: 012112.

Zanatta, M., L. Tavagnacco, E. Buratti, M. Bertoldo, F. Natali, E. Chiessi et al. 2018. Evidence of a low-temperature dynamical transition in concentrated microgels. Science advances 4: eaat5895.

Zanotti, J. M., M. C. Bellissent-Funel and S. H. Chen. 1999. Relaxational dynamics of supercooled water in porous glass. Phys. Rev. E 59: 3084.

Zhang, Y., M. Lagi, E. Fratini, P. Baglioni, E. Mamontov and S.-H. Chen. 2009. Dynamic susceptibility of supercooled water and its relation to the dynamic crossover phenomenon. Phys. Rev. 79: 040201.

Chapter 3

The Franzese-Stanley Coarse Grained Model for Hydration Water

Luis Enrique Coronas,[1] *Oriol Vilanova,*[1] *Valentino Bianco,*[2]
Francisco de los Santos[3] *and Giancarlo Franzese*[1,*]

Introduction

Water has more than 60 thermodynamic, dynamic, and structural anomalies (Chaplin 2006) whose origin is largely debated (Amann-Winkel et al. 2016, Franzese and Stanley 2010, Gallo et al. 2016, Handle et al. 2017). For example, in contrast with normal fluids, water has the property of *polyamorphism*, that is, it has at least three amorphous solids (Amann-Winkel et al. 2016, Mishima 1994, Mishima et al. 1984, 1985), whose formation depends on the preparation route (Handle et al. 2017), and a large number of (crystal) ice polymorphs: 17 have been confirmed experimentally and other are predicted computationally (Salzmann 2019). Water can be supercooled in its liquid state almost 50 degrees below its melting temperature (Kim et al. 2017). Ice has a lower density than liquid water at ambient pressure and its density decreases below 4°C. Experiments (Angell et al. 1973, 1982, Speedy and Angell 1976) have shown that water's isobaric specific heat C_P and isothermal compressibility K_T have a non-monotonic behavior with minima at approximately 35°C and 46°C, respectively, at ambient pressure, while is obaricthermal expansivity α_P turns negative at approximately 4°C. All the anomalies of water become more relevant in the supercooled region, where the fluctuations increase upon cooling (Speedy and Angell 1976, Stanley et al. 1981, Gallo et al. 2016), instead of decreasing as in normal liquids.

---

[1] Departament de Física de la Matèria Condensada & Institut de Nanociència i Nanotecnologia (IN2UB), Universitat de Barcelona, Carrer Martí i Franquès 1, 08028 Barcelona (Spain).
[2] Departamento de Química Física, Universidad Complutense de Madrid, Plaza de las Ciencias, Ciudad Universitaria, 28040 Madrid (Spain).
[3] Departamento de Electromagnetismo y Física de la Materia, Universidad de Granada, Fuentenueva s/n, 18071 Granada (Spain).
* Corresponding author: G.F.gfranzese@ub.edu

Several thermodynamic scenarios have been proposed to explain the origin of the anomalies and polyamorphism. The *singularity free* (SF) scenario hypothesizes that the uncommon volume-entropy and volume-energy anticorrelations of water, due to the water hydrogen bonds (HBs) properties, are responsible for the increase of fluctuations in the supercooled region with no singular behavior (Sastry et al. 1996, Stanley and Teixeira 1980). Three other scenarios, the *stability limit* (SL) conjecture (Speedy 1982), the *liquid-liquid critical point* (LLCP) scenario (Poole et al. 1992), and the *critical pointfree* (CPF) hypothesis (Angell 2008, Poole et al. 1994), in stead postulate a singular behavior that enhances the fluctuations at low temperature: are entrant spinodal for the first, a critical point for the second, and a first-order phase transition for the third scenario. Stokely et al. demonstrated that all the scenarios belong to the same theoretical framework and that it is possible to go from one to another by tuning a single parameter related to the water's cooperativity (Stokely et al. 2010a), as we will discuss in the following sections. Nevertheless, which of these scenarios hold for water is still a matter of debate, because, so far, no definitive experimental evidence has been found, although many recent experiments have contributed to enrich our insight (Caupin et al. 2018, Kim et al. 2017, 2018, Woutersen et al. 2018). One of the issues is that water freezes before experimental measurements are made in the region conventionally called 'no-man's land' (Handle et al. 2017, Mishima and Stanley 1998, Stanley et al. 2005), where the different scenarios predict different behaviors. Several strategies have been explored to overcome the inevitable crystallization of supercooled water, including strong confinement (Liu et al. 2007, Mallamace et al. 2007a,b) and anti-freezing solutions (Murata and Tanaka 2012, 2013, Woutersen et al. 2018) with results possibly related to the bulkcase (Leoni and Franzese 2016a, Mancinelli et al. 2009, Soper 2012, Wang et al. 2016).

In our research, we use all-atom simulations to study systems where water is at an interface, forming a monolayer or a few confined layers. Experiments (Paul 2012, Zhang et al. 2011) and simulations on confined water (Camisasca et al. 2017, Han et al. 2010, Zangi and Mark 2003) show controversial phenomena, as water under confinement can have properties significantly different from those found in bulk water. For instance, experiments on water confined between flat crystals of graphite and hexagonal boron nitride find a decay of two orders of magnitude in the dielectric constant of water as a function of the sample thickness (down to 1 nanometer) compared to the bulkcase (Fumagalli et al. 2018). The mobility of water is also strongly affected by confinement, as different regimes can be observed depending on the structure, composition, and geometry of the confining surfaces, going from subdiffusive to superdiffusive (Gallo et al. 2010, 2000, Leoni and Franzese 2016b). Moreover, confinement also modifies the phase diagram, as crystallization can be avoided, at least partially, at temperatures down to 160 K for water confined *in silica* MCM-41 nanopores (Faraone et al. 2004, Leoni and Franzese 2016a, Stefanutti et al. 2019). Confined water is also of great importance in biological systems. For example, hydration water plays an important role in protein denaturation (Bianco and Franzese 2015, Bianco et al. 2017a,b, Piana and Shaw 2018) and protein aggregation (Bianco et al. 2019a,b). Water between cell membranes is responsible for the self-assembly of phospholipids into bilayers (Nagle and Tristram-Nagle 2000, Zhao et al. 2008) and stabilizes the membrane structure (Calero and Franzese 2019, Martelli et al. 2017,

Samatas et al. 2019). Therefore, it is clear that a deeper understanding of confined water properties is essential to develop new applications in bionanotechnology.

We study hydrated systems over timescales going from 10 fs to 1 μs and length-scales from 1 Å to 100 nm, adopting atomistic water models such as SCP/E, ST2, TIP3P, TIP4P/2005, and TIP5P with numbers of water molecules upto 7040 (Calero and Franzese 2019, Calero et al. 2016, Kesselring et al. 2012, 2013, Kumar et al. 2006a, Martelli et al. 2017, Martí et al. 2017, Samatas et al. 2019). Nevertheless, these models have extremely large equilibration times at very low temperatures or in under extreme pressure (González et al. 2016, Vega et al. 2009, Yagasaki et al. 2014) and have free-energy minima that differ from those of more realistic polarizable models (Hernández-Rojas et al. 2010, James et al. 2005). We, therefore, use the all-atoms results to define coarse-grained models for water monolayers (Franzese et al. 2003, Franzese and Stanley 2002a,b, 2007, 2010, Franzese et al. 2008, 2000, Kumar et al. 2008a,b,c, Mazza et al. 2009, Stokely et al. 2010a,b) hydrating proteins or nanoparticles (Bianco and Franzese 2014, 2015, Bianco et al. 2017a, 2012b, 2013a, 2017b, 2013b, Franzese and Bianco 2013, Franzese et al. 2011, 2010, Mazza et al. 2011, 2012, Strekalova et al. 2011, 2012b). With the coarse-grained models, we can study the large-scale dynamics of hydration water at extreme conditions or the protein folding, spanning scales from 1 nm to 1 μ min space and from 10 ns to 0.1 sin time.

This chapter reviews our contribution to the ongoing debate on the origin of the water properties, adopting the Franzese-Stanley coarse-grained model for a water monolayer (Franzese and Stanley 2002a,b). The model offers a rationale for the thermodynamic and dynamic anomalies of water, relating the cooperativity of the hydrogen bond network to the occurrence of a possible liquid-liquid phase transition (LLPT) in the supercooled water region.

Computational Models: All-Atom Simulations

Given the experimental difficulties to avoid crystallization of water at extremely low temperatures, it is interesting to resort to computational models to understand if the different theoretical scenarios are, at least, thermodynamically consistent. However, modeling water is a difficult problem (Barnes et al. 1979, Finney 2001), in particular, because it is not settled how to include the quantum many-body, or cooperative, effects in water interactions. In literature, there are more than a hundred water models (Ouyang and Bettens 2015), from those parametrized based on experimental data to those fitting *ab initio* calculations, and each model is coarse-grained at a different level, from atomistic non-polarizable and rigid, to polarizable or flexible, from spatially resolved to spatially coarse-grained. For a number of these models, it is possible to explore the *no-man's land*, although it could require extremely large computational times (Kesselring et al. 2012) and elaborate analysis (Chandler 2016, Limmer and Chandler 2013, Palmer et al. 2018, 2014, 2016).

Specifically, for those models belonging to the family of ST2, TIP4P, and TIP5P potentials (Gallo et al. 2016), the LLCP hypothesis is, among the different scenarios, the one that better adjusts to the low-temperature phase diagram. In particular, we have shown that the ST2 model has a LLCP between two metastable liquid phases, low-density liquid (LDL) and high-density liquid (HDL), that belongs to the 3D Ising Universality class (Kesselring et al. 2012, Lascaris et al. 2013). To achieve this goal, we have performed all-atom simulations and a detailed finite-size scaling analysis based on an appropriate order parameter, defined as a combination of energy and density (Wilding and Binder 1996). Furthermore, we have shown that structural parameters that can quantify the amount of diamond structure in the first shell and the amount of hcp structure in the second shell (Kesselring et al. 2013) discriminate better between the different local structures of LDL and HDL.

Similar conclusions have been reached for different models, for example recently in a work by Shi et al. (2018), with results that are in principle, model-dependent. Therefore, understanding which feature of these models regulates the occurrence of anomalies of water and its peculiar properties is a task that requires a detailed analysis for each of them (Shi et al. 2018).

In the following, we will describe our approach to define a model of water that

- is suitable for large-scale, long-time simulations as needed in biologically relevant problems;
- is manageable for theoretical calculations, essential for understanding general properties;
- includes many-body interactions, necessary for a proper water model;
- equilibrates at extreme conditions.

The Coarse-grained Franzese-Stanley Water Model

In their seminal paper about computer simulations of liquid water, Barnes et al. (1979) stated that the use of pair-additive interactions for water is a "serious oversimplification" that "is necessary to abandon" because it does not account for many-body forces, to which "solution and interfacial properties of aqueous systems are particularly sensitive". For example, they reviewed quantum calculations for small water clusters, showing that HB energy in trimers and tetramers is 20–30% stronger than in dimers. More recently, James et al. (2005) and Hernández-Rojas et al. (2010) studied the configurations that minimize the energy of water clusters made of upto 21 molecules adopting non-polarizable and polarizable models. The comparison shows structural differences for clusters with more than five molecules, in particular with six or more than ten. This observation emphasizes that many-body effects in water are especially important when there are at least five molecules.

To account for these many-body effects, in 2002 Franzese and Stanley (FS), proposed a coarse-grained Hamiltonian model[1] that is analytically tractable (Franzese and Stanley 2002a,b, 2007) and is suitable for Monte Carlo (MC) calculations at constant pressure P and temperature T (de los Santos and Franzese 2009, Franzese and de los Santos 2009, Franzese et al. 2003, 2008, Kumar et al. 2008a,b,c). Thanks to a percolation mapping (Bianco and Franzese 2019) and a very efficient *cluster* MC algorithm (Franzese et al. 2010, Mazza et al. 2009, Stokely et al. 2010a), the model can be equilibrated at extreme $T < 125$ K and P ranging from negative to more

---

[1] A preliminary version of the model and its mean-field solution at zeroth order was proposed in 2000 (Fanzese et al. 2000).

than 10 GPa (Bianco et al. 2012a, 2013a, de los Santos and Franzese 2011, 2012, Mazza et al. 2012, Strekalova et al. 2011, 2012a,b). Furthermore, the coarse-graining allows the simulation of large systems, with more than 160.000 molecules (Bianco and Franzese 2014), a size that is challenging for atomistic models.

On the other hand, by adopting *local* MC dynamics (Kumar et al. 2008c) and a rescaling of time units based on experimental data, it is possible to simulate the model for times upto 100 s (Mazza et al. 2011). This timescale is out of reach for any atomistic simulation and is accessible only for coarse-grained models.

Furthermore, the use of *diffusive* MC dynamics allows evaluating transport properties (de los Santos and Franzese 2009, Franzese and de los Santos 2009). The results have offered a new interpretation of the diffusion anomaly (de los Santos and Franzese 2011, 2012, Franzese et al. 2010).

The FS model is defined as follows. For N water molecules distributed in a volume V, we partition V into N equal cells, each with volume $v \equiv V/N$ that depends on P and T. If v_0 is the van der Waals volume of a water molecule, then $v \geq v_0$, and v_0/v is the cell density when there are no HBs. For example, near the liquid-gas transition, the number of HBs is negligible and it is convenient to associate, to each cell and its corresponding molecule $i \in [i, ..., N]$, an index n_i, with $n_i = 0$ if $v_0/v \leq 0.5$, and $n_i = 1$ otherwise. Because the two-states variable n_i does not include the volume variation due to the HBs, we can consider it as a discretized *density field* due to the van der Waals interaction. However, as we will discuss in the following, the formation of HBs leads to density heterogeneities, encoded within the model in a discretized density field with many states and the same resolution as the cell's grid.

The thermodynamic average $\langle n_i \rangle$ reminds of the order parameter for the liquid-gas phase transition of the lattice-gas model. However, here is always $\langle n_i \rangle = n_i$, and the order parameter, as we will discuss, is a more complex function of the molecular configuration.

An essential feature of water is its ability to form HBs between neighboring molecules. The HB has a strong directional component due to the dipole-dipole interaction between the highly concentrated positive charge on each H and each of the two excess negative charges concentrated on the O of another water molecule. The features of the HB in water have been studied over the last decades with a variety of models and computational techniques, from *ab initio* calculations to classical molecular dynamics simulations. All these studies emphasize that, at each pressure and temperature, the HB has a strong directional (covalent) component (Galkina et al. 2017) that qualitatively can be described by the $\widehat{OOH}$ angle of the O–H···O bond (see, for example, Ceriotti et al. 2013, Schran and Marx 2019). Recent calculations suggest that $\widehat{OOH} < 24°$ with ~ 50% probability at 250 K (Schran and Marx 2019). However, consistent with Debye-Waller factors estimates (Teixeira and Bellissent-Funel 1990), previous calculations (Ceriotti et al. 2013) showed that $\widehat{OOH} < 30°$ can be associated to a HB, confirming a previous assumption (Luzar and Chandler 1996). Hence, the HB must have $-30° < \widehat{OOH} < 30°$, that is, only 1/6 of the entire range of possible values $[0, 360°]$ of the $\widehat{OOH}$ is associated with a bonded state.

In the FS model, this condition is accounted for by introducing a *bonding variable* $\sigma_{ij} = 1,..., q$ for each molecule i facing the n.n. molecule j. By definition, a HB between i facing and j is possible only if $\delta_{\sigma_{ij},\sigma_{ji}} = 1$, with $\delta_{ab} = 1$ if $a = b$, 0 otherwise, where the *bonding variables* have $q = 6$ possible states. Therefore, only 1/6 of all the possible $(\sigma_{ij}, \sigma_{ji})$, configurations correspond to a HB.

Furthermore, calculations show that the larger the O-O distance, r, between the two molecules, the smaller is the HB strength (see, for example, Huš and Urbic 2012) in such a way that it is possible to define a limiting value, r_{max}, above which the HB can be considered broken. Different authors have proposed different values for r_{max}, ranging from 3.1 Å (Schran and Marx 2019) to 3.5 Å (Luzar and Chandler 1996) in bulk water.

If we use as reasonable condition $r_{max} \simeq 3.65$ Å, considering that the van der Waals diameter of a water molecule is $r_0 \simeq 2.9$ Å, we find that, if $r > r_{max}$, then $r_0^3/r^3 \equiv v_0/v < 0.5$, that is, $n_i = 0$ in the FS model. Hence, two nearest neighbors (n.n.) molecules i and j can form a HB if $n_i n_j = 1$. Because by definition $n_i = n_j$, $\forall i \neq j$, then $n_i n_j = n_i = n_j = 1$, and n_i can be considered as a *bonding index*.

Therefore, in the FS model, the geometrical definition of the HB is granted when two n.n. molecules (1) have the facing *bonding variables* in the same state, and (2) their *bonding indices* are set to 1. As a consequence, the number of HBs is given by

$$N_{HB} = \sum_{\langle i,j \rangle} n_i n_j \, \delta_{\sigma_{ij},\sigma_{ji}}, \tag{1}$$

where the sum is performed over n.n. water molecules.

FS, following previous works (Sastry et al. 1996), make the reasonable assumption that the formation of HBs is the primary source of local density fluctuations. A water molecule fully bonded to its hydration shell, formed by other four water molecules in a tetrahedral configuration, occupies approximately the same volume of a hydrated water molecule with no HBs and a larger coordination number (Soper and Ricci 2000). Therefore, to each HB we can associate a proper volume v_{HB} given by 1/4 of the proper volume fluctuation in the hydration shell. A reasonable choice for this parameter is $v_{HB}/v_0 = 0.5$, equal to the average volume increase between (high density) ices VI and VIII and (tetrahedral, low density) ice Ih. Hence, the total volume V_{tot} occupied by the system increases linearly with the number of HBs, N_{HB}, that is,

$$V_{tot} \equiv V + N_{HB} v_{HB} \tag{2}$$

While V is the volume without HBs, used for partitioning the system, and is homogeneously distributed among the N water molecules, V_{tot} includes the local heterogeneities in the density field due to the HBs.

The FS Hamiltonian is by definition

$$\mathcal{H} \equiv \mathcal{H}_{vdW} + \mathcal{H}_{HB} + \mathcal{H}_{Coop} \tag{3}$$

where the first term accounts for the van der Waals (dispersive) attraction and hard core (electron) repulsion between water molecules and is

$$\mathcal{H}_{vdW} \equiv \sum_{i,j} U(r_{ij}) \tag{4}$$

summed over all the water molecules i and j at O-O distance r_{ij}, with

$$U(r) \equiv \begin{cases} \infty & \text{if } r \leq r_0 \\ 4\,\epsilon \left[\left(\dfrac{r_0}{r} \right)^{12} - \left(\dfrac{r_0}{r} \right)^{6} \right] & \text{if } r_0 \leq r \leq 25\,r_0 \\ 0 & \text{if } r > 25\,r_0 \end{cases} \tag{5}$$

a (double) truncated Lennard-Jones (LJ) interaction with $\epsilon \equiv 5.8$ kJ/mol, close to the estimate based on isoelectronic molecules at optimal separation $\simeq 5.5$ kJ/mol (Henry 2002).

The LJ potential is a function of the continuous variable r and it is truncated at both a large distance and at a short distance for numerical efficiency. Our previous analysis (de los Santos and Franzese 2011) shows that both truncations, introduced to simplify the implementation of the model, do not affect the results. This interaction regulates the liquid-gas phase transition and dominates above the liquid-gas spinodal temperature.

Because the formation of HBs does not affect the distance r between a molecule and those in its hydration shell (Soper and Ricci 2000), the van der Waals interaction is not affected by the HBs. This observation is crucial to state that the FS model is not mean field.[2]

The second term is proportional to N_{HB} and accounts for the additive (two-bodies) component of the HB,

$$\mathcal{H}_{HB} = -JN_{HB} \tag{6}$$

with an energy decrease for each HB given by $J = 0.5 \times 4\,\epsilon \simeq 11$ kJ/mol, close to the estimate from the optimal HB energy and a HB cluster analysis (Stokely et al. 2010a).

Each new HB leads to an entropy decrease equal to $-k_B \ln 6$ (k_B is the Boltzmann constant), due to the selection of one of the six possible states for the bonding variable. This entropy loss for increasing N_{HB} is consistent with the anti correlation between entropy and volume that, for the Clausius-Clapeyron relation, is responsible for the negative slope of the melting line of water near ambient pressure. The total number of accessible states for a free water molecule with four bonding variables is $q^4 = 1296$, for $q = 6$, and for a system of N molecules at high temperature is 1296^N.

The third term of the Hamiltonian accounts for the HB cooperativity due to many-body correlations (Barnes et al. 1979). The many-body interactions lead to the local order in the hydration shell (Soper and Ricci 2000) and are a consequence of the quantum nature of the HB (Barnes et al. 1979, Hernández de la Peña and Kusalik 2005, Ludwig 2001). This term is modeled in classical atomistic potentials with a long-range dipolar interaction, as for example in Dang and Chang (1997). In

---

[2] We acknowledge the late Professor David Chandler for discussing this point.

the FS Hamiltonian, it is modeled as a five-bodies term in which each molecule i is interacting with its first coordination shell

$$\mathcal{H}_{Coop} = -J_\sigma \sum_i n_i \sum_{(k,l)_i} \delta_{\sigma_{ik},\sigma_{il}} \tag{7}$$

where $(l,k)_i$ indicates each of the six different pairs of the four variables σ_{ij} of the molecule i and J_σ is the extra energy-gain provided by each cooperative HB. A direct experimental evaluation of such a term is not available, however it can be estimated by attributing to it (Heggie et al. 1996) the 3 kJ/mol increase in strength of the HBs in ice Ih with respect to liquid water (Eisenberg and Kauzmann 1969). This value is consistent by order of magnitude with the 6 kJ/mol estimated for D_2O (Shimoaka et al. 2012), for which the effect is expected to be stronger. Considering that each variable σ_{ij} appears in three terms in the sum of Eq. (7) at fixed index i, we can estimate the value of J_σ to be ~ 1.0 kJ/mol. Because $J_\sigma \ll J$, this term is relevant only when $N_{HB} \gg 1$. This asymmetry between the two components of the HB interaction is necessary for the model to represent water.

In the *NPT* ensemble, the partition function of the system is

$$Z(T, P) \equiv \sum_{\{\sigma\}\{v\}} e^{-H/k_B T} \tag{8}$$

where the sum is over all the possible configurations of bonding variables $\{\sigma\}$ and cell volumes $\{v\}$, and

$$H \equiv \mathcal{H} + P \, V_{tot} \equiv \mathcal{H}_{vdW} - J_{eff} \sum_{\langle i,j \rangle} n_i n_j \delta_{\sigma_{ij},\sigma_{ji}} + \mathcal{H}_{Coop} + P \, Nv \tag{9}$$

is the enthalpy, with $J_{eff} \equiv J - P \, v_{HB}$, the effective interaction between σ-variables of n.n. molecules, that depends on P.

Eq. (8) can be rewritten as

$$Z(T, P) = \sum_{\{v\}} e^{-[U(v) + PNv]/k_B T} \times Z\_\{\sigma\} \tag{10}$$

where

$$Z_{\{\sigma\}} \equiv \sum_{\{\sigma\}} e^{(J_{eff}/k_B T)\sum_{\langle i,j\rangle} n_i n_j \delta_{\sigma_{ij},\sigma_{ji}}} \times e^{\left(\frac{J_\sigma}{k_B T}\right)\sum_i n_i \sum_{(k,l)} \delta_{\sigma_{ik},\sigma_{il}}}$$

$$= \sum_{\{\sigma\}} \prod_{\langle i,j\rangle} \left[1 + \left(e^{\left(\frac{J_{eff}}{k_B T}\right)} - 1 \right) n_i n_j \delta_{\sigma_{ij},\sigma_{ji}} \right] \tag{11}$$

$$\times \prod_{i=1}^{N} \prod_{(k,l)_i} \left[1 + \left(e^{\left(\frac{J_\sigma}{k_B T}\right)} - 1 \right) n_i \delta_{\sigma_{ik},\sigma_{il}} \right]$$

Here $\prod_{\langle i,j\rangle}$ runs over all the n.n. molecules, $\prod_{i=1}^{N}$ runs over all molecules, and $\prod_{(k,l)_i}$ extends over all the six pairs of the bonding variables of a specific molecule i.

Due to its simplicity, it is possible to perform a thorough analysis of how the macroscopic properties of the model depend on its limited number of parameters, each describing a molecular mechanism, both with theoretical calculations with the *cavity method* and with MC simulations (Bianco et al. 2013a, Franzese et al. 2010, Franzese and Stanley 2002a,b, 2007, Franzese et al. 2008, Kumar et al. 2008c, Mazza et al. 2011, 2012, 2009, Stanley et al. 2010, 2011, 2009, Stokely et al. 2010a,b). The model, initially defined for monolayers with height $h = 0.5$ nm and $v \equiv hr^2$, has recently been extended to bulkwater (Coronas et al. 2016).

Dynamic Behavior of the FS Water Model

Two Dynamic Crossovers in FS Water

Data on protein hydration water and confined waters how the presence of a dynamical crossover from a non-Arrhenius to an Arrhenius regime (Franzese et al. 2008) at ~ 220 K. For example, this crossover is found in the translational correlation time of water hydrating lysozyme proteins (Chen et al. 2006b) or in the structural relaxation time of water confined *in silica* pores (Faraone et al. 2004, Liu et al. 2005). Also, simulations of the TIP5P water models how a dynamic crossover from a non-Arrhenius to an Arrhenius regime in the diffusivity of water hydrating lysozyme and DNA (Kumar et al. 2006b). As this crossover takes place at much higher temperatures than T_G, the so-called *glass transition temperature*, Kumar et al. discard any relation with the glass state. According to this numerical and experimental evidence, a possible hypothesis, among others (Cerveny et al. 2006, Swenson 2006), for the origin of this crossover is the local rearrangement of the HB network at low temperatures (Chen et al. 2006, Kumar et al. 2008c).

In particular, MC simulations of the FS model display a dynamic crossover from a non-Arrhenius to an Arrhenius regime, which is a consequence of the local rearrangement of the water HBs (Kumar et al. 2008b,c). Kumar et al. by mean-field calculations and MC simulations of the FS water monolayer, estimate the orientational correlation time, that is, the relaxation time τ of the auto correlation $C_S(t) \equiv < S_i(t)S_i(0) >/(S_i^2)$, where $S_i \equiv \sum_j \sigma_{ij}/4$, for a FS monolayer. The physical meaning of S_i is the total bond ordering of the i-th water molecule, and τ is defined as the time at which $C_S(t)$ decays by a factor $1/e$. Kumar et al. find a non-Arrhenius regime at high-T upon cooling at constant pressure, where τ can be fitted with the Vogel-Fulcher-Tamman (VTF) function

$$\tau^{\mathrm{VTF}} \equiv \tau_0^{\mathrm{VTF}} \exp\left[\frac{T_1}{T - T_0}\right] \tag{12}$$

with τ_0^{VTF}, T_0, and T_1 fitting parameters. At low-T, τ displays Arrhenius behavior, $\tau = \tau_0 \exp[E_A/k_B T]$, where τ_0 is the limiting time at high-T, and E_A is the T-independent activation energy. The crossover occurs at the same temperature where the specific heat displays a maximum. The authors discuss that they interpret this dynamic crossover as an effect of the breaking and reorientation of HBs, leading to a more tetrahedrally structure for the HB network at low-T.

In Kumar et al. (2008c), the authors also investigate how the cooperative term affects the dynamics. At constant $J/\epsilon = 0.5$, they compare the results for $J_\sigma/\epsilon = 0.05$ and $J_\sigma = 0$, corresponding to the LLCP scenario and the SF scenarios, respectively, as we will discuss in the following. In both scenarios, they find that (i) the crossover time τ_C is approximately P-independent, (ii) the Arrhenius activation energy $E_A(P)$ decreases upon increasing P and, (iii) the temperature $T_A(P)$, at which τ reaches a fixed macroscopic time $\tau_A \geq \tau_C$, decreases upon increasing P. Furthermore, they show that (iv) $E_A/(k_B T_A)$ increases upon pressurization in the LLCP scenario, but it remains constant in the SF scenario.

These new predictions have been tested in experiments in a protein hydration layer (Chu et al. 2009, Franzese et al. 2008). In particular, quasi-elastic neutron scattering (QENS) (Chu et al. 2009, Franzese et al. 2008) verifies that the predictions (i)–(iii) are correct for a water monolayer hydrating lysozyme. Nevertheless, the resolution of these experiments does not allow to finalize, on the basis of the prediction (iv), which among the LLCP and SF scenarios is satisfied by protein hydration water (Franzese et al. 2008).

Further research on water monolayers at lower T, surprisingly, has shown the presence of no tone, but two dynamic crossovers (Mazza et al. 2011). Mazza et al. measured the dielectric relaxation time of water protons τ_{WP}, as it is sensitive to breaking and formation of HBs (Peyrard 2001), in a water monolayer hydrating lysozyme protein, and compare it with the MC simulation of a FS water monolayer. In both cases, the authors found two crossovers at ambient pressure. The first was at $T \sim 252$ K, as previously reported by Kumar and coworkers (2008b). The second is at $T \sim 181$ K.

They found that the crossover at $T \sim 252$ K was between two VTF behaviors. They interpreted this result as a change in the diffusion of water protons, between a high-T diffusive regime and a low-T sub-diffusive regime.

The crossover at $T \sim 181$ K is between a VTF and an Arrhenius regime, corresponding to the rearrangement of the HB network structure. The experimental measurements of τ_{WP} compare well with FS water calculations of the S_i relaxation time from MC simulations τ_{MC}. Mazza and coworkers consider the auto correlation function

$$C_M(t) \equiv \frac{1}{N}\sum_i \frac{\langle S_i(t_0 + t)S_i(t_0)\rangle - \langle S_i\rangle^2}{\langle S_i^2\rangle - \langle S_i\rangle^2} \tag{13}$$

which decays to 0 as $t \to \infty$. By definition, $C_M(0) = 1$. Following (Kumar et al. 2008c), τ_{MC} is by definition such that $C_M(\tau_{MC}) = 1/e$. They find two dynamic crossovers and relate them to the presence of two specific heat maxima: the high-T weak maximum and the low-T strong maximum. At high-T, the C_p weak maximum occurs when the fluctuations of N_{HB} are maximum. At low-T, the C_p strong maximum is due to the maximum in the fluctuations of the cooperative term (Eq. 7) of the Hamiltonian. Hence, the high-T crossover is associated with the formation of the HB network. Instead, the low-T crossover is due to the rearrangement of the HBs in an ordered structure in the monolayer.

Effect of Pressure and Temperature on the Dynamics

Protein hydration water undergoes a liquid-glass transition (LGT) with the glass state characterized by a huge increment of viscosity and the freezing of long-range translational diffusion (Doster 2010). The macroscopic structural arrest of the glass state emerges from the slowing down of its nearest neighbor HBs' dynamics. Doster and coworkers performed neutron scattering experiments on myoglobin at hydration level $h = 0.35$ g_{H_2O}/g, as this technique allows for monitoring displacements at the microscopic scale. They measured the incoherent intermediate scattering functions $I(q,t)$ at $T = 320$ K (above the LTG but relevant for translational diffusion) and for scattering vector 0.4 Å^{-1} ≤ 2 (Settles and Doster 1996). They found a two-step time-decay in $I(q, t)$ that, at high-q and long times t, can be fitted to a stretched exponential

$$C(t) = C_0 \exp[-(t/\tau_0)^\beta] \tag{14}$$

where τ_0 is the correlation time, $0 < \beta \leq 1$ is the stretched exponent, and C_0 is a normalization factor. For large q, their results show that $0.3 \leq \beta \leq 0.4$. When they measure $I(q, t)$ at constant q = 1.8 Å^{-1} and $180 \leq T/K \leq 320$ as the temperature decreases, the two-step decay turns into a plateau, which leads to a relaxation time that exceeds the observation time (Doster 2010).

Calculations from MC simulations of the FS model are consistent with these two experimental results. In particular, to study the microscopic origins of the complex dynamics of water on low-hydrated proteins, de los Santos and Franzese consider a monolayer of FS water adsorbed on a generic inert substrate at 75% hydration (de los Santos and Franzese 2011, 2012, Franzese and de los Santos 2009). At such a hydration level, adsorbed water molecules are restricted to diffuse on a surface geometry with an upto four coordination number. They calculate $C_M(t)$ at different T and P (Franzese and de los Santos 2009) finding that at high pressure, $P \geq 1$ ϵ/v_0, the correlation function decays exponentially for any T allowing the system to equilibrate easily. At these pressures, the HB network is inhibited, inducing rapid dewetting and large dry cavities with decreasing temperature.

At lower pressure, $P = 0.7$ ϵ/v_0, and low T, the behavior of $C_M(t)$ can be fitted with a stretched exponential function, with no strong increase of the correlation time as T decreases (Fig. 1). The authors associate this behavior with (i) the rapid ordering of the HBs that generates heterogeneities and (ii) with the lack of a single timescale due to the vicinity of the liquid-liquid critical point, as we will discuss in the next section.

At even lower pressures, the gradual formation of the HB network, starting at higher T, is responsible for the dynamics lowing down as T decreases and for the dynamical arrest at $(P,T) = (0.1\epsilon/v_0, 0.05\epsilon/k_B)$, with an increase in τ_0 of more than four orders of magnitude, as in glass. Under these conditions, the dewetting process is strikingly different, with the formation of many small cavities.

Comparison between the FS results and experiments (Doster 2010, Settles and Doster 1996) show that the complex dynamic behavior of protein hydration water at low hydration can be well reproduced by solely taking into account the dynamics of the HB network. The LGT emerges as a consequence of the slowing of the HBs dynamics by decreasing T. At extremely low T, this results in a dynamic arrest of the system.

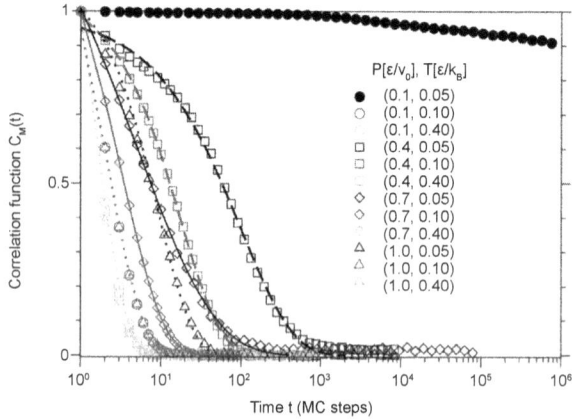

Fig. 1: The correlation function $C_M(t)$ for pressures $P \le 1.0$ ϵ/v_0 and temperatures $T \le 0.4$ ϵ/k_B. $C_M(t)$ decays exponentially (dotted lines)at $P = 1.0$ ϵ/v_0 for all T (triangles) and at $P < 1.0$ ϵ/v_0 for any $T \ge 0.4$ ϵ/k_B (not shown). For $T \le 0.1$ ϵ/k_B and $P = 0.7$ ϵ/v_0 (diamonds, continuous line) or $P = 0.4$ ϵ/v_0 (squares, dashed line), $C_M(t)$ can be described by a stretched exponential, with an exponent β that decreases as T is lowered. At low pressure, $P = 0.1$ ϵ/v_0 (circles), $C_M(T)$ is exponential for $T \ge 0.1$ ϵ/k_B, and non-exponential at $T = 0.05$ ϵ/k_B (solid circle). (Figure reprinted from (Franzese and de los Santos 2009) "Dynamically slow processes in supercooled water confined between hydrophobic plates". Copyright (2009) IOP Publishing. Reproduced with permission. All rights reserved.)

The Diffusion Anomaly

Experiments and simulations of the diffusion of confined water show controversial results. For water confined in NaX and NaA zeolites and for T between 310 K and 260 K, experiments observe a reduction of two orders of magnitude of the translational diffusion coefficient D respect to the bulkcase (Kamitakahara and Wada 2008). Other results show that D decreases upon increasing the confinement in either hydrophilic (Takahara et al. 1999) or hydrophobic conditions (Naguib et al. 2004, Takahara et al. 1999). However, for confined water in carbon nanotubes with a diameter smaller than 2 nm, the experiments find that the transport of mass is extremely fast (Holt et al. 2006). Furthermore, water confined in smooth graphene capillaries shows a fast flow (of ~ 1 m/s) that is enhanced if the height of the channel can accommodate only a few water layers (Radha et al. 2016). The authors associate the fast flow to the great capillary pressure and relate its enhancement to the increased structural order of nanoconfined water.

Several models, for example (Netz et al. 2001, Szortyka and Barbosa 2007), can reproduce numerically the diffusion a nomaly, but they display a variety of different results still controversial. Classical molecular dynamics simulations of SPC/E water in hydrophilic MCM-41or Vycor show that the mobility of water molecules decreases as the hydration level is lowered (Gallo et al. 2010). This is interpreted as due to the greater proportion of molecules bonded to the surface at low hydration.

A diffusion decrease, by two orders of magnitude compared to the bulk, is found in simulations of TIP5P water between hydrophobic, smooth, planar plates. In particular, the diffusion is anomalous in the direction parallel to the walls, while it is not for the orthogonal direction (Han et al. 2008, Kumar et al. 2005). However,

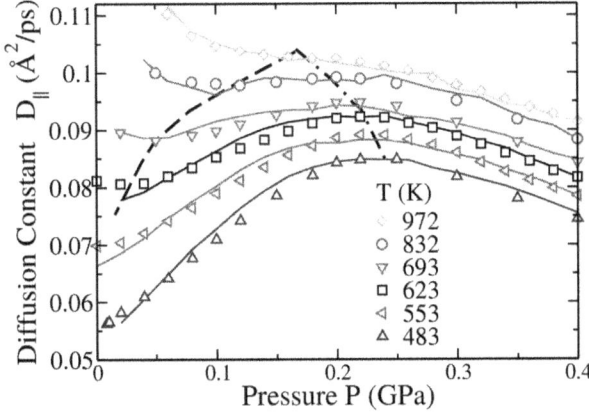

Fig. 2: Diffusion coefficient $D_\parallel$ from MC simulations (symbols) as a function of pressure along isotherms. For $T < 972$ K, $D_\parallel$ has maxima (dotted-dashed line) and minima (dashed line). Solid lines are from $W_{v,\mu}$ calculations (defined in the text). (Reprinted figure with permission from (de los Santos and Franzese 2012). Copyright (2012) by the American Physical Society.)

first-principle molecular dynamics simulations of SPC/E water confined in graphene sheets and carbon nanotubes show a faster diffusion compared to the bulkcase (Cicero et al. 2008), possibly due to weaker H-bonding at the interface.

To shed light on this controversy, de los Santos and Franzese perform MC simulations of a water monolayer confined in a smooth slit pore at partial hydration. Their results describe the origins of the diffusion anomaly (de los Santos and Franzese 2011, 2012) in terms of *cooperative rearranging regions* (CRR) of water. They calculate the diffusion coefficient $D_\parallel$ parallel to the walls using Einstein's formula

$$D_\parallel = \lim_{t \to \infty} \frac{\langle |\vec{r}_i(t + t_0) - \vec{r}_i(t)|^2 \rangle}{4t} \tag{15}$$

where $\vec{r}_i$ is the project ion of the position of molecule i on to the plates. The average $<\cdot>$ is calculated over all the molecules and different times t_0. The analysis of $D_\parallel$ shows the presence of maxima and minima along isotherms at high temperature (Fig. 2).

The authors describe the anomaly in terms of the joint probability

$$W_{v,\mu}(P, T) \equiv \mathcal{P}_F^v \mu \mathcal{P}_b \frac{1}{Z} \exp[-H(T,P)/k_B T] \tag{16}$$

of finding v molecules with a free cell available for diffusion with in a region with a number $\mu \mathcal{P}_b$ of HBs, where Z is the partition function, $\mathcal{P}_F$ and $\mathcal{P}_b$ are the probability for each cell to have a free n.n., and the probability for each HB to be formed, respectively, with $\mathcal{P}_F \equiv <n_F>/4$, $\mathcal{P}_b \equiv <n_{HB}>/4$, $<n_F>$ the average number of free n.n. cells per molecule and $<n_{HB}>$ the average number of HBs permolecule.

The authors find that $D_\parallel$ is proportional to $W_{v,\mu}$, implying that the diffusion is dominated by the cooperativity of water (Fig. 3 main panel). The resulting value for $v = 12.5 \pm 0.5$ suggests that diffusion requires a CRR that reaches ~ 3.5 molecules (Fig. 3 lower inset). Thus, the FS model clarifies that the diffusion anomaly, at

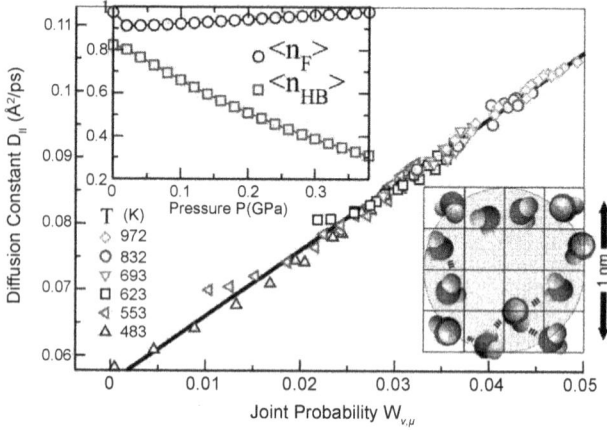

Fig. 3: Upper inset: Average number offreen.n.cells around a molecule $< n_F >$ and average number of HBs formed by a molecule $< n_{HB} >$ at $T = 693$ K. There is a discontinuity in $< n_F >$ at low P corresponding to the gas-liquid phase transition. At high P they are both monotonic. Main panel: Linear dependency of $D_\parallel$ vs. $W_{v,\mu}$ along the isotherms represented in Fig. 2. Lower inset: Example of a CRR (shaded) of about 1 nm size, with $v = 12$ molecules and $\mu < n_{HB} >/4 = 5$ HBs. (Reprinted figure with permission from (de los Santos and Franzese 2012). Copyright (2012) by the American Physical Society.)

constant T by increasing P, originates from the competition between, on the one hand, the increase of free volume $< n_F >$ and the decrease of the energy cost for a molecule to move due to the reduction of $< n_{HB} >$ (Fig. 3 upper inset), and, on the other hand, the decrease of free volume due to the increase of density. The diffusion is favored by pressurizing at constant T until the HB formation is unfavorable for enthalpic reasons, giving origin to the $D_\parallel$ maxima. The diffusion coefficient $D_\parallel$ correlates to the phase diagram of FS water (Fig. 4). The loci of D^{max} and D^{min} along isotherms lie between the temperature of maximum density (TMD) line and the liquid-gas spinodal. The constant $D_\parallel$ lines resemble the melting line of bulk water. In the deeply supercooled region, there is a sub diffusive regime due to the increment of the relaxation time of the HB network, which at low P and T leads to the dynamical arrest and to the amorphous glassy water (Handle et al. 2017).

FS-water Phase Diagram

The phase diagram of the FS model has been extensively studied in the case of a water monolayer by analytic (Franzese and Stanley 2002a,b, Franzese et al. 2000) and numerical methods (Franzese et al. 2003, de los Santos and Franzese 2009). By changing the model's parameters, Stokely et al. (2010a) reveal the relations among the different scenarios for the anomalies of water.

The T-P plane (Fig. 5) has several features (Bianco and Franzese 2014).

- The isobaric TMD line
    - ○ (i) at high P, has a negative slope, as in the experiments, while
    - ○ (ii) at negative P, it has a positive slope and asymptotically approaches the liquid-gas (LG) spinodal.

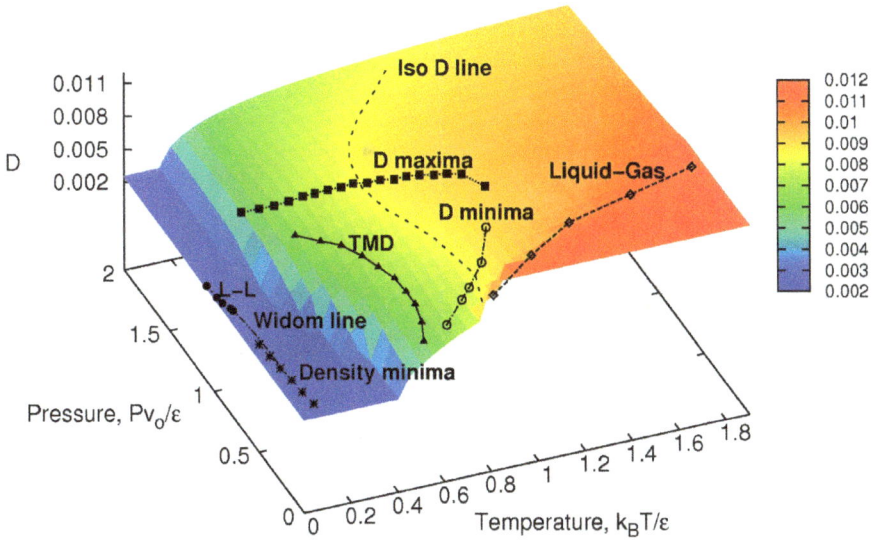

Fig. 4: Phase diagram of a water monolayer nanoconfined between hydrophobic plates. The Z-axis and color scale represent the diffusion constant $D_\parallel$ for values $0.12 \geq D_\parallel$ (Å/ps^2) ≥ 0.03. At high T and low P there is the liquid-gas phase transition line (open diamonds) ending in the liquid-gas critical point (diamond symbol at highest T), as discussed in Section "FS-water Phase Diagram". The loci of isothermal $D_\parallel$ extrema, D^{max} (solids quares), and minima, D^{min} (open circles), envelope the TMD line (solid triangles). Loci at constant $D_\parallel$ (e.g., the dashed line marked as "Iso-D") resemble in their reentrant behavior the water melting line. (Reprinted with permission from (de los Santos and Franzese 2011). Copyright (2011) American Chemical Society.)

- o (iii) At the turning point,it crosses the line of minima of the isothermal compressibility along isobars, $K_T^{min}(T)$ (Fig. 5a).

- o (iv) It avoids crossing the LG spinodal line at low T and turns into an isobaric line of the temperature of minimum density with a negative slope, as suggested in the experiments by Mallamace et al. (2007a).

- o (v) Its low-T turning point occurs where the line of (weak) minima of α_p along isobars, $\alpha_p^{wmin}(T)$, crosses it (Fig. 5c).

- o (vi) This point is also where the line of specific heat (weak) maxima along isotherms, $C_P^{wMax}(P)$, is aiming for (Fig. 5b).

- o (vii) The locus of (weak) maxima of compressibility along isobars, $K_T^{wMax}(T)$ turns, at its minimum P, into the locus of minima of K_T along isobars, $K_T^{min}(T)$ (Fig. 5a).

- o (viii) The resulting line of extrema of isobaric K_T coincides within the error bar (not shown) with the locus of (weak) minima of α_p along isotherms, $\alpha_p^{wmin}(P)$ (Fig. 5c).

All the findings (i)–(viii) are consistent with thermodynamic relations and atomistic models (Poole et al. 2005), confirming the correctness of the numerical calculations performed in the study by Bianco and Franzese (2014).

Thanks to its mapping onto a percolation formulation (Bianco and Franzese 2019), the FS water can be easily simulated adopting a fast (cluster) MC algorithms

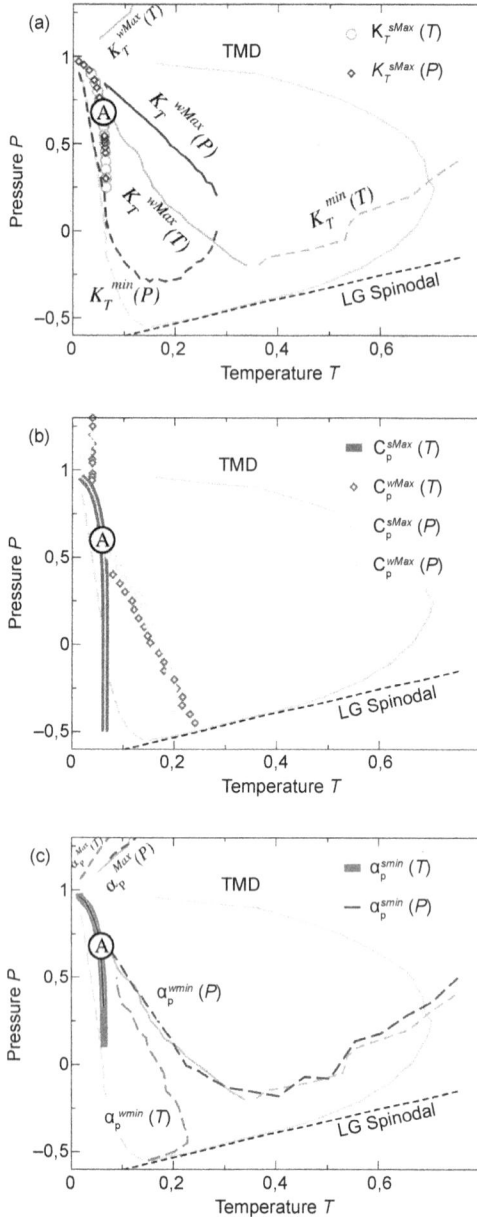

Fig. 5: Phase diagram of a water monolayer. In all panels, the black dashed line represents the LG spinodal, while all the colored dashed lines turning into solid lines of the same color represent loci of minima and maxima, respectively, of different quantities: for example, in green the temperature of minimum density and the TMD. Each panel focuses on one response function: (a) K_T along isobars (isotherms) in orange (blue), with dashed (continuous) lines for weak minima (maxima) and symbols for strong maxima; (b) C_P along isobars (isotherms) in red (turquoise), with lines (symbols) for strong (weak) maxima; (c) α_p along isobars (isotherms) in magenta (violet), with dashed (continuous) lines for weak (strong) minima. All the loci of extrema of response functions converge to ward the large circle with label A corresponding to the LLCP. (Reprinted with permission from (Bianco and Franzese 2014.))

(Mazza et al. 2009) that allows us to explore in detail the deeply supercooled states (Franzese and Stanley 2007) and the high-pressure region (Franzese et al. 2010, 2008, Stokely et al. 2010b). In particular, in the deeply supercooled region, Mazza et al. (2011, 2012) and Bianco et al. (Bianco and Franzese 2014, Bianco et al. 2013a) reveal the presence of strong extrema, for the response functions C_P, K_T, and α_P, occurring at temperatures lower than those for the weak extrema (Fig. 5). While the weak extrema occur both in the FS model and in atomistic water models, the strong extrema have been found only in the FS model for the monolayer. Their presence in bulk models, both FS and atomistic, for water is still under investigation. It is important to observe here that the FS model predicts the strong extrema in a region in which the atomistic models freeze into a glassy state (Kesselring et al. 2012, 2013, Palmer et al. 2014).

Mazza and coauthors (2011) show that the two (weak and strong) maxima of the response functions correspond to the two dynamic crossovers, discussed in Section 4.1, found in experiments and the FS model for a hydration monolayer. In this way, they establish a connection between thermodynamics and dynamics that extends up to moderate supercooling at low P.

As initially observed by Franzese and coworkers (Bianco and Franzese 2014, Franzese and Stanley 2007, Franzese et al. 2008, Kumar et al. 2008b,c, Mazza et al. 2012, Stokely et al. 2010b), all the loci of the extrema of response functions converge toward a point (large circle with label A in Fig. 5). At this point A, the extrema reach their maximum values, as expected at a critical point in a finite-size system. This point A corresponds to the LLCP, as shown by analyzing the fluctuations around it (Mazza et al. 2011, Mazza et al. 2012). In particular, as described with more detail in the next section, a finite-size scaling analysis shows that the LLCP in the FS monolayer belongs to the 2D Ising Universality class (Bianco and Franzese 2014).

This result is consistent with those for ST2 water (Kesselring et al. 2012, Palmer et al. 2014, Smallenburg and Sciortino 2015), TIP4P/2005 water (Abascal and Vega 2010, González et al. 2016), but not with those for the three-body interactions mW model (Limmer and Chandler 2013). Works by Anisimov and coauthors show that, in the case of the mW model, the mixed-field order parameter is entropy-driven, instead of energy-driven as in the models with the LLCP, being not strong enough to induce the liquid-liquid phase separation (Holten et al. 2013, Singh et al. 2016).

Extrapolations from the experiments show large fluctuations for the response functions (Fuentevilla and Anisimov 2006, Holten and Anisimov 2012, Kim et al. 2017) as calculated in the FS-water, at variance with the atomistic models. This significant difference could be a consequence of the explicit many-body interactions included in the FS model, but not in the (non-polarizable) atomistic models. Further investigation regarding this difference is necessary.

The FS water monolayer presents the first-order LLPT, between two liquids with different energies and structures, at pressures above (and T below) the LLCP. Other (atomistic) models adopting a two-state description of bulk water show the same property (Gallo et al. 2016, Holten et al. 2013, 2014, Russo and Tanaka 2014, Shi et al. 2018, Singh et al. 2016). In the FS model, the phase transition occurs between the HDL, with a disordered HB network and high energy, and the LDL, with an ordered HB network and low energy (Franzese et al. 2003).

The Critical Point Analysis

Bianco and Franzese study the universality class of the LLCP (Bianco and Franzese 2014). To define the correct order parameter M for the LLCP, they first analyze the free-energy landscape. The LLCP in a finite-size system, and the landscape has two

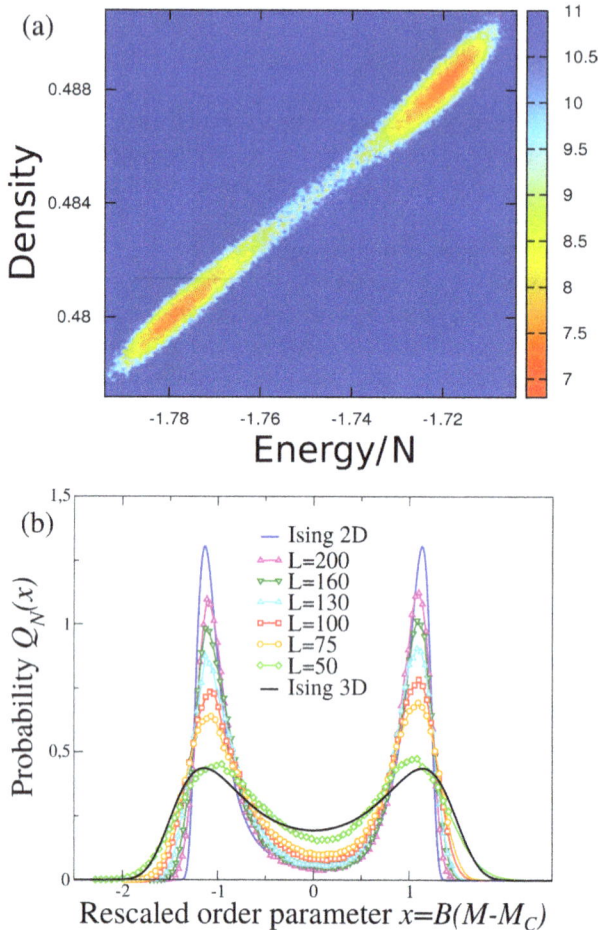

Fig. 6: The free-energy landscape in the vicinity of the LLCP. (a) The Gibbs free energy, G in units of $k_B T$, is represented as a heat-map in the energy-density plane, showing two basins separated by a barrier $\Delta G \sim k_B T$, as expected near a critical point. The phase transition is described by the order parameter M given by the reaction coordinate (an imaginary straight line in the plane, not shown) between the two minima. (Reprinted with permission from Springer Nature Customer Service Centre GmbH: Springer Food Biophys (Franzese and Bianco 2013), Copyright (2013)). (b) At the LLCP, as the size L of the system increases, the M probability distribution, Q_N, approximates the 2D-Ising universality-class (blue solid line). Here, B is a scaling factor, and the black reference distribution holds for the 2D-Ising model. (Reprinted with permission from (Bianco and Franzese 2014.))

equivalent minima, each with a basin of attraction, with different energy and density, corresponding to two coexisting HDL and LDL phases (Fig. 6a). They found that a free energy barrier $\sim k_B T$ separates the two basins. Hence, as expected near a critical point,the system has enough thermal energy to cross the barrier between the two basins.

The order parameter M is, by definition, the reaction coordinate between the two minima, that is, the linear combination of energy and density $M \equiv \rho + sE$, as in the mixed-field approach (Bruce and Wilding 1992, Franzese and Bianco 2013, Wilding 1995). M is related to the diagonal connecting the centers of the two wells (Fig. 6a), as its probability distribution must be symmetric at the LLCP. By performing a finite-size scaling analysis at fixed sample thickness $h = 0.5$ nm, and by varying the number of water molecules from 2500 to 40000 at fixed density, Bianco and Franzese find that the FS monolayer in the thermodynamic limit displays a LLCP belonging to the 2D-Ising universality-class (Fig. 6b).

The probability distribution of M shows a crossover from 3D to 2D-Ising Universality class as the number of water molecules in the monolayer increases (Fig. 6.b). The system crosses from 3D to 2D behavior for $L/h > 50$, where L is the system lateral size, while for normal liquids this crossover takes place for $L/h > 5$ (Liu et al. 2010). This can be interpreted as a consequence of the high cooperativity, and the low coordination number, of the water molecules. For small L, the strong cooperativity at the LLCP increases the HBs fluctuations, resulting in a (3D-like) probability distribution for M broader than the one in 2D. The water coordination number–four both in 3D and 2D–emphasizes the effect because it reduces the fluctuations differences between the two cases when the system is small (Bianco and Franzese 2014, Franzese and Bianco 2013).

The Widom Line

The Widom line (WL) is defined as the locus of maxima of the statistical correlation length ξ, emanated from the critical point as the analytic continuation of the first-order transition line, and spanning into the supercritical region (Franzese and Stanley 2007, Holten et al. 2012, Kumar et al. 2007). The expression "Widom line" was first used by Stanley and coworkers in 2005 as the locus where the lines of the maxima for different response functions asymptotically converge approaching the critical point from the supercritical region (Xu et al. 2005b, Kumar et al. 2007, Franzese and Stanley 2007).

Close to the critical point, the thermodynamic response functions can be expressed as power-law functions of ξ. Since ξ by definition is maximum along the WL, the loci of maxima of thermodynamic response functions converge toward the WL on approaching the critical point and, near (T_c, P_c), they are often used as *proxies* for the Widom line (Abascal and Vega 2010, Simeoni et al. 2010).

Following the work of Bianco and Franzese (2014), it is possible to calculate the correlation length ξ using the spatial correlation function $G(r)$, that, within the FS model, is defined as

$$G(r) \equiv \frac{1}{4N} \sum_{|(\vec{r}_i) - (\vec{r}_l)| = r} [\langle \sigma_{ij}(\vec{r}_i) \sigma_{lk}(\vec{r}_l) \rangle - \langle \sigma_{ij} \rangle^2] \tag{17}$$

Far from the critical region, $G(r)$ decays exponentially, and ξ is by definition the characteristic length of the decay, $G(r) \sim e^{-r/\xi}$. By approaching the critical point, the correlation function can be written as $G(r) \sim e^{-r/\xi}/r^{d-2+\eta}$, being d the system dimension and η a (critical) positive exponent (Stanley 1971).

In the FS model for a water monolayer, the correlation length can be calculated along isobars down to deeply supercooled temperatures, showing that it exhibits a maximum, ξ^{Max} at all the P explored. The $T - P$ locus of ξ^{Max} overlaps (i) with the LLPT at high P and (ii) with the locus of strong maxima of the specific heat at lower P (Mazza et al. 2011, 2012). This line, at low P, identifies the WL (Mazza et al. 2012, Bianco and Franzese 2014), with a large slope in the T–P plane, consistent with extrapolations from experiments (Fuentevilla and Anisimov 2006, Taschin et al. 2013). This finding, however, is at variance with previous works identifying the WL line in supercooled water with the higher–T weak maxima of the response function (Xu et al. 2005a, Kumar et al. 2007, Franzese and Stanley 2007, Franzese et al. 2008, Mallamace et al. 2008, Abascal and Vega 2010).

Nevertheless, the FS water monolayer is the only model with the LLCP where ξ-maxima have been found. Furthermore, it is the only model that can be equilibrated so deeply into the supercooled region. Therefore, the prediction of a WL below the temperatures accessible to other models or experiments is consistent with the experimental observation of an increasing isobaric ξ for decreasing T but no maxima (Huang et al. 2010).

In a recent work, Bianco and Franzese use a percolation approach (Coniglio and Klein 1980, Kasteleyn and Fortuin 1969) to identify the regions (clusters) of statistically correlated water molecules (Bianco and Franzese 2019). According to this mapping, the FS model under goes a percolation transition along a T–P locus that is numerically consistent with the line of extrema of ξ^{Max}. A detailed cluster analysis reveals that at high P, along the LLPT line, the percolation transition is due to the building up of the HB network. At lower P, the origin of the percolation transition is related to the local tetrahedral reordering of the water molecules.

This percolation approach allows us to compute the connectivity length ξ_C of clusters of molecules, which, by definition, diverges along the percolation line at any P. Hence, as pointed out by the authors, the rigorous equivalence $\xi \sim \xi_C$ holds only at the critical pressure, but not at other pressures where ξ does not diverge, while ξ_C does (Bianco and Franzese 2019, Coniglio and Klein 1980, Kasteleyn and Fortuin 1969). Nevertheless, it shows that ξ diverges at the LLCP and grows approaching the line of diverging ξ_C for both decreasing and increasing T at low P.

The Effect of Cooperativity in the Phase Diagram

The FS phase diagram depends on the ratio between the directional and cooperative components of the HB, encoded by the parameters J and J_σ, respectively. Stokely et al. (2010a) show that tuning J_σ/J of the FS model accounts for all the scenarios proposed for the origin of the water anomalies.

By setting $J_\sigma = 0$, that is, with no cooperativity among the HBs, the FS model reproduces the *SF scenario* for any J. For $J_\sigma/J > 0$, but close to zero, the model presents a LLPT ending in a LLCP at $P > 0$, as in the *LLCP scenario*. By increasing the ratio $J_\sigma/J > 0$, the LLCP moves from positive pressures and low temperatures to negative pressures and larger temperatures, until it crosses the limit of stability of the liquid phase respect to the gas phase, as in the *CPF scenario*. In this last case, the (positively sloped) liquid-to-gas spinodal merges with the (negatively sloped) liquid-to-liquid spinodal, determining the reentrance of the resulting spinodal for the liquid, as in the *SL conjecture* (Speedy 1982). Stokely et al. (2010a) show that the hypothesis of the LLCP at positive pressure is the only scenario consistent with estimates made from experimental data on the structural and dynamical properties of liquid water (Chumaevskii and Rodnikova 2003, Eisenberg and Kauzmann 1969, Heggie et al. 1996, Henry 2002, Suresh and Naik 2000).

Summary and Future Perspectives

In this chapter, we presented the Hamiltonian coarse-grained approach proposed by Franzese and Stanley (FS) in 2002 (Franzese and Stanley 2002a,b) for water under confinement. The FS model coarse-grains the configurations of water molecules and describes the water hydrogen bonding (HB) using two terms, accounting for (i) the directional and (ii) the cooperative components of the HB interaction. A percolation mapping allows the implementation of a cluster Monte Carlo algorithm for an efficient equilibration of a water monolayer at extreme temperatures and pressures, in regions where atomistic water models freeze in a glassy state.

First, were viewed the dynamical properties of FS water monolayers. The model displays two dynamic crossovers for the HB time-correlation function: the first at $T \sim 252$ K between two non-Arrhenius regimes, and the second at $T \sim 181$ K from non-Arrhenius to Arrhenius behavior (Kumar et al. 2008c, Mazza et al. 2011). These dynamical crossovers, experimentally observed in protein hydration water (Mazza et al. 2011), originate from the maxima of the fluctuations in the HB network, due to the directional and cooperative HB components. Moreover, the FS model reproduces the anomalous diffusion upon pressurization (de los Santos and Franzese 2012).

Second, were viewed the phase diagram of the FS water (Bianco and Franzese 2014). The model reproduces the increase of density and energy fluctuations upon cooling and the existence of temperatures of maximum and minimum density along isobars. Thermodynamic response functions—the isobaric specific heat, the isothermal compressibility, and the thermal expansivity—display loci of strong and weak maxima. These two sets of maxima correspond to the two dynamic crossovers discussed above. In the supercooled region of the phase diagram, a first-order phase transition separates a high-density liquid (HDL) at higher P from a low-density liquid (LDL) at lower P. This transition line end sin a critical region (where the loci of thermodynamic maxima converge), where the finite-size scaling analysis of the proper order parameter indicates the presence of a liquid-liquid critical point (LLCP) belonging to the 2D-Ising universality-class. From the LLCP stems the Widom line (WL) calculated as the locus of maxima of the water spatial correlation length ξ. The WL coincides with the locus of strong maxima of the response functions in a region that is not accessible to atomistic models. Its geometrical description provides further insight into the building up of the HB network (Bianco and Franzese 2019).

Interestingly, by tuning the ratio between the direction a land cooperative components of the HB, the FS model reproduces the proposed scenarios to explain the origin of the water anomalies (Stokely et al. 2010a). When the ratio is zero, no singularity is observed in the thermodynamic quantities at low temperatures, as in the *Singularity Free* scenario. By increasing the ratio, a transition line ending in a critical point appears, as in the LLCP scenario. The larger the ratio, the lower the LLCP pressure (and the larger the LLCP temperature), which passes from positive to negative until the liquid-to-gas stability limit is reached, as in the *Critical Point Free* scenario, with a retracing liquid spinodal, as in the *Stability Limit* conjecture.

These findings, confirmed also by preliminary results for the bulk FS model (Coronas et al. 2016), offer valuable contributions to the on going debate on the thermodynamic scenarios for supercooled water and the characterization of the HB network. We believe that the approach of the FS model, by keeping a molecular description of the HB dynamics but strongly reducing the computational cost to sample huge systems, could represent a valuable tool to tackle problems of biological relevance (Bianco et al. 2019a, Bianco and Franzese 2015, Bianco et al. 2019b, 2017a,b) and to explore the design of water-adapted bio-materials (Bianco et al. 2017a, Cardelli et al. 2017, 2019, 2018, Nerattini et al. 2019).

Acknowledgements

We thank C. A. Angell, M. Anisimov, M. Bernabei, F. Bruni, C. Calero, D. Chandler, I. Coluzza, F. Coupin, P. Debenedetti, C. Dellago, P. Gallo, P. Kumar, M. Mazza, F. Martelli, P. Poole, S. Sastry, F. Sciortino, F. Starr, K. Stokely, E. Strekalova, A. Zantop, and L. Xu for their helpful contributions to our discussions. In particular, we thank H. E. Stanley for introducing us to this field and for his many contributions to it. L. C. acknowledges the support provided by grant n. 5757200 (APIF_18_19-University of Barcelona). O.V. acknowledges support provided by the Barcelona University Institute of Nanoscience and Nanotechnology (IN2UB). V. B. acknowledges the support from the European Commission through the Marie Sklodowska-Curie Fellowship No. 748170 ProFrost. F. S. acknowledges the support offered by Consejería de Conocimiento, Investigación y Universidad, Junta de Andalucía and European Regional Development Found (ERDF), ref. SOMM17/6105/UGR and Spanish Ministry MINECO project FIS2017-84256-P. G. F. acknowledges the support of the ICREA Foundation (ICREA Academia prize) and the Spanish grant PGC2018-099277-B-C22 (MCIU/AEI/ERDF).

References

Abascal, J. L. F. and C. Vega. 2010. Widom line and the liquid-liquid critical point for the TIP4P/2005 water model. J. Chem. Phys. 133: 234502.

Amann-Winkel, K., R. Böhmer, F. Fujara, C. Gainaru, B. Geil and T. Loerting. 2016. Colloquium: Water's controversial glass transitions. Rev. Mod. Phys. 88: 011002.

Angell, C. A., J. Shuppert and J. C. Tucker. 1973. Anomalous properties of supercooled water. Heat capacity, expansivity, and proton magnetic resonance chemical shift from 0 to -38%. J. Phys. Chem. 77: 3092–3099.

Angell, C. A., W. J. Sichina and M. Oguni. 1982. Heat capacity of water at extremes of supercooling and super heating. J. Phys. Chem. 86: 998–1002.

Angell, C. A. 2008. Insights into phases of liquid water from study of its unusual glass-forming properties. Science 319: 582–587.

Barnes, P., J. L. Finney, J. D. Nicholas and J. E. Quinn. 1979. Cooperative effects in simulated water. Nature 282: 459–464.

Bianco, V., G. Franzese, R. Ruberto and S. Ancherbak. 2012a. Water and anomalous liquids. pp. 113–128. *In*: Mallamace, F. and H. E. Stanley (eds.). Complex Materials in Physics and Biology, volume 176 of Proceedings of the International School of Physics "Enrico Fermi". IOS Press.

Bianco, V., S. Iskrov and G. Franzese. 2012b. Understanding the role of hydrogen bonds in water dynamics and protein stability. J. Biol. Phys. 38: 27–48.

Bianco, V., M. G. Mazza, K. Stokely, F. Bruni, H. E. Stanley and G. Franzese. 2013a. Prediction and observation of two dynamic crossovers in hydration water allows the reconciliation of simulations and experiments. *In*: Ganapathy, R., A. L. Greer, K. F. Kelton and S. Sastry (eds.). Proceedings of "Symposium on the Fragility of Glass-formers: A Conference in Honor of C. Austen Angell".

Bianco, V., O. Vilanova and G. Franzese. 2013b. Polyamorphism and polymorphism of a confined water monolayer: Liquid-liquid critical point, liquid-crystal and crystal-crystal phase transitions. pp. 126–149. *In:* Proceedings of Perspectives and Challenges in Statistical Physics and Complex Systems for the Next Decade: A Conference in Honor of Eugene Stanley and Liacir Lucen.

Bianco, V. and G. Franzese. 2014. Critical behavior of a water monolayer under hydrophobic confinement. Sci. Rep. 4: 4440.

Bianco, V. and G. Franzese. 2015. Contribution of water to pressure and cold denaturation of proteins. Phys. Rev. Lett. 115: 108101.

Bianco, V., G. Franzese, C. Dellago and I. Coluzza. 2017a. Role of water in the selection of stable proteins at ambient and extreme thermodynamic conditions. Phys. Rev. X. 7: 021047.

Bianco, V., N. Pagès-Gelabert, I. Coluzza and G. Franzese. 2017b. How the stability of a folded protein depends on interfacial water properties and residue-residue interactions. Journal of Molecular Liquids 245: 129–139.

Bianco, V. and G. Franzese. 2019. Hydrogen bond correlated percolation in a supercooled water monolayer as a hallmark of the critical region. J. Mol. Liq. 285: 727–739.

Bianco, V., M. Alonso-Navarro, D. Di Silvio, S. Moya, A. L. Cortajarena and I. Coluzza. 2019a. Proteins are solitary! Pathways of protein folding and aggregation in protein mixtures. J. Phys. Chem. Lett. 10: 4800–4804.

Bianco, V., G. Franzese and I. Coluzza. 2019b. *In silico* evidence that protein unfolding is as a precursor of the protein aggregation. Chem. Phys. Chem. 21: 1–9.

Bruce, A. D. and N. B. Wilding. 1992. Scaling fields and universality of the liquid-gas critical point. Phys. Rev. Lett. 68: 193–196.

Calero, C., H. E. Stanley and G. Franzese. 2016. Structural interpretation of the large slowdown of water dynamics at stacked phospholipid membranes for decreasing hydration level: All-atom molecular dynamics. Materials 9: 319.

Calero, C. and G. Franzese. 2019. Membranes with different hydration levels: The interface between bound and unbound hydration water. J. Mol. Liq. 273: 488–496.

Camisasca, G., M. De Marzio, M. Rovere and P. Gallo. 2017. Slow dynamics and structure of supercooled water in confinement. Entropy 19: 185.

Cardelli, C., V. Bianco, L. Rovigatti, F. Nerattini, L. Tubiana, C. Dellago et al. 2017. The role of directional interactions in the designability of generalized heteropolymers. Sci. Rep. 7: 4986.

Cardelli, C., L. Tubiana, V. Bianco, F. Nerattini, C. Dellago and I. Coluzza. 2018. Heteropolymer design and folding of arbitrary topologies reveals an unexpected role of alphabet size on the knot population. Macromolecules 51: 8346–8356.

Cardelli, C., F. Nerattini, L. Tubiana, V. Bianco, C. Dellago, F. Sciortino et al. 2019. General methodology to identify the minimum alphabet size for heteropolymer design. Adv. Theory Simul. 2: 1900031.

Caupin, F., V. Holten, C. Qiu, E. Guillerm, M. Wilke, M. Frenz et al. 2018. Comment on "maxima in the thermodynamic response and correlation functions of deeply supercooled water". Science 360: eaat1634.

Ceriotti, M., J. Cuny, M. Parrinello and D. E. Manolopoulos. 2013. Nuclear quantum effects and hydrogen bond fluctuations in water. Proc. Natl. Acad. Sci. U.S.A. 110: 15591.

Cerveny, S., J. Colmenero, A. Alegria and J. Swenson. 2006. Comment on "Pressure dependence of fragile-to-strong transition and a possible second critical point in supercooled confined water". Phys. Rev. Lett. 97: 189802.

Chandler, D. 2016. Metastability and no criticality. Nature 531: E1–E2.

Chaplin, M. 2006. Do we underestimate the importance of water in cell biology? Nat. Rev. Mol. Cell Biol. 7: 861–866.

Chen, S.-H., L. Liu and A. Faraone. 2006a. Chen, Liu, and Faraone reply. Phys. Rev. Lett. 97: 189803.

Chen, S.-H., L. Liu, E. Fratini, P. Baglioni, A. Faraone and E. Mamontov. 2006b. Observation of fragile-to-strong dynamic crossover in protein hydration water. Proc. Natl. Acad. Sci. U.S.A. 103: 9012–9016.

Chu, X.-q., A. Faraone, C. Kim, E. Fratini, P. Baglioni, J. B. Leão et al. 2009. Proteins remain soft at lower temperatures under pressure. J. Phys. Chem. B. 113: 5001–5006.

Chumaevskii, N. A. and M. N. Rodnikova. 2003. Some peculiarities of liquid water structure. J. Mol. Liq. 106: 167–177.

Cicero, G., J. C. Grossman, E. Schwegler, F. Gygi and G. Galli. 2008. Water Confined in Nanotubes and between Graphene Sheets: A First Principle Study. J. Am. Chem. Soc. 130: 1871–1878.

Coniglio, A. and W. Klein. 1980. Clusters and Ising critical droplets: a renormalisation group approach. J. Phys. A Math. Gen. 13: 2775.

Coronas, L. E., V. Bianco, A. Zantop and G. Franzese. 2016. Liquid-liquid critical point in 3d many-body water model. ArXiv e-prints arXiv:1610.00419.

Dang, L. X. and T.-M. Chang. 1997. Molecular dynamics study of water clusters, liquid, and liquid-vapor interface of water with many-body potentials. J. Chem. Phys. 106: 8149–8159.

de los Santos, F. and G. Franzese. 2009. Influence of intramolecular couplings in a model for hydrogen bonded liquids. pp. 185–197. In: Marro, J., P. L. Garrido and P. I. Hurtado (eds.). Modeling and Simulation of New Materials: Proceedings of Modeling and Simulation of New Materials: Tenth Granada Lectures. AIP, Granada, Spain.

de los Santos, F. and G. Franzese. 2011. Understanding diffusion and density anomaly in a coarse-grained model for water confined between hydrophobic walls. J. Phys. Chem. B. 115: 14311–14320.

de los Santos, F. and G. Franzese. 2012. Relations between the diffusion anomaly and cooperative rearranging regions in a hydrophobically nanoconfined water monolayer. Phys. Rev. E. 85: 010602.

Doster, W. 2010. The protein-solvent glass transition. Biochim. Biophys. Acta. 1804: 3–14.

Eisenberg, D. and W. Kauzmann. 1969. The Structure and Properties of Water. p. 139. Oxford University Press.

Faraone, A., L. Liu, C.-Y. Mou, C.-W. Yen and S.-H. Chen. 2004. Fragile-to-strong liquid transition in deeply supercooled confined water. J. Chem. Phys. 121: 10843–10846.

Finney, J. L. 2001. The water molecule and its interactions: The interaction between theory, modelling, and experiment. J. Mol. Liq. 90: 303–312.

Franzese, G., M. Yamada and H. E. Stanley. 2000. Hydrogen-bonded liquids: Effects of correlations of orientational degrees of freedom. In: Statistical Physics 519: 281–287.

Franzese, G. and H. E. Stanley. 2002a. Liquid-liquid critical point in a Hamiltonian model for water: Analytic solution. J. Phys. Condens. Matter. 14: 2201–2209.

Franzese, G. and H. E. Stanley. 2002b. A theory for discriminating the mechanism responsible for the water density anomaly. Physica A. 314: 508–513.

Franzese, G., M. I. Marqués and H. E. Stanley. 2003. Intramolecular coupling as a mechanism for a liquid-liquid phase transition. Phys. Rev. E. 67: 11103.

Franzese, G. and H. E. Stanley. 2007. The Widom line of supercooled water. J. Phys. Condens. Matter. 19: 205126.

Franzese, G., K. Stokely, X. Q. Chu, P. Kumar, M. G. Mazza, S.-H. Chen et al. 2008. Pressure effects in supercooled water: Comparison between a 2D model of water and experiments for surface water on a protein. J. Phys. Condens. Matter 20: 494210.

Franzese, G. and F. de los Santos. 2009. Dynamically slow processes in supercooled water confined between hydrophobic plates. J. Phys. Condens. Matter 21: 504107.

Franzese, G., A. Hernando-Martínez, P. Kumar, M. G. Mazza, K. Stokely, E. G. Strekalova et al. 2010. Phase transitions and dynamics of bulk and interfacial water. J. Phys. Condens. Matter 22: 284103.

Franzese, G. and H. E. Stanley. 2010. Understanding the Unusual Properties of Water. Chapter 7. CRC Press.

Franzese, G., V. Bianco and S. Iskrov. 2011. Water at Interface with Proteins. Food Biophys. 6: 186–198.

Franzese, G. and V. Bianco. 2013. Water at biological and inorganic interfaces. Food Biophys. 8: 153–169.

Fuentevilla, D. A. and M. A. Anisimov. 2006. Scaled equation of state for supercooled water near the liquid-liquid critical point. Phys. Rev. Lett. 97: 195702.

Fumagalli, L., A. Esfandiar, R. Fabregas, S. Hu, P. Ares, A. Janardanan et al. 2018. Anomalously low dielectric constant of confined water. Science 360: 1339–1342.

Galkina, Y. A., N. A. Kryuchkova, M. A. Vershinin and B. A. Kolesov. 2017. Features of strong O-H-O and N-H-O hydrogen bond manifestation in vibrational spectra. J. Struct. Chem. 58: 911–918.

Gallo, P., M. Rovere and E. Spohr. 2000. Supercooled confined water and the mode coupling crossover temperature. Phys. Rev. Lett. 85: 4317–4320.

Gallo, P., M. Rovere and S.-H. Chen. 2010. Dynamic crossover in supercooled confined water: understanding bulk properties through Con_nement. J. Phys. Chem Lett. 1: 729–733.

Gallo, P., K. Amann-Winkel, C. A. Angell, M. A. Anisimov, F. Caupin, C. Chakravarty et al. 2016. Water: A tale of two liquids. Chem. Rev. 116: 7463–7500.

González, M. A., C. Valeriani, F. Caupin and J. L. F. Abascal. 2016. A comprehensive scenario of the thermodynamic anomalies of water using the TIP4P/2005 model. J. Chem. Phys. 145.

Han, S., P. Kumar and H. E. Stanley. 2008. Absence of a diffusion anomaly of water in the direction perpendicular to hydrophobic nanoconfining walls. Phys. Rev. E. 77: 30201.

Han, S., M. Choi, P. Kumar and H. E. Stanley. 2010. Phase transitions in confined water nanofilms. Nat. Phys. 6: 685.

Handle, P. H., T. Loerting and F. Sciortino. 2017. Supercooled and glassy water: Metastable liquid(s), amorphous solid(s), and a no-man's land. Proc. Natl. Acad. Sci. U.S.A. 114: 13336–13344.

Heggie, M. I., C. D. Latham, S. C. P. Maynard and R. Jones. 1996. Cooperative polarisation in ice I h and the unusual strength of the hydrogen bond. Chem. Phys. Lett. 249: 485–490.

Henry, M. 2002. Nonempirical quantification of molecular interactions in supramolecular assemblies. Chem. Phys. Chem. 3: 561–569.

Hernández de la Peña, L. and P. G. Kusalik. 2005. Temperature dependence of quantum effects in liquid water. J. Am. Chem. Soc. 127: 5246–5251.

Hernández-Rojas, J., F. Calvo, F. Rabilloud, J. Bretón and J. M. Gómez Llorente. 2010. Modeling water clusters on cationic carbonaceous seeds. J. Phys. Chem. A. 114: 7267–7274.

Holt, J. K., H. G. Park, Y. Wang, M. Stadermann, A. B. Artyukhin, C. P. Grigoropoulos et al. 2006. Fast mass transport through sub-2-nanometer carbon nanotubes. Science 312: 1034.

Holten, V. and M. A. Anisimov. 2012. Entropy-driven liquid-liquid separation in supercooled water. Sci. Rep. 2: 713.

Holten, V., C. E. Bertrand, M. A. Anisimov and J. V. Sengers. 2012. Thermodynamics of supercooled water. J. Chem. Phys. 136: 94507–94518.

Holten, V., D. T. Limmer, V. Molinero and M. A. Anisimov. 2013. Nature of the anomalies in the supercooled liquid state of the mW model of water. J. Chem. Phys. 138: 174501.

Holten, V., J. C. Palmer, P. H. Poole, P. G. Debenedetti and M. A. Anisimov. 2014. Two-state thermodynamics of the ST2 model for supercooled water. J. Chem. Phys. 140: 104502.

Huang, C., T. M. Weiss, D. Nordlund, K. T. Wikfeldt, L. G. M. Pettersson and A. Nilsson. 2010. Increasing correlation length in bulk supercooled H2O, D2O, and NaCl solution determined from small angle x-ray scattering. J. Chem. Phys. 133: 134504.

Huš, M. and T. Urbic. 2012. Strength of hydrogen bonds of water depends on local environment. Journal of Chemical Physics 136: 144305.

James, T., D. J. Wales and J. Hernández-Rojas. 2005. Global minima for water clusters (h2o)n, n21, described by a five-site empirical potential. Chem. Phys. Lett. 415: 302–307.

Kamitakahara, W. A. and N. Wada. 2008. Neutron spectroscopy of water dynamics in NaX and NaA zeolites. Phys. Rev. E. 77: 41503–41510.

Kasteleyn, P. and C. M. Fortuin. 1969. Phase transitions in lattice systems with random local properties. Physical Society of Japan Journal Supplement, Proceedings of the International Conference on Statistical Mechanics held 9–14 September, 1968 in Koyto. 26: 11.

Kesselring, T., G. Franzese, S. Buldyrev, H. Herrmann and H. E. Stanley. 2012. Nanoscale Dynamics of Phase Flipping in Water near its Hypothesized Liquid-Liquid Critical Point. Sci. Rep. 2: 474.

Kesselring, T., E. Lascaris, G. Franzese, S. V. Buldyrev, H. Herrmann and H. E. Stanley. 2013. Finite-size scaling investigation of the liquid-liquid critical point in ST2 water and its stability with respect to crystallization. J. Phys. Chem. 138: 244506.

Kim, K. H., A. Späh, H. Pathak, F. Perakis, D. Mariedahl, K. Amann-Winkel et al. 2017. Maxima in the thermodynamic response and correlation functions of deeply supercooled water. Science 358: 1589.

Kim, K. H., A. Späh, H. Pathak, F. Perakis, D. Mariedahl, K. Amann-Winkel et al. 2018. Response to comment on "Maximain the thermodynamic response and correlation functions of deeply supercooled water". Science 360: eaat1729.

Kumar, P., S. V. Buldyrev, F. W. Starr, N. Giovambattista and H. E. Stanley. 2005. Thermodynamics, structure, and dynamics of water confined between hydrophobic plates. Phys. Rev. E. 72: 51503.

Kumar, P., G. Franzese, S. V. Buldyrev and H. E. Stanley. 2006a. Molecular dynamics study of orientational cooperativity in water. Phys. Rev. E. 73: 41505.

Kumar, P., Z. Yan, L. Xu, M. G. Mazza, S. V. Buldyrev, S.-H. Chen et al. 2006b. Glass transition in biomolecules and the liquid-liquid critical point of water. Phys. Rev. Lett. 97: 177802.

Kumar, P., S. V. Buldyrev, S. R. Becker, P. H. Poole, F. W. Starr and H. E. Stanley. 2007. Relation between the Widom line and the breakdown of the Stokes-Einstein relation in supercooled water. Proc. Natl. Acad. Sci. U.S.A. 104: 9575–9579.

Kumar, P., G. Franzese, S. V. Buldyrev and H. E. Stanley. 2008a. Dynamics of water at low temperatures and implications for biomolecule. pp. 3–22. *In*: Volume 752 of Lecture Notes in Physics. Springer Berlin Heidelberg.

Kumar, P., G. Franzese and H. E. Stanley. 2008b. Dynamics and thermodynamics of water. J. Phys. Condens. Matter. 20: 244114.

Kumar, P., G. Franzese and H. E. Stanley. 2008c. Predictions of dynamic behavior under pressure for two scenarios to explain water anomalies. Phys. Rev. Lett. 100: 105701.

Kumar, P. and H. E. Stanley. 2011. Thermal conductivity minimum: A new water anomaly. J. Phys. Chem. B. 115: 14269–14273.

Lascaris, E., T. A. Kesselring, G. Franzese, S. V. Buldyrev, H. J. Herrmann and H. E. Stanley. 2013. Response functions near the liquid-liquid critical point of st2 water. AIP Conf. Proc. 1518: 520–526.

Leoni, F. and G. Franzese. 2016a. Effects of confinement between attractive and repulsive walls on the thermodynamics of an anomalous liquid. Phys. Rev. E. 94: 062604.

Leoni, F. and G. Franzese. 2016b. Structural behavior and dynamics of an anomalous liquid between attractive and repulsive walls: Templating, molding, and superdiffusion. J. Chem. Phys. 141: 174501.

Limmer, D. T. and D. Chandler. 2013. The putative liquid-liquid transition is a liquid-solid transition in atomistic models of water. II. J. Chem. Phys. 138: 214504.

Liu, D., Y. Zhang, C.-C. Chen, C.-Y. Mou, P. H. Poole and S.-H. Chen. 2007. Observation of the density minimum in deeply supercooled confined water. Proc. Natl. Acad. Sci. U.S.A. 104: 9570–9574.

Liu, L., S.-H. Chen, A. Faraone, C.-W. Yen and C.-Y. Mou. 2005. Pressure dependence of fragile-to-strong transition and a possible second critical point in supercooled confined water. Phys. Rev. Lett. 95: 117802.

Liu, Y., A. Z. Panagiotopoulos and P. G. Debenedetti. 2010. Finite-size scaling study of the vapor-liquid critical properties of confined fluids: Crossover from three dimensions to two dimensions. J. Chem. Phys. 132: 144107.

Ludwig, R. 2001. Water: From Clusters to the Bulk. Angew. Chem. Int. Ed. Engl. 40: 1808–1827.

Luzar, A. and D. Chandler. 1996. Effect of environment on hydrogen bond dynamics in liquid water. Phys. Rev. Lett. 76: 928–931.

Mallamace, F., C. Branca, M. Broccio, C. Corsaro, C.-Y. Mou and S.-H. Chen. 2007a. The anomalous behavior of the density of water in the range $30\ K < T < 373\ K$. Proc. Natl. Acad. Sci. U.S.A. 104: 18387–18391.

Mallamace, F., M. Broccio, C. Corsaro, A. Faraone, D. Majolino, V. Venuti et al. 2007b. Evidence of the existence of the low-density liquid phase in supercooled, confined water. Proc. Natl. Acad. Sci. U.S.A. 104: 424–428.

Mallamace, F., C. Corsaro, M. Broccio, C. Branca, N. González-Segredo, J. Spooren et al. 2008. NMR evidence of a sharp change in a measure of local order in deeply supercooled confined water. Proc. Natl. Acad. Sci. U.S.A. 105: 12725–12729.

Mancinelli, R., S. Imberti, A. K. Soper, K. H. Liu, C. Y. Mou, F. Bruni et al. 2009. Multiscale Approach to the Structural Study of Water Confined in MCM41. J. Phys. Chem. B. 113: 16169–16177.

Martelli, F., H.-Y. Ko, C. Calero Borallo and G. Franzese. 2017. Structural properties of water confined byphospholipid membranes. Front. Phys. 13: 136801.

Martí, J., C. Calero and G. Franzese. 2017. Structure and dynamics of water at carbon-based interfaces. Entropy 19: 135.

Mazza, M. G., K. Stokely, E. G. Strekalova, H. E. Stanley and G. Franzese. 2009. Cluster monte-carlo and numerical mean field analysis for the water liquid-liquid phase transition. Comput. Phys. Commun. 180: 497–502.

Mazza, M. G., K. Stokely, S. E. Pagnotta, F. Bruni, H. E. Stanley and G. Franzese. 2011. More than one dynamic crossover in protein hydration water. Proc. Natl. Acad. Sci. U.S.A. 108: 19873–19878.

Mazza, M. G., K. Stokely, H. E. Stanley and G. Franzese. 2012. Effect of pressure on the anomalous response functions of a confined water monolayer at low temperature. J. Chem. Phys. 137: 204502.

Mishima, O., L. D. Calvert and E. Whalley. 1984. 'Melting ice'l at 77 K and 10 kbar: A new method of making amorphous solids. Nature 310: 393–395.

Mishima, O., L. D. Calvert and E. Whalley. 1985. An apparently 1st-order transition between 2 amorphous phases of ice induced by pressure. Nature 314: 76–78.

Mishima, O. 1994. Reversible first-order transition between two H2O amorphs at ~ 0.2 GPa and ~ 135 K. J. Chem. Phys. 100: 5910–5912.

Mishima, O. and H. E. Stanley. 1998. The relationship between liquid, supercooled and glassy water. Nature 396: 329–335.

Murata, K.-I. and H. Tanaka. 2012. Liquid-liquid transition without macroscopic phase separation in a water-glycerol mixture. Nat. Matter. 11: 436–443.

Murata, K.-I. and H. Tanaka. 2013. General nature of liquid-liquid transition in aqueous organic solutions. Nat. Commun. 4.

Nagle, J. F. and S. Tristram-Nagle. 2000. Structure of lipid bilayers. Biochim. Biophys. Acta Rev. Biomembranes. 1469: 159–195.

Naguib, N., H. Ye, Y. Gogotsi, A. G. Yazicioglu, C. M. Megaridis and M. Yoshimura. 2004. Observation of water confined in nanometer channels of closed carbon nanotubes. Nano Lett. 4: 2237–2243.

Nerattini, F., L. Tubiana, C. Cardelli, V. Bianco, C. Dellago and I. Coluzza. 2019. Design of protein-protein binding sites suggests a rationale for naturally occurring contact areas. J. Chem. Theory Comput. 15: 1383–1392.

Netz, P. A., F. W. Starr, H. E. Stanley and M. C. Barbosa. 2001. Static and dynamic properties of stretched water. J. Chem. Phys. 115: 344–348.

Ouyang, J. F. and R. P. A. Bettens. 2015. Modelling water: A lifetime enigma. CHIMIA Int. J. Chem. 69: 104–111.

Palmer, J. C., F. Martelli, Y. Liu, R. Car, A. Z. Panagiotopoulos and P. G. Debenedetti. 2014. Metastable liquid-liquid transition in a molecular model of water. Nature 510: 385–388.

Palmer, J. C., F. Martelli, Y. Liu, R. Car, A. Z. Panagiotopoulos and P. G. Debenedetti. 2016. Palmer et al. reply. Nature 531: E2–E3.

Palmer, J. C., A. Haji-Akbari, R. S. Singh, F. Martelli, R. Car, A. Z. Panagiotopoulos et al. 2018. Comment on "The putative liquid-liquid transition is a liquid-solid transition in atomistic models of water" [I and II: J. Chem. Phys. 135, 134503(2011); J. Chem. Phys. 138, 214504(2013)]. J. Chem. Phys. 148: 137101.

Paul, D. R. 2012. Creating new types of carbon-based membranes. Science 335: 413–414.

Peyrard, M. 2001. Glass transition in protein hydration water. Phys. Rev. E Stat. Nonlin. Soft. Matter. Phys. 64: 011109.

Piana, S. and D. E. Shaw. 2018. Atomic-level description of protein folding inside the groEL cavity. J. Phys. Chem B. 122: 11440–11449.

Poole, P., F. Sciortino, U. Essmann and H. Stanley. 1992. Phase-behavior of metastable water. Nature 360: 324–328.

Poole, P. H., F. Sciortino, T. Grande, H. E. Stanley and C. A. Angell. 1994. Effect of hydrogen bonds on the thermodynamic behavior of liquid water. Phys. Rev. Lett. 73: 1632–1635.

Poole, P. H., I. Saika-Voivod and F. Sciortino. 2005. Density minimum and liquid-liquid phase transition. J. Phys. Condens. Matter. 17: L431–L437.

Radha, B., A. Esfandiar, F. C. Wang, A. P. Rooney, K. Gopinadhan, A. Keerthi et al. 2016. Molecular transport through capillaries made with atomic-scale precision. Nature 538: 222.

Russo, J. and H. Tanaka. 2014. Understanding water's anomalies with locally favoured structures. Nat. Commun. 5: 3556.

Salzmann, C. G. 2019. Advances in the experimental exploration of water's phase diagram. J. Chem. Phys. 150: 060901.

Samatas, S., C. Calero, F. Martelli and G. Franzese. 2019. Water between membranes: structure and dynamics. *In*: Berkowitz, M. L. (ed.). Biomembrane Simulations: Computational Studies of Biological Membranes. CRC Press. Chap. 4.

Sastry, S., P. G. Debenedetti, F. Sciortino and H. E. Stanley. 1996. Singularity-free interpretation of the thermodynamics of supercooled water. Phys. Rev. E. 53: 6144–6154.

Schran, C. and D. Marx. 2019. Quantum nature of the hydrogen bond from ambient conditions down to ultra-low temperatures. Phys. Chem. Chem. Phys. 21: 24967–24975.

Settles, M. and W. Doster. 1996. Anomalous diffusion of adsorbed water: A neutron scattering study of hydrated myoglobin. Faraday Discuss. 103: 269–279.

Shi, R., J. Russo and H. Tanaka. 2018. Common microscopic structural origin for water's thermodynamic and dynamic anomalies. J. Chem. Phys. 149: 224502.

Shimoaka, T., T. Hasegawa, K. Ohno and Y. Katsumoto. 2012. Correlation between the local OH stretching vibration wavenumber and the hydrogen bonding pattern of water in a condensed phase: Quantum chemical approach to analyze the broad OH band. Journal of Molecular Structure 1029: 209–216.

Simeoni, G. G., T. Bryk, F. A. Gorelli, M. Krisch, G. Ruocco, M. Santoro et al. 2010. The Widom line as the crossover between liquid-like and gas-like behaviour in supercritical liquids. Nat. Phys. 6: 503–507.

Singh, R. S., J. W. Biddle, P. G. Debenedetti and M. A. Anisimov. 2016. Two-state thermodynamics and the possibility of a liquid-liquid phase transition in supercooled TIP4P/2005 water. J. Chem. Phys. 144: 144504.

Smallenburg, F. and F. Sciortino. 2015. Tuning the liquid-liquid transition by modulating the hydrogen-bond angular flexibility in a model for water. Phys. Rev. Lett. 115: 015701.

Soper, A. K. and M. G. Phillips. 1986. A new determination of the structure of water at 25C. Chem. Phys. 107: 47–60.

Soper, A. K. and M. A. Ricci. 2000. Structures of high-density and low-density water. Phys. Rev. Lett. 84: 2881–2884.

Soper, A. K. 2012. Density profile of water confined in cylindrical pores in MCM-41 silica. J. Phys. Condens. Matter. 24: 64107.

Speedy, R. J. and C. A. Angell. 1976. Isothermal compressibility of supercooled water and evidence for a thermodynamic singularity at –45C. J. Phys. Chem. 65: 851–858.

Speedy, R. J. 1982. Limiting forms of the thermodynamic divergences at the conjectured stability limits in superheated and supercooled water. J. Phys. Chem. 86: 3002–3005.

Stanley, H. E. 1971. Introduction to Phase Transitions and Critical Phenomena. Volume 40 of International Series of Monographs on Physics. Oxford University Press.

Stanley, H. E. and J. Teixeira. 1980. Interpretation of the unusual behavior of H2O and D2O at low temperatures: Tests of a percolation model. J. Chem. Phys. 73: 3404–3422.

Stanley, H. E., J. Teixeira, A. Geiger and R. L. Blumberg. 1981. Interpretation of the unusual behavior of H2O and D2O at low temperature: Are concepts of percolation relevant to the puzzle of liquid water? Physica A. 106: 260–277.

Stanley, H. E., S. V. Buldyrev, G. Franzese, N. Giovambattista and F. W. Starr. 2005. Static and dynamic heterogeneities in water. Philos. Trans. A Math. Phys. Eng. Sci. 363: 509–523.

Stanley, H. E., P. Kumar, S. Han, M. G. Mazza, K. Stokely, S. V. Buldyrev et al. 2009. Heterogeinties in Confined Water and Protein Hydration Water. J. Phys. Condens. Matter 21: 504105.

Stanley, H. E., S. V. Buldyrev, G. Franzese, P. Kumar, F. Mallamace, M. G. Mazza et al. 2010. Liquid polymorphism: Water in nanoconfined and biological environments. J. Phys. Condens. Matter. 22: 284101.

Stanley, H. E., S. V. Buldyrev, P. Kumar, F. Mallamace, M. G. Mazza, K. Stokely et al. 2011. Water in nanoconfined and biological environments: (Plenary Talk, Ngai-Ruocco 2009 IDMRCSConf.). J. Non Cryst. Solids. 357: 629–640.

Stefanutti, E., L. E. Bove, G. Lelong, M. A. Ricci, A. K. Soper and F. Bruni. 2019. Ice crystallization observed in highly supercooled confined water. Phys. Chem. Chem. Phys. 21: 4931–4938.

Stokely, K., M. G. Mazza, H. E. Stanley and G. Franzese. 2010a. Effect of hydrogen bond cooperativity on the behavior of water. Proc. Natl. Acad. Sci. U.S.A. 107: 1301–1306.

Stokely, K., M. G. Mazza, H. E. Stanley and G. Franzese. 2010b. Metastable systems under pressure. Chapter metastable water under pressure. pp. 197–216. *In:* NATO Science for Peace and Security SeriesA: Chemistry and Biology. Springer.

Strekalova, E. G., M. G. Mazza, H. E. Stanley and G. Franzese. 2011. Large decrease of fluctuations for supercooled water in hydrophobic nanoconfinement. Phys. Rev. Lett. 106: 145701.

Strekalova, E., D. Corradini, M. Mazza, S. Buldyrev, P. Gallo, G. Franzese et al. 2012a. Effect of hydrophobic environments on the hypothesized liquid-liquid critical point of water. J. Biol. Phys. 38: 97–111.

Strekalova, E. G., M. G. Mazza, H. E. Stanley and G. Franzese. 2012b. Hydrophobic nanoconfinement suppresses fluctuations in supercooled water. J. Phys. Condens. Matter. 24: 064111.

Suresh, S. J. and V. M. Naik. 2000. Hydrogen bond thermodynamic properties of water from dielectric constant data. J. Chem. Phys. 113: 9727–9732.

Swenson, J. 2006. Comment on "Pressure dependence of fragile-to-strong transition and a possible second critical point in supercooled confined water". Phys. Rev. Lett. 97: 189801.

Szortyka, M. M. and M. C. Barbosa. 2007. Diffusion anomaly in an associating lattice gas model. Physica A. 380: 27–35.

Takahara, S., M. Nakano, S. Kittaka, Y. Kuroda, T. Mori, H. Hamano et al. 1999. Neutron scattering study on dynamics of water molecules in MCM-41. J. Phys. Chem. B. 103: 5814–5819.

Taschin, A., P. Bartolini, R. Eramo, R. Righini and R. Torre. 2013. Evidence of two distinct local structures of water from ambient to supercooled conditions. Nat. Commun. 4: 2401.

Teixeira, J. and M.-C. Bellissent-Funel. 1990. Dynamics of water studied by neutron scattering. J. Phys. Condens. Matter. 2: SA105–SA108.

Vega, C., J. L. F. Abascal, M. M. Conde and J. L. Aragonés. 2009. What ice can teach us about water interactions: A critical comparison of the performance of different water models. Faraday Discuss. 141: 251–276.

Wang, Z., K. Ito and S.-H. Chen. 2016. Detection of the liquid-liquid transition in the deeply cooled water confined in MCM-41 with elastic neutron scattering technique. Nuovo Cimento C. 39: 299.

Wilding, N. B. 1995. Critical-point and coexistence-curve properties of the Lennard-Jones liquid: A finite-size scaling study. Phys. Rev. E. 52: 602–611.

Wilding, N. B. and K. Binder. 1996. Finite-size scaling for near-critical continuum liquids at constant pressure. Physica A. 231: 439–447.

Woutersen, S., B. Ensing, M. Hilbers, Z. Zhao and C. A. Angell. 2018. A liquid-liquid transition in supercooled aqueous solution related to the HDA-LDA transition. Science 359: 1127–1131.

Xu, L., P. Kumar, S. V. Buldyrev, S.-H. Chen, P. H. Poole, F. Sciortino et al. 2005a. Relation between the Widom line and the dynamic crossover in systems with a liquid-liquid phase transition. Proc. Natl. Acad. Sci. U.S.A. 102: 16558–16562.

Xu, W.-X., J. Wang and W. Wang. 2005b. Folding behavior of chaperonin-mediated substrate protein. Proteins: Struct. Funct. Genet. 61: 777–794.

Yagasaki, T., M. Matsumoto and H. Tanaka. 2014. Spontaneous liquid-liquid phase separation of water. Phys. Rev. E. 89: 020301.

Zangi, R. and A. E. Mark. 2003. Monolayer Ice. Phys. Rev. Lett. 91: 25502.

Zhang, Y., A. Faraone, W. A. Kamitakahara, K.-H. Liu, C.-Y. Mou, J. B. Leão et al. 2011. Density hysteresis of heavy water confined in a nanoporous silica matrix. Proc. Natl. Acad. Sci. U.S.A. 108: 12206–12211.

Zhao, W., D. E. Moilanen, E. E. Fenn and M. D. Fayer. 2008. Water at the surfaces of aligned phospholipid multibilayer model membranes probed with ultrafast vibrational spectroscopy. J. Am. Chem. Soc. 130: 13927–13937.

CHAPTER 4

Nineteen Phases of Ice and Counting

*Alfred Amon, Bharvi Chikani, Siriney O. Halukeerthi,
Carissa Ponan, Alexander Rosu-Finsen, Zainab Sharif,
Rachael L. Smith, Sukhpreet K. Talewar* and
*Christoph G. Salzmann**

Introduction

H_2O is the thermodynamically stable combination of hydrogen and oxygen, and exists in vast quantities in our Universe (Tielens 2013). From the oceans on Earth and the poles of Mars to the icy moons of the gas giants and the distant mountain ranges of Pluto, H_2O is a constant companion throughout our solar system and a critical ingredient for the evolution and support of life (Ball 1999). Although often depicted as a V-shaped molecule, the van der Waals volume of the water molecule is dominated by the oxygen atom with two 'bulges' where the two hydrogen atoms are located (Finney 2001). Unlike any of the other dihydrogen chalcogenides, H_2O is capable of forming up to four hydrogen bonds with neighboring molecules which has a pronounced impact on the physical properties of its condensed phases including the temperatures of melting and boiling (Petrenko and Whitworth 1999).

The familiar hexagonal phase of ice is called ice Ih. Based on X-ray diffraction, it was established that the oxygen atoms in ice Ih are arranged in tetrahedral coordination environments (Dennison 1921, Bragg 1921, Barnes and Bragg 1929). However, the positions of the hydrogen atoms remained unresolved and even controversial until the late 1940s when neutron diffraction became available (Wollan et al. 1949). Four of the debated structural models are shown in Fig. 1. According to the Barnes model (Barnes and Bragg 1929), the water molecules lose their molecular character in ice with hydrogen atoms located halfway between the oxygen atoms. The Bernal-Fowler model on the other hand suggested intact and orientationally ordered H_2O molecules (Bernal and Fowler 1933). Pauling proposed a structural

Department of Chemistry, University College London, 20 Gordon Street, London WC1H 0AJ, UK.
* Corresponding author: c.salzmann@ucl.ac.uk

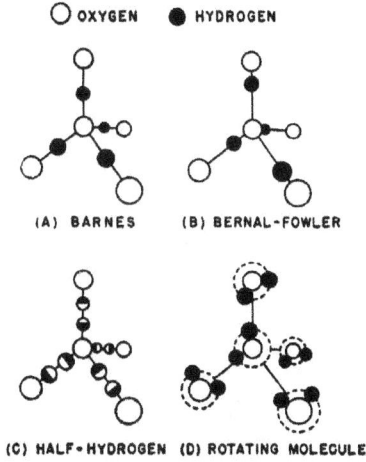

Fig. 1: The four structural models used to analyze the first powder neutron diffraction data of D_2O ice I*h* (Barnes and Bragg 1929, Bernal and Fowler 1933, Pauling 1935, Wollan et al. 1949). Reproduced with permission from Wollan et al. (1949).

model with two half-occupied hydrogen sites along each of the hydrogen bonds (Pauling 1935). Finally, rotating water molecules, in what would now be considered a plastic phase of ice, were considered as well (Wollan et al. 1949). The analysis of the powder neutron diffraction of data of D_2O ice I*h* revealed that Pauling's model provided the best match to the recorded intensities of the Bragg peaks. The half-occupied hydrogen sites thereby reflect the average structure of ice in which the orientations of the fully hydrogen-bonded water molecules are random. Such an ice structure is commonly described as hydrogen disordered. The molar configurational entropy of such a phase of ice was estimated by Pauling as $R \ln (3/2)$ (Pauling 1935). The frequently used term "proton disordered" is chemically incorrect since the water molecule contains hydrogen atoms and not H^+ ions.

In addition to ice I*h*, several other phases of ice exist that have been labelled with increasing Roman numerals in chronological order of their discoveries. The up-to-date phase diagram of ice is shown in Fig. 2. Like ice I*h*, all other phases of ice that share phase boundaries with liquid water including ices III, V, VI, and VII are hydrogen disordered and hence follow Pauling's model. Despite not being able to reproduce the diffraction pattern of ice I*h*, the other three structural models shown in Fig. 1 still turned out to be highly relevant for ice research. Upon cooling, the hydrogen-disordered ices III and VII transform to their hydrogen-ordered counterparts, ices IX and VIII with defined orientations in line with the Bernal-Fowler model (La Placa and Hamilton 1973, Whalley et al. 1968, Whalley et al. 1966). Compression of ice VII beyond 60 GPa leads to the formation of ice X as the hydrogen bonds become symmetric like in the Barnes model (Stillinger and Schweitzer 1983, Polian and Grimsditch 1984). The formation of a plastic state of ice VII at high temperatures has been suggested (Takii et al. 2008, Aragones and Vega 2009, Hernandez and Caracas 2018) and the recently discovered superionic ice XVIII (Millot et al. 2018, Millot et al. 2019) can even be seen as a state beyond plastic rotations with completely dissociated water molecules.

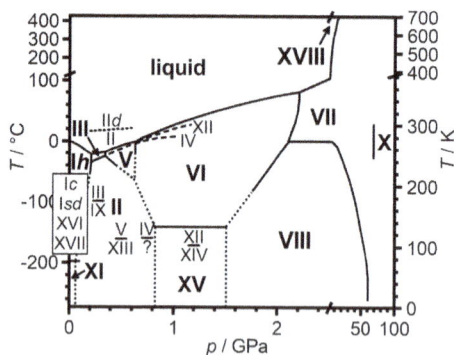

Fig. 2: The phase diagram of ice. Phases in their regions of stability are shown as bold Roman numerals whereas a smaller font size is used for metastable phases. Dashed lines indicate metastable melting lines whereas dotted lines are either extrapolated or estimated computationally (Nakamura et al. 2016).

The hydrogen-bonded networks of all phases of ice discovered so far are shown in Fig. 3. A complex array of structures exists, ranging from ices Ih/XI, II, and XVII with large open channels to the densely knit interpenetrating networks of ices

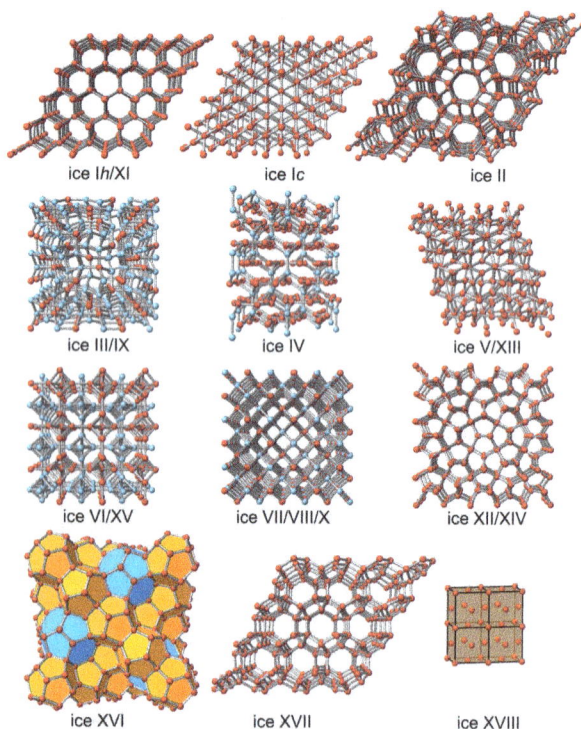

Fig. 3: The hydrogen-bonded networks of the various phases of ice. Oxygen atoms are shown as spheres and hydrogen bonds as grey lines. Blue spheres indicate the oxygen atoms belonging to 4-fold spirals in ice III/IX, interpenetrating hydrogen bonds in ice IV, and the second independent networks in ices VI/XV/XIX and VII/VIII/X. The structure of ice Ic is shown with the corresponding supercell of ice Ih. The orange and blue polyhedra in ice XVI indicate the two different types of cages. The unit cell of the superionic ice XVIII is shown with black lines.

VI/XV/XIX and VII/VIII/X (Salzmann 2019). In the following sections, the various phases of ice will be introduced by describing their structures, physical properties, and the most recent experimental advances in understanding these phases.

The Ice I Family of Polytypes

Hexagonal ice Ih is the most ubiquitous, naturally occurring form of ice on Earth which forms below 0°C at atmospheric pressure (Kuhs and Lehmann 1983, Röttger et al. 1994). The crystal structure of ice Ih with space group $P6_3/mmc$ contains puckered layers of six-membered rings of water molecules in the chair conformation as shown in Fig. 4(a). These layers extend in the a-b-plane and are stacked along the c-axis where each layer is the mirror image of the previous layer. The hexagonal stacking of layers is indicated by 'h' in Fig. 4(b). Due to the stacking, six-membered rings in the boat conformation are formed so that the ice Ih structure contains equal amounts of six-membered rings in the chair and boat conformations. The hexagonal unit cell of ice Ih contains 4 water molecules and one distinct oxygen site. Although of very similar length, two different types of hydrogen bonds exist in ice Ih: those within the layers and those connecting them. Ice Ih is fully hydrogen disordered and it has been discussed that the hydrogen disorder induces positional disorder of the oxygen atoms (Bertie and Whalley 1964a, Kuhs and Lehmann 1983). Highly accurate measurements of the lattice constants of ice Ih as a function of temperature have recently been reported (Fortes 2018). Furthermore, insights into the reorientation dynamics and weak ordering processes at low temperatures have been gained from measurement of the lattice constants (Fortes 2019). Below ~ 60 K, ice Ih displays negative thermal expansion (Röttger et al. 1994, Gupta et al. 2018).

The crystal structure of ice Ic with space group Fd-$3m$ comprises layers identical to those in ice Ih but they are stacked on top of one another by shifting halfway

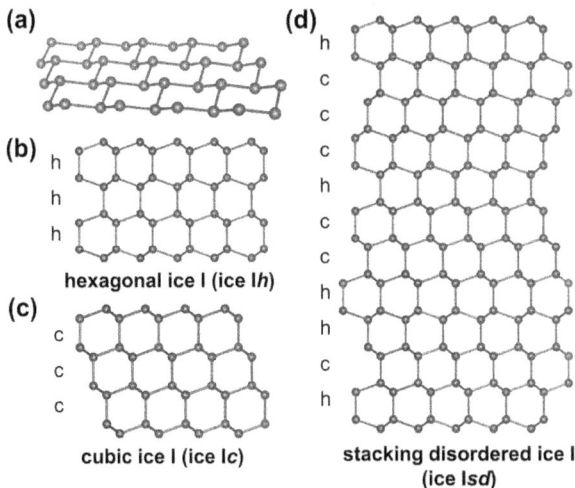

Fig. 4: Stacking of layers in ice I. (a) Single puckered ice I layer. Side views of the stacking of layers in (b) ice Ih, (c) ice Ic, and (d) stacking disordered ice I (ice Isd). Oxygen atoms of water molecules are shown as spheres connected by hydrogen bonds drawn as lines. Hexagonal and cubic stacking is indicated as 'h' and 'c', respectively.

across the diagonal of the six-membered rings (Kuhs et al. 1987). Cubic stacking is indicated by 'c' in Fig. 4(c). The arrangement of oxygen atoms in ice Ic is identical to the cubic diamond structure. Hence, the overall structure contains only six-membered rings in the chair conformation. The cubic unit cell contains eight water molecules with one distinct oxygen site and one type of hydrogen bond. Since the structural difference between ice Ih and Ic lies only in the way that identical layers are stacked, these two forms of ice are not considered to be distinct polymorphs but different polytypes of ice I.

Pure cubic ice Ic has been prepared for the first time in two recent independent studies. The first route involves heating the low-density ice XVII under vacuum to ~ 160 K (del Rosso et al. 2020). Alternatively, ice Ic can be prepared by decompressing the C$_2$ high-pressure hydrogen hydrate to ambient pressure at ~ 100 K (Komatsu et al. 2020b). The host framework of the C$_2$ hydrate is isostructural with ice Ic which may provide a templating effect for its formation. The recent discoveries of ice Ic have now paved the way for detailed investigations of its physical properties and how they differ from those of ice Ih. A rare 28° halo from the sun, called Scheiner's halo, has been suggested to arise from the presence of octahedral ice Ic crystal in the atmosphere (Whalley 1983).

Before the recent discovery of ice Ic, diffraction data of so-called 'cubic ice' samples always showed the evidence of mixing of cubic and hexagonal stacking in the form of diffuse diffraction features (Kuhs et al. 1987, Hansen et al. 2008, Malkin et al. 2012, Kuhs et al. 2012, Malkin et al. 2015, Amaya et al. 2017). The name 'stacking disordered' ice (ice Isd) was therefore suggested for this type of ice I, which allows varying fractions of cubic and hexagonal stacking ranging between pure ice Ih and pure ice Ic (Malkin et al. 2012). A possible stacking sequence of ice Isd is shown in Fig. 4d. The space group of ice Isd is $P3m1$ which could explain the rare occurrence of snowflakes with 3-fold rotational symmetry (Murray et al. 2015). Computational fitting of the diffraction data of ice Isd enables the determination of its cubicity (that is, the percentage of cubic stacking). Furthermore, information on memory effects in the stacking sequences can be obtained, which is most usefully visualized with the help of the 'stackogram' plot (Hansen et al. 2008, Malkin et al. 2015, Salzmann et al. 2015). Independent of their cubicities, all ice Isd samples so far have displayed a greater tendency to stay with a given type of stacking rather than switching. In addition to diffraction, stacking disorder in ice I can also be identified from its Raman spectrum (Carr et al. 2014).

Following the preparation of ice Isd using cryogenic vapor deposition (König 1943), subsequent routes included the recrystallisation of high-pressure phases, heating of low-density amorphous ice, and freezing of water and aqueous solutions (Mayer and Hallbrucker 1987, Kuhs et al. 2012, Malkin et al. 2015). The enthalpies of transformation to ice Ih range from 23 to 160 J mol^{-1} (Malkin et al. 2015, Handa et al. 1986). It seems possible that changes in the interfacial energies may contribute to the measured enthalpies (Handa et al. 1988).

Ice XI is the hydrogen-ordered counterpart of ice Ih and forms below 72 K with the help of a base dopant (Tajima et al. 1982). Out of all alkali hydroxides, KOH was found to be the most effective dopant (Tajima et al. 1984, Matsuo and Suga 1987). The space group of ice XI is $Cmc2_1$, which makes it the only ferroelectric phase of ice with a net dipole moment (Leadbetter et al. 1985, Howe and Whitworth 1989, Line and Whitworth 1996, Jackson et al. 1997). Its unit cell contains eight water molecules with two different oxygen sites and three types of hydrogen bonds. Complete hydrogen ordering of ice XI has so far not been achieved. For example, annealing a KOD-doped D_2O ice Ih sample at 70 K for 135 hours leads to a 58% weight fraction of ice XI (Fukazawa et al. 2005). The ice Ih to XI phase transition has been shown to exhibit memory effects which means that once ice XI has been formed, it is easier to form it again from ice Ih (Arakawa et al. 2011). The problems in achieving full hydrogen order in ice XI have been attributed to the strain of the ice Ih/XI interface (Johari 1998). Acid dopants have been shown to be ineffective for preparing ice XI (Ueda et al. 1982, Matsuo and Suga 1987).

The possible existence of ice XI in space has been discussed (Fukazawa et al. 2006) and the ferroelectric nature of the ground state structure of ice Ih has recently been challenged (Parkkinen et al. 2014). Furthermore, now that ice Ic has been made, future work can focus on isolating its hydrogen-ordered counterpart for which several different structures have been discussed (Raza et al. 2011). In fact, hydrogen-ordered ice Ic could potentially be the most polar state of condensed H_2O with all dipole moments perfectly aligned.

In addition to ice I structures with extended hexagonal and cubic stacking, high-order polytypes are possible as well in principle. According to the Ramsdell notation, the 4H, 6H, and 9R polytype include periodic sequences of (hc), (hcc), and (hhc) blocks. If and how such polytypes of ice I can be prepared is unclear at present. However, it may well be that the ice I family of polytypes will continue to grow in the future (Salzmann 2019).

ICe II: The Odd One Out?

Ice II was the first of the high-pressure phases of ice to be discovered (Tammann 1900, Bridgman 1912). Its rhombohedral unit cell with R-3 space group symmetry (Kamb 1964, Finch et al. 1968, Kamb et al. 1971) contains open channels akin to nanotubes that are hydrogen-bonded to one another (*cf.* Fig. 3) (Salzmann et al. 2011). The rhombohedral unit cell contains 12 water molecules, two distinct oxygen sites and four types of hydrogen bonds. The corresponding hexagonal cell has three times the volume and hence contains 36 molecules. The channels consist of two types of six-membered rings, one significantly puckered and the other one almost flat, that are stacked in an alternating fashion. Eight and ten-membered rings exist between adjacent channels. Upon compression, ice II displays anisotropic changes in its structure (Fortes et al. 2003).

The open channels of ice II can be filled with hydrogen (Dyadin et al. 1999), helium (Londono et al. 1992, Lobban et al. 2002), and neon (Yu et al. 2014) which stabilizes the ice II structure with respect to ices III and V. In contrast to its neighboring phases in the phase diagram, ice II is fully hydrogen ordered (Kamb et al. 1971, La Placa and Hamilton 1973, Arnold et al. 1971, Lobban et al. 2002). This is remarkable considering that its triple point with ices III and V is only ~ 5 K away from the melting line (Journaux et al. 2020). The complete hydrogen order of ice II was also confirmed spectroscopically (Bertie and Whalley 1964b, Bertie et al.

1968a, Bertie and Francis 1980, 1982, Minceva-Sukarova et al. 1985, Li et al. 1991, Tran et al. 2017). Its compressibility and thermal expansion under *in situ* conditions has been studied using neutron diffraction (Fortes et al. 2005).

The thermal stability of the hydrogen-ordered ice II means that it is the only phase of ice for which the hydrogen-disordered counterpart is unknown. Upon heating ice II, it transforms to ices I, III, V, or VI depending on the pressure and not to its hydrogen-disordered counterpart (*cf*. Fig. 2). It has been predicted computationally that the phase transition from ice II to its disordered counterpart ice II*d* takes place at *p*/*T* conditions where liquid water is the stable phase (Nakamura et al. 2016). The space group of ice II*d* has been suggested as *R*-3c (Kamb 1973) although it should be noted that ice II can also appear to have this space group because of twinning (Kamb 1964). In addition to a perfectly hydrogen-ordered structure, dielectric spectroscopy suggests a complete lack of defects and a static structure in stark contrast to the hydrogen-disordered ices (Wilson et al. 1965). The lack of defects could also explain why ice II is much harder compared to its neighbors in the phase diagram which could have an impact on the geophysics of icy moons (Echelmeyer and Kamb 1986, Durham et al. 1997). Upon heating at ambient pressure, H_2O ice II is the only phase of ice displaying an endothermic phase transition to ice I*sd*, again illustrating its stability (Handa et al. 1988, Fuentes-Landete et al. 2020). Out of all the high-pressure phases, it also gives the ice I*sd* with the highest cubicity (Malkin et al. 2015). Whalley et al. (1965) stated that "the complexity of the phase diagram of ice is due largely to ice II being ordered at all temperatures".

Recently it was discovered that doping ice with ammonium fluoride (NH_4F)leads to the disappearance of ice II from the phase diagram (Shephard et al. 2018).NH_4F has a hydrogen-disordering effect on ice (Salzmann et al. 2019), which means that in the case of ice II, the free energy is raised by such an amount that competing hydrogen-disordered phases become more stable. The NH_4F-doping experiment also illustrates that ice II is strictly topologically constrained since small amounts of dopant have such a dramatic effect. Since ice II occupies an important central region of the phase diagram, it has been argued that its unique properties, including its exceptionally long correlation length, could be regarded as the original anomaly in the H_2O system (Shephard et al. 2018). In any case, the doping-induced disappearance of ice II was the first instance of a dopant being used to remove a phase from the phase diagram.

If ice II should be regarded as the 'odd one out' is ultimately a matter of perspective. Considering its stability as well as its hydrogen ordered and defect-free structure, it could be argued that it is an 'ideal' phase of ice and that the hydrogen-disordered phases with their dynamic and defective structures should instead be regarded as 'odd'.

Ices III and IX: Chiral H_2O

Ice III occupies the smallest region of thermodynamic stability in the phase diagram between 0.22 and 0.34 GPa (Tammann 1900, Bridgman 1912, 1937). Its tetragonal unit cell with $P4_12_12$ space group symmetry contains 12 water molecules, two different oxygen sites, and three types of hydrogen bonds (McFarlan 1936, Kamb and Datta 1960, Kamb and Prakash 1968, Rabideau et al. 1968, La Placa and Hamilton

1973). The crystal structure includes five-, seven-, and eight-membered rings. An important structural feature are 4-fold spirals of hydrogen-bonded water molecules running in the c direction (*cf.* Fig. 3). These spirals can be either left- or right-handed which makes ice III a chiral phase of ice (Salzmann 2019). The enantiomorphic counterpart of $P4_12_12$ is $P4_32_12$. Ice III is less dense than ice II and it is isostructural with the SiO_2 polymorph keatite. Originally, ice III was believed to be completely hydrogen-disordered (Wilson et al. 1965, Whalley and Davidson 1965). However, *in situ* neutron diffraction showed that ice III is actually partially ordered with the occupancies of the hydrogen sites deviating from ½ (Londono et al. 1993, Kuhs et al. 1998, Lobban et al. 2000).

Upon cooling under pressure, ice III transforms to its antiferroelectric hydrogen-ordered counterpart ice IX between 208 and 165 K (Whalley et al. 1968). The ice III to IX phase transition is the only hydrogen-ordering phase transition in ice observed so far where the space group symmetry does not change. Unlike ice III, the dimensions of the unit cell of ice IX are close to cubic (La Placa and Hamilton 1973, Londono et al. 1993). In stark contrast to the "ice-nine" in Kurt Vonnegut's novel *Cat's Cradle* (Vonnegut 1963), which is a fictional stable form of ice at room temperature and ambient pressure, the real ice IX is metastable with respect to ice II below the region of stability of ice III (Petrenko and Whitworth 1999). Due to its metastability, ice IX is often found in sequences of phase transitions at low temperatures following Ostwald's rule of stages up to ~ 0.7 GPa (Salzmann et al. 2004a, Salzmann et al. 2008). Using single-crystal neutron diffraction, D_2O ice IX was shown to contain a small amount of residual hydrogen disorder (La Placa and Hamilton 1973, Londono et al. 1993). Spectroscopic studies of ice IX samples at low temperatures and ambient pressure confirmed its highly hydrogen-ordered nature (Bertie and Bates 1977, Bertie and Francis 1982, Bertie et al. 1968a). The experimental ordered structure was confirmed to be the lowest energy structure with DFT calculations (Knight and Singer 2006).

The phase transition from ice III to ice IX upon cooling under pressure was followed with dielectric spectroscopy (Whalley et al. 1968), calorimetry (Nishibata and Whalley 1974), and Raman spectroscopy (Minceva-Sukarova et al. 1984). It was found that cooling rates greater than 1–2 K min^{-1} are needed in order to suppress its transformation to the stable ice II (Minceva-Sukarova et al. 1984, Arnold et al. 1971). Upon heating, ice IX transforms to ice II, which means that the ice IX to ice III phase transition has not been observed for pure ice so far (Whalley et al. 1968, Nishibata and Whalley 1974, Minceva-Sukarova et al. 1984). Using ND4F doping to prevent the formation of ice II, the reversible ice III to ice IX phase transition was recently observed with *in situ* neutron diffraction (Sharif et al. 2021).

Ices V and XIII: Structural Complexity at Its Most Extreme

Ice V is denser than ice II and its region of thermodynamic stability is found between 0.35 and 0.60 GPa (Tammann 1900, Bridgman 1912). Its monoclinic unit cell with $C2/c$ space group symmetry contains 28 water molecules, four distinct oxygen sites, and seven types of hydrogen bonds (Kamb et al. 1967, Hamilton et al. 1969). Four-, five-, six-, eight-, nine-, ten-, and twelve-membered rings exist within its structure illustrating the remarkable structural complexity of this phase of ice (*cf.* Fig. 3). The structure can be described with two types of zigzag chains that run parallel to the *a* axis and form two layers within the unit cell. The hydrogen-disordered nature of ice V was demonstrated using dielectric spectroscopy (Wilson et al. 1965) and thermodynamic considerations (Whalley and Davidson 1965). The structural complexity of ice V also manifests in vibrational spectroscopy (Bertie and Whalley 1964b, Minceva-Sukarova et al. 1986, Tran et al. 2017). Using *in situ* neutron diffraction, it was shown that ice V is not fully hydrogen disordered with some of the fractional occupancies of the hydrogen sites deviating substantially from ½ (Kuhs et al. 1998, Lobban et al. 2000). Two of the hydrogen sites in $C2/c$ are restricted to ½ occupancy which means that this space group does not allow full hydrogen order. Pure H_2O ice V shows an orientational glass transition at ~ 130 K upon heating at ambient pressure at 30 K min^{-1} corresponding to the unfreezing of reorientation dynamics of the water molecules (Salzmann et al. 2003a, Salzmann et al. 2011).

Weak hydrogen ordering upon cooling pure ice V has been suggested in several studies (Kamb and La Placa 1974, Minceva-Sukarova et al. 1988, Handa et al. 1988). Fully hydrogen-ordered ice V, which was named ice XIII, was obtained using hydrochloric acid (HCl) as a dopant and slow-cooling at ambient pressure (Salzmann et al. 2006c). The space group symmetry of the antiferroelectric ice XIII is $P2_1/a$ with seven distinct oxygen positions and 14 different hydrogen bonds which makes it by far the most structurally complicated phase of ice. During the ice V to ice XIII phase transition, the *a* lattice constant and the monoclinic angle *β* increase whereas the *b* and *c* lattice constants contract (Salzmann et al. 2007).

Using calorimetry, it was shown that the ice V to ice XIII phase transition at ~ 113 K goes along with a 66% loss of Pauling entropy upon slow-cooling at ambient pressure (Salzmann et al. 2008). Considering the highly ordered nature of ice XIII, this value reflects the partially ordered nature of ice V. Using neutron diffraction, it has been shown that the partial order in ice V is consistent with the way hydrogen order is established in ice XIII (Salzmann 2021). According to calorimetry, the ice V to ice XIII phase transition takes place in at least two overlapping stages (Salzmann et al. 2008). Slow-cooling under pressure gives a slightly more disordered ice XIII. The highly hydrogen-ordered nature of ice XIII can also be seen from its Raman spectrum (Salzmann et al. 2006a). The ice V to ice XIII phase transition has also been reproduced computationally (Knight and Singer 2008).

The role of the acid dopant is to produce mobile H_3O^+ point defects that speed up molecular reorientation dynamics and hence enable the hydrogen ordering phase transition to take place at low temperatures. The effect of HCl doping on the dielectric relaxation times was shown (Köster et al. 2016). Comparative studies of the effects of different acid dopants, such as HCl, HBr, $HClO_4$ and HF, on the hydrogen-ordering process during the ice V/XIII transition found that the interplay between acid strength and its solubility in ice determines the effectiveness of dopants with respect to enabling the hydrogen-ordering transition (Salzmann et al. 2008, Rosu-Finsen and Salzmann 2018). HF doping has an ordering effect on ice V but to a lesser extent than HCl which is the most effective hydrogen-ordering agent for ice V/XIII. Further studies into base dopants showed that LiOH facilitates the transition of ice

V to ice XIII to the same extent as HF (Rosu-Finsen and Salzmann 2018, Salzmann et al. 2021). Therefore, ice XIII is the first hydrogen-ordered phase of ice that can be prepared with the help of both acid and base dopants.

Ice IV: The Rulebreaker

The rule of numbering the phases of ice chronologically according to their dates of discovery was only broken once. Tammann suspected a metastable phase of ice to exist below the melting line of ice Ih (Tammann 1910) which led Bridgman to skip the number IV when he discovered ice V (Bridgman 1912). Investigating the phase diagram of D$_2$O, Bridgman then identified a metastable phase which he named ice IV (Bridgman 1935). As far as we know, Bridgman's ice IV is different from the metastable phase originally suspected by Tammann. The metastable nature of ice IV is reflected in its challenging synthesis. After crystallization of liquid water to ice VI, ice IV is formed in the narrow window of 258.6–259.6 K and 0.500–0.535 GP a upon slow decompression of ice VI, taking care not to approach the melting pressure of ice VI (Nishibata 1972). The formation of ice IV following this route occurs only sporadically and in an unpredictable manner. However, organic nucleation agents have been successfully used to aid its formation (Evans 1967, Engelhardt and Whalley 1972). Due to its transient nature, ice IV has been called a "will-o'-the-wisp, ghostly form of ice" (Ball 1999). In a new approach, ice IV was prepared in a reproducible fashion by slowly heating high-density amorphous ice (HDA) at 0.81 GPa (Salzmann et al. 2002b, Salzmann et al. 2003c, Salzmann et al. 2004a). The importance of applying a smooth heating curve for making ice IV has recently been pointed out (Rosu-Finsen and Salzmann 2022).

Ice IV forms in a rhombohedral structure with 16 water molecules per unit cell (space group R-$3c$), two distinct oxygen sites, and four types of hydrogen bonds (Engelhardt and Kamb 1981, Klotz et al. 2003). The corresponding hexagonal cell contains 48 water molecules. The crystal structure features layers of puckered six-membered rings in a chair conformation that are interpenetrated by hydrogen bonds as shown in Fig. 3. The structure is in some sense similar to ice I but with hydrogen bonding between the second-nearest layers. In addition to the six-membered rings, eight- and ten-membered rings exist as well. Infrared and Raman spectroscopy suggest ice IV is hydrogen-disordered (Engelhardt and Whalley 1979, Salzmann et al. 2003b). At ambient pressure, H$_2$O ice IV displays an orientational glass transition at ~ 140 K upon heating at 30 K min^{-1} (Salzmann et al. 2004b). Recent molecular dynamics simulations of HDA identified local environments similar to those of ice IV and experimental comparisons with the isostructural NH$_4$F phases suggested that HDA is a "derailed state" forming during the transition from ice Ih to ice IV (Martelli et al. 2018, Shephard et al. 2017). Doping ice IV with HCl leads to the appearance of a weak endotherm in calorimetry around 113 K (Salzmann et al. 2011) and this feature increases in area as the sample is annealed at pressures up to 2.5 GPa (Rosu-Finsen and Salzmann 2022). Whether this feature arises as part of the kinetic unfreezing of molecular reorientation dynamics, release of stress/strain after the preparation under pressure or from the transition from a weakly hydrogen-ordered counterpart of ice IV is unclear (see question mark in Fig. 2).

Ices XII and XIV: Large Rings and Extreme Bending of Hydrogen Bonds

Another metastable polymorph, ice XII, forms in the stability domains of ices V and VI and was first prepared through slow crystallization of liquid water at 0.55 GPa (Lobban et al. 1998). The tetragonal ice XII structure (space group *I*-42*d*) contains 12 water molecules per unit cell with two distinct oxygen sites and two types of hydrogen bonds (Lobban et al. 1998, Koza et al. 1999). In the *a-b* projection, the ice XII structure resembles the Cairo tiling, which has five-membered rings only (Salzmann 2019). While such a two-dimensional phase of ice has been identified computationally (Chen et al. 2016), the ice XII structure actually contains seven- and eight-membered rings (O'Keeffe 1998) and is the densest known phase of ice without interpenetrating structural features. Instead, the high density is achieved by significant bending of the hydrogen bonds. In addition to the crystallization from the liquid, ice XII has been identified as an accidental byproduct during the preparation of HDA from ice I*h* at 77 K (Koza et al. 1999, Koza et al. 2000). Later, it was suggested that the ice XII formation is due to shock-wave heating (Kohl et al. 2001). Isobaric heating of HDA in the 0.7 to 1.4 GPa pressure range then revealed ice XII in mixtures with other crystalline polymorphs (Loerting et al. 2002). To prepare pure ice XII by heating HDA, it was shown that controlling the heating rate is crucial (for example > 11 K min^{-1} at 0.8 GPa) (Salzmann et al. 2003c, Salzmann et al. 2004a).

Ice XII is fully hydrogen disordered as shown by neutron diffraction and Raman spectroscopy (Lobban et al. 1998, Salzmann et al. 2002a). Spectroscopically, ice XII appears to be very similar to a metastable phase discovered by Chou et al. (1998). However, if the Chou phase is indeed ice XII is still debated (Salzmann et al. 2004b, Yoshimura et al. 2007). Upon heating at ambient pressure, H$_2$O ice XII displays an orientational glass transition at ~ 131 K upon heating at 30 K min^{-1} (Salzmann et al. 2004b). Annealing below the glass transition temperature has been shown to produce kinetic overshoot effects (Salzmann et al. 2003a).

Doping ice XII with HCl leads to the formation of its hydrogen-ordered counterpart ice XIV below ~ 103 K (Salzmann et al. 2006c). Ice XIV is antiferroelectric and orthorhombic with $P2_12_12_1$ space group symmetry. Its unit cell contains three distinct oxygen sites and four types of hydrogen bonds. The reversibility of the ice XII to ice XIV phase transition at ambient pressure has been demonstrated (Salzmann et al. 2007, Salzmann et al. 2006b). However, cooling under pressure leads to a much more hydrogen-ordered ice XIV, which has been attributed to the pressure helping to overcome orthorhombic strain. The phase transition from ice XII to ice XIV goes along with expansions in the *a* and *c* lattice constants and a contraction in *b* (Salzmann et al. 2007). Doping with HF leads to less-ordered ice XIV (Köster et al. 2015), whereas KOH seems to be ineffective (Salzmann et al. 2006c). Following an initial claim of a complete release of Pauling entropy during the ice XII to ice XIV transition under pressure using HCl doping (Köster et al. 2015), an integration mistake of the calorimetric data was conceded and the value was corrected to 60% (Köster et al. 2018). If the effect of temperature is considered correctly upon integration, it has been argued that the actual value should be 51%, which means that ice XIV still contains significant amounts of hydrogen disorder (Rosu-Finsen and

Salzmann 2018). The energetics of the ordering of ice XIV are consistent with DFT calculations (Tribello et al. 2006).

Ices VI, XV and XIX: Self-clathrates

Ice VI is stable between 0.6 and 2.2 GPa and the first of the high-pressure phases of ice to display a melting point greater than 0°C (Bridgman 1912). For this reason, it was nick-named "hot ice" (Walter 1990). Its tetragonal crystal structure with $P4_2/nmc$ space group symmetry contains 10 water molecules, two distinct oxygen sites, and three types of hydrogen bonds (*cf.* Fig. 3) (Kamb 1965). Furthermore, ice VI is the first of the high-pressure phases of ice to contain two independent and interpenetrating hydrogen-bonded networks. Ice VI has therefore been called a "self-clathrate" (Kamb 1965). The networks consist of hexameric units with structures like the $(H_2O)_6$ 'cage-like' cluster in the gas phase (Liu et al. 1996, Ludwig 2001). To build up a network, the hexameric units share corners in the c direction and are hydrogen-bonded to one another in the a and b directions. Ice VI is isostructural with the edingtonite SiO_2 polymorph (Kamb 1965). The $P4_2/nmc$ space group requires ice VI to be completely hydrogen disordered and its crystal structure was confirmed using *in situ* neutron diffraction (Kuhs et al. 1984, Kuhs et al. 1989). The hydrogen-disordered nature of ice VI is also consistent with dielectric (Wilson et al. 1965) and calorimetric measurements (Hobbs 1974). The *in situ* equation of state of ice VI has been measured using neutron diffraction (Fortes et al. 2012). The crystallization of ice VI from the liquid has been shown to be affected by the presence of sodium halides (Zeng et al. 2017, Zeng et al. 2016). Upon heating H_2O ice VI at ambient pressure, an orientational glass transition is found at ~ 134 K (Shephard and Salzmann 2016).

The search for the hydrogen-ordered counterpart of ice VI began quite early. Kamb (1965) already mentioned the observation of a reflection in the X-ray diffraction of ice VI at 95 K and ambient pressure which is inconsistent with the reflection conditions of $P4_2/nmc$ (Kamb 1965) and he later assigned *Pmmn* space group symmetry to what he called ice VI' (Kamb 1973). *Pmmn* allows partial hydrogen order and implies antiferroelectric hydrogen ordering. However, this was not reproduced in later neutron diffraction work (Kuhs et al. 1984, Kuhs et al. 1989). Weak hydrogen ordering upon cooling ice VI under pressure has been suggested from dielectric (Johari and Whalley 1976, 1979), thermal conductivity (Ross et al. 1978), and thermal expansion measurements (Mishima et al. 1979). In contrast to Kamb's ice VI' (Kamb 1973), dielectric measurements implied ferroelectric ordering of ice VI upon cooling under pressure (Johari and Whalley 1976, 1979). Raman and FT-IR spectroscopy of ice VI at low temperatures did not identify any signs of hydrogen order (Bertie et al. 1968b, Minceva-Sukarova et al. 1986), even when the sample was doped with KOH (Minceva-Sukarova et al. 1988).

Using HCl as a dopant led to the discovery of the antiferroelectric ice XV, the hydrogen-ordered counterpart of ice VI (Salzmann et al. 2009). The most ordered ice XV was obtained after slow-cooling at ambient pressure. Its pseudo-orthorhombic unit cell with P-1 space group symmetry contains five distinct oxygen sites and ten types of hydrogen bonds (Salzmann et al. 2009, Salzmann et al. 2016). There are three different ways in which the hexameric units can become hydrogen ordered.

The experimental structure contains the most polar individual networks. However, the polarity of the networks is cancelled by the centrosymmetric *P*-1 space group symmetry (Salzmann et al. 2016). The antiferroelectric nature of ice XV is consistent with its Raman spectrum (Whale et al. 2013).

The phase transition from ice VI to ice XV starting at ~ 130 K at ambient pressure goes along with contractions in *a* and *b* whereas there is a significant expansion in *c* which leads to an overall volume expansion. The increase in volume upon hydrogen-ordering explains why the most ordered states of ice XV are obtained upon cooling at ambient pressure. As can be seen from the changes in lattice constants but also calorimetric data, the ice VI to ice XV phase transition is quite fast at first, followed by a long tail so that the entire phase transition takes place over a ~ 30 K window (Shephard and Salzmann 2015, Salzmann et al. 2016). This rather large window has been explained with the complication that the ice VI to ice XV phase transitions must include the hydrogen-ordering of the individual networks as well as establishing ordering of the two networks with respect to one another. Even slow-cooling at ambient pressure only achieves a ~ 50% loss of the Pauling entropy. Recently, HCl was confirmed as the most effective dopant for making ice XV (Rosu-Finsen and Salzmann 2018). In addition to HF, doping with HBr also proved to enable hydrogen ordering. This was somewhat surprising since HBr doping is ineffective for preparing ice XIII. It was speculated that the large bromide anions could be accommodated within the ice VI/XV lattice by partially substituting one of the hexameric units and that this is not possible for the more 'densely-knit ice' V/XIII network (Rosu-Finsen and Salzmann 2018).

In contrast to the experimental *P*-1 structure, DFT calculations suggested that the lowest energy ice XV structure is ferroelectric with *Cc* space group symmetry (Knight and Singer 2005, Kuo and Kuhs 2006). A later study using fragment-based 2nd order perturbation and coupled cluster theory then suggested the antiferroelectric *P*-1 structure as the lowest energy structure (Nanda and Beran 2013). This result was subsequently contested using the fully periodic 2nd order Møller-Plesset perturbation theory and the DFT random phase approximation reconfirming the original *Cc* structure (Del Ben et al. 2014). However, in this study it was also shown that the various possible ferroelectric structures can be destabilized by the dielectric properties of the surrounding medium. Hence, a mechanism by which antiferroelectric domains could become stabilized was identified providing an explanation for the apparent discrepancy between theory and experiment.

Upon increasing the pressure, the formation of ice XV becomes more and more difficult (Komatsu et al. 2016, Salzmann et al. 2016). For example, D_2O samples either quenched at 1.0 GPa or slow-cooled at 1.4 GPa were identified as ice VI in neutron diffraction, and hence, lacking any significant hydrogen ordering. Heating such samples at ambient pressure revealed a 'transient' ordering feature both in neutron diffraction as well as in calorimetry which was followed by a phase transition to ice VI (Shephard and Salzmann 2015, Salzmann et al. 2016). The transient ordering feature can be explained by the high free energy of the essentially ice VI states at low temperatures and hence the strong tendency to hydrogen order at low pressures.

Slow-cooling HCl-doped ice VI at pressures greater than 1 GPa then brought about an irreversible endotherm upon heating at ~ 105 K preceding the transient ordering. This new low-temperature endotherm was initially attributed to a new hydrogen ordered phase of ice that was suggested to be more and differently hydrogen-ordered than ice XV (Gasser et al. 2018, Thoeny et al. 2019). Furthermore, a strong isotope effect was suggested that prevents the formation of a corresponding D_2O phase. In response to this, it was suggested that the new low-temperature endotherm is related to the glass transition of 'deep-glassy' states of ice VI (Rosu-Finsen and Salzmann 2019, Rosu-Finsen et al. 2020). It was shown that the low-temperature endotherms displayed all the characteristics of endothermic overshoot effects related to a glass transition. This included a typical dependence on pressure and cooling rate, the fact that they could be produced by sub-T_g annealing at ambient pressure and that they could be made to appear or disappear depending on the heating rate and the initial extent of relaxation. Such deep glassy states have been observed for a variety of other materials (Moynihan et al. 1976, Zhao et al. 2013). Furthermore, deep-glassy states of ice VI were also identified for corresponding D_2O samples and the proposed deep-glassy ice VI scenario was shown to be consistent with X-ray diffraction recorded by Gasser et al. (Rosu-Finsen and Salzmann 2019). Further to this, using a combination of neutron spectroscopy and diffraction, an HCl-doped H_2O deep-glassy ice VI sample was compared with pure ice VI and ice XV (Rosu-Finsen et al. 2020). This showed that deep-glassy ice VI is very closely related to ice VI from both the spectroscopic as well as the diffraction point of view (Rosu-Finsen et al. 2020). Additionally, in line with expectation, it has been shown that the hydrogen-ordered structure of D_2O ice XV can be used to describe the hydrogen order in the corresponding H_2O phase. Most recently, new Bragg peaks were discovered by three independent groups upon cooling DCl-doped ice VI at pressures greater than 1.6 GPa (Yamane et al. 2021, Gasser et al. 2021, Salzmann et al. 2021). In line with the deep-glassy ice VI scenario, these are consistent with the formation of a structurally distorted ice VI structure combined with potentially weak hydrogen ordering (Salzmann et al. 2021). Alternatively, it has been suggested that what is now called ice XIX displays considerable hydrogen order (Yamane et al. 2021, Gasser et al. 2021).

Ices VII, VIII, X, and XVIII: the High-pressure Frontier

Ice VII is a stable phase of ice above 2.1 GPa (Bridgman 1937, Pistorius et al. 1968). Like ice VI, it contains two interpenetrating networks that are isostructural with ice Ic (Kuhs et al. 1984, Jorgensen and Worlton 1985). Its cubic unit cell with $Pn\text{-}3m$ space group symmetry contains two water molecules, one distinct oxygen site, and one type of hydrogen bond (*cf.* Fig. 3). The $Pn\text{-}3m$ space group requires ice VII to be completely hydrogen disordered. In addition to ice Ih, ice VII is the second phase of ice considered to be a mineral following its discovery as an inclusion compound inside diamonds (Tschauner et al. 2018). Due to its simple structure, ice VII has been a testbed for probing the positional disorder of the oxygen atoms induced by the hydrogen-disorder (Kuhs et al. 1984, Jorgensen and Worlton 1985, Nelmes et al. 1998, Knight and Singer 2009, Bellin et al. 2011). In a recent neutron diffraction study, ice VII has been compressed beyond 60 GPa (Guthrie et al. 2019). In contrast to ice Ih, ice VII can accommodate lithium chloride and bromide at ~ 1:6 LiX:H_2O

molar ratios (Klotz et al. 2016, Klotz et al. 2009). The unit cell volume of 'salty' ice VII is strongly increased and the concomitant positional disorder of H_2O makes it a "plastic" phase of ice. Compared to the lithium halides, the solubility of NaCl in ice VII is much smaller (Ludl et al. 2017).

Upon cooling, ice VII transforms to its hydrogen-ordered counterpart ice VIII within a narrow temperature range around 273 K over quite a large pressure range (Whalley et al. 1966, Pistorius et al. 1968, Johari et al. 1974). In addition to the ice III to ice IX phase transition, this is the only other hydrogen ordering phase transition that takes place without the help of dopants. The ice VIII unit cell is tetragonal with $I4_1/amd$ space group symmetry ($a_{VIII} = \sqrt{2}a_{VII}$ and $c_{VIII} = 2a_{VII}$) (Jorgensen et al. 1984, Kuhs et al. 1984). Similar to the situation for ice XV, the two networks in ice VIII are as polar as they can be but because of the $I4_1/amd$ space group symmetry, the polarities of the individual networks are cancelled so that ice VIII is antiferroelectric overall. Recent *in situ* neutron diffraction work has shown that the ice VII to ice VIII phase transitions slows down around 10 GPa but then becomes fast again at higher pressures (Komatsu et al. 2020a). These findings have been attributed to a crossover in the hydrogen dynamics and several of the previously observed anomalies of ice VII in this pressure range (Okada et al. 2014, Noguchi and Okuchi 2016, Hirai et al. 2014) have been attributed to this phenomenon. Furthermore, the hydrogen ordering of ice VII can be slowed down and partially prevented by using NH_4F as a dopant (Salzmann et al. 2019). Close to completely hydrogen-disordered ice VII can be obtained by low temperature compression of HDA or ice VI to make a phase of ice called ice VII' (Hemley et al. 1989, Klotz et al. 1999).

Recovered at ambient pressure, ices VII' and VIII behave differently to the other high-pressure phases of ice upon heating. Instead of transforming to ice Isd, they form low-density amorphous ice (LDA) first (Klug et al. 1989, Klotz et al. 1999, Klotz et al. 2005). Using Raman spectroscopy and X-ray diffraction, the LDA from ice VIII was shown to be different compared to other types of LDA formed through, for example, vapor deposition onto a cold substrate (Shephard et al. 2016). Using ultrafast diffraction, it has recently been shown that HDA forms before LDA upon heating ice VIII (Lin et al. 2020).

Compression of ice VII beyond 60 GPa leads to the formation of cubic ice X with $Pn\text{-}3m$ space group symmetry (Stillinger and Schweitzer 1983, Polian and Grimsditch 1984, Aoki et al. 1996, Teixeira 1998, Goncharov et al. 1996, Goncharov et al. 1999). The hydrogen atoms in ice X are located halfway between the oxygen atoms which means that the water molecules lose their molecular character as shown in Fig. 1b. Experimentally, it has been shown that ice X persists up to at least 210 GPa (Goncharov et al. 1996). However, recent *in situ* Raman measurements have suggested that the transition from ice VII to ice X may not actually occur below 120 GPa (Zha et al. 2016). Furthermore, incorporating LiCl into ice VII has been shown to shift the phase transition from ice VII to ice X towards higher pressures (Bove et al. 2015). A variety of post-ice X phases have been predicted computationally (Benoit et al. 1996, Militzer and Wilson 2010, Pickard et al. 2013).

Subjecting ice VII to sophisticated laser-driven shock-compression, superionic ice XVIII with $Fm\text{-}3m$ space group symmetry forms above 100 GPa and 2000 K (Millot et al. 2018, Millot et al. 2019). The hydrogen atoms in ice XVIII are free to move between the densely packed oxygen atoms. Ice XVIII has a melting point near 5000 K at 190 GPa which means that it may be present in the interiors of Neptune and Uranus. A variety of other superionic ices have been predicted computationally (Sun et al. 2015).

Ices XVI and XVII: The Empty Clathrate Hydrates

Two low-density ices can be accessed from the corresponding gas clathrate hydrates. Several hundreds of hypothetical structures have been explored by computational efforts in the search for new ice modifications, but hitherto, only two have been realized experimentally (Kosyakov and Shestakov 2001, Matsui et al. 2017, Huang et al. 2016, Huang et al. 2017).

A neon clathrate hydrate with the cubic structure II can be formed by pressurizing ice Ih to 0.35 GPa at 244 K. By pumping the material to low pressures, the neon atoms escape the cages in the clathrate structure, leaving behind the empty clathrate hydrate structure, which was named ice XVI (Falenty et al. 2014). Hydrogen-disordered ice XVI contains 136 water molecules and three distinct oxygen sites in its cubic unit cell with Fd-$3m$ space group symmetry. The hydrogen bonds form two types of empty cages as shown in Fig. 3. These cages consist of five- and six-membered rings. The removal of neon from the cages during formation leads to a volume increase, resulting in a density of 0.81 g cm^{-3} (Falenty et al. 2014). Recently, refilling of the clathrate cages in ice XVI with helium led to the first reported helium clathrate hydrate (Kuhs et al. 2018).

In a similar approach, a hydrogen-filled ice structure was first prepared by exposing ice to ~ 0.4 GPa of H$_2$ gas at 255 K (Strobel et al. 2016). By pumping the material at 120 K, hydrogen can be reversibly extracted from the clathrate hydrate with C$_0$-type structure to obtain the empty clathrate named ice XVII (del Rosso et al. 2016a). Reducing the H$_2$ content causes a decrease of the lattice parameter c, but an increase of a. The water molecules in the hexagonal unit cell of ice XVII with space group $P6_122$ (or $P6_522$) and one distinct oxygen site form a network of five-membered rings. This gives rise to interconnected spiral chains around spacious hexagonal channels in the overall hydrogen-disordered structure (del Rosso et al. 2016b). In addition to ices III/IX, ice XVII displays a second type of chiral ice structure. This type of structure was also recently found for a CO$_2$ clathrate hydrate (Amos et al. 2017).

Conclusions

The exploration of water's phase diagram started more than a century ago with the work of Tammann and Bridgman (Tammann 1900, 1910, Bridgman 1912). Starting with the development of piston cylinders to the modern-day laser shock-wave heating experiments (Lin and Tse 2021), new experimental techniques have always been tested with respect to exploring the phase diagram of H$_2$O to ever higher pressures and temperatures, and there is no doubt that this trend will continue in the future. In addition to exploring extreme pressure and temperature conditions, the last decade has seen a new trend emerging with a large number of studies exploring the 'chemical' dimension of ice research. This has included work on enabling hydrogen ordering with the help of dopants, using small gas species as removable templates for growing low-density ice phases, the formation of 'salty' ices with very high concentrations of dissolved species and the selective disappearance of a phase of ice with the help of an impurity. Knowing how ice behaves when mixed with other chemical species is of course highly relevant for environmental research and the vast research area of clathrate hydrates (Ripmeester et al. 2006, Loveday and Nelmes 2008). Interestingly, working on this review we also realized that the intermediate pressure range, which includes ices III, V, and II and their (dis)ordered counterparts, has actually not received as much attention as the exploration of the higher-pressure phases. In any case, the H_2O system will without doubt reveal many more secrets in the future and we can only wonder what the next phase of ice will be.

References

Amaya, A. J., H. Pathak, V. P. Modak, H. Laksmono, N. D. Loh, J. A. Sellberg et al. 2017. How cubic can ice be? J. Chem. Phys. Lett. 8: 3216–3222.

Amos, D. M., M.-E. Donnelly, P. Teeratchanan, C. L. Bull, A. Falenty, W. F. Kuhs et al. 2017. A chiral gas–hydrate structure common to the carbon dioxide–water and hydrogen–water systems. J. Phys. Chem. Lett. 8: 4295–4299.

Aoki, K., H. Yamawaki, M. Sakashita and H. Fujihisa. 1996. Infrared absorption study of the hydrogen-bond symmetrization in ice to 110 GPa. Phys. Rev. B 54: 15673–15677.

Aragones, J. L. and C. Vega. 2009. Plastic crystal phases of simple water models. J. Chem. Phys. 130: 244504.

Arakawa, M., H. Kagi, J. A. Fernandez-Baca, B. C. Chakoumakos and H. Fukazawa. 2011. The existence of memory effect on hydrogen ordering in ice: the effect makes ice attractive. Geophys. Res. Lett. 38: L16101.

Arnold, G. P., R. G. Wenzel, S. W. Rabideau, N. G. Nereson and A. L. Bowman. 1971. Neutron diffraction study of ice polymorphs under helium pressure. J. Chem. Phys. 55: 589–595.

Ball, P. 1999. H_2O A Biography of Water: Weidenfeld & Nicolson.

Barnes, W. H. and W. H. Bragg. 1929. The crystal structure of ice between 0°C and –183°C. Proc. Roy. Soc. A125: 670–693.

Bellin, C., B. Barbiellini, S. Klotz, T. Buslaps, G. Rousse, T. Strässle et al. 2011. Oxygen disorder in ice probed by x-ray Compton scattering. Phys. Rev. B 83: 094117.

Benoit, M., M. Bernasconi, P. Rocher and M. Parrinello. 1996. New high-pressure phase of ice. Phys. Rev. Lett. 76: 2934–2936.

Bernal, J. D. and R. H. Fowler. 1933. A theory of water and ionic solution, with particular reference to hydrogen and hydroxyl ions. J. Chem. Phys. 1: 515–549.

Bertie, J. E. and E. Whalley. 1964a. Infrared Spectra of Ices Ih and Ic in the Range 4000 to 350 cm^{-1}. J. Chem. Phys. 40: 1637–1645.

Bertie, J. E. and E. Whalley. 1964b. Infrared spectra of ices II, III, and V in the range 4000 to 350 cm^{-1}. J. Chem. Phys. 40: 1646–1659.

Bertie, J. E., H. J. Labbé and E. Whalley. 1968a. Far-infrared spectra of ice II, V, and IX. J. Chem. Phys. 49: 775–780.

Bertie, J. E., H. J. Labbé and E. Whalley. 1968b. Infrared Spectrum of Ice VI in the Range 4000–50 cm^{-1}. J. Chem. Phys. 49: 2141–2144.

Bertie, J. E. and F. E. Bates. 1977. Mid-infrared spectra of deuterated ices at 10°K and interpretation of the OD stretching bands of ices II and IX. J. Chem. Phys. 67: 1511–1518.

Bertie, J. E. and B. F. Francis. 1980. Raman spectra of the O-H and O-D stretching vibrations of ice II and IX to 25°K at atmospheric pressure. J. Chem. Phys. 72: 2213–2221.

Bertie, J. E. and B. F. Francis. 1982. Raman spectra of ices II and IX above 35 K at atmospheric pressure: translational and rotational vibrations. J. Chem. Phys. 77: 1–15.

Bove, L. E., R. Gaal, Z. Raza, A.-A. Ludl, S. Klotz, A. M. Saitta et al. 2015. Effect of salt on the H-bond symmetrization in ice. Proc. Natl. Acad. Sci. USA 112: 8216–8220.

Bragg, S. W. H. 1921. The crystal structure of ice. Proc. Phys. Soc. 34: 98–103.

Bridgman, P. W. 1912. Water, in the liquid and five solid forms under pressure. Proc. Am. Acad. Arts Sci. 47: 441–558.

Bridgman, P. W. 1935. The pressure-volume-temperature relations of the liquid, and the phase diagramm of heavy water. J. Chem. Phys. 3: 597–605.

Bridgman, P. W. 1937. The phase diagramm of water to 45000 kg/cm². J. Chem. Phys. 5: 964–966.

Carr, T. H. G., J. J. Shephard and C. G. Salzmann. 2014. Spectroscopic signature of stacking disorder in ice I. J. Phys. Chem. Lett. 5: 2469–2473.

Chen, J., G. Schusteritsch, C. J. Pickard, C. G. Salzmann and A. Michaelides. 2016. Two dimensional ice from first principles: Structures and phase transitions. Phys. Rev. Lett. 116: 025501.

Chou, I.-M., J. G. Blank, A. F. Goncharov, H.-K. Mao and R. J. Hemley. 1998. *In situ* observations of a high-pressure phase of H_2O ice. Science 281: 809–812.

Del Ben, M., J. VandeVondele and B. Slater. 2014. Periodic MP2, RPA, and boundary condition assessment of hydrogen ordering in ice XV. J. Phys. Chem. Lett. 5: 4122–4128.

del Rosso, L., M. Celli and L. Ulivi. 2016a. New porous water ice metastable at atmospheric pressure obtained by emptying a hydrogen-filled ice. Nat. Comm. 7: 13394.

del Rosso, L., F. Grazzi, M. Celli, D. Colognesi, V. Garcia-Sakai and L. Ulivi. 2016b. Refined structure of metastable ice XVII from neutron diffraction measurements. J. Phys. Chem. C 120: 26955–26959.

del Rosso, L., M. Celli, F. Grazzi, M. Catti, T. C. Hansen, A. D. Fortes et al. 2020. Cubic ice Ic without stacking defects obtained from ice XVII. Nat. Mat.: https://doi.org/10.1038/s41563-020-0606-y.

Dennison, D. M. 1921. The crystal structure of ice. Phys. Rev. 17: 20–22.

Durham, W. B., S. H. Kirby and L. A. Stern. 1997. Creep of water ices at planetary conditions: A compilation. J. Geophys. Res. 102: 16293–16302.

Dyadin, Y. A., E. G. Larionov, E. Y. Aladko, A. Y. Manakov, F. V. Zhurko, T. V. Mikina et al. 1999. Clathrate formation in water-noble gas (Hydrogen) systems at high pressures. J. Struct. Chem. 40: 790–795.

Echelmeyer, K. and B. Kamb. 1986. Rheology of ice II and ice III from high-pressure extrusion. Geophys. Res. Lett. 13: 693–696.

Engelhardt, H. and E. Whalley. 1972. Ice IV. J. Chem. Phys. 56: 2678–2684.

Engelhardt, H. and E. Whalley. 1979. The infrared spectrum of ice IV in the range 4000–400 cm⁻¹. J. Chem. Phys. 71: 4050–4051.

Engelhardt, H. and B. Kamb. 1981. Structure of Ice IV, a Metastable High-pressure Phase. J. Chem. Phys. 75: 5887–5899.

Evans, L. F. 1967. Selective nucleation of the high-pressure ices. J. Appl. Phys. 38: 4930–4932.

Falenty, A., T. C. Hansen and W. F. Kuhs. 2014. Formation and Properties of Ice XVI Obtained by Emptying a Type sII Clathrate Hydrate. Nature 516: 231–233.

Finch, E. D., S. W. Rabideau, R. G. Wenzel and N. G. Nereson. 1968. Neutron-diffraction study of ice polymorphs. II. Ice II. J. Chem. Phys. 49: 4361–4365.

Finney, J. L. 2001. The water molecule and its interactions: the interaction between theory, modelling, and experiment. J. Mol. Liq. 90: 303–312.

Fortes, A. D., I. G. Wood, J. P. Brodholt and L. Vočadlo. 2003. *Ab initio* simulation of the ice II structure. J. Chem. Phys. 119: 4567–4572.

Fortes, A. D., I. G. Wood, M. Alfredsson, L. Vocadlo and K. S. Knight. 2005. The incompressibility and thermal expansivity of D_2O ice II determined by powder neutron diffraction. J. Appl. Crystal. 38: 612–618.

Fortes, A. D., I. G. Wood, M. G. Tucker and W. G. Marshall. 2012. The P-V-T equation of state of D_2O ice VI determined by neutron powder diffraction in the range 0 < P < 2.6 GPa and 120 < T < 330 K, and the isothermal equation of state of D_2O ice VII from 2 to 7 GPa at room temperature.

Fortes, A. D. 2018. Accurate and precise lattice parameters of H_2O and D_2O ice Ih between 1.6 and 270 K from high-resolution time-of-flight neutron powder diffraction data. Acta Cryst. B74: 196–216.

Fortes, A. D. 2019. Structural manifestation of partial proton ordering and defect mobility in ice Ih. Phys. Chem. Chem. Phys. 21: 8264–8274.

Fuentes-Landete, V., S. Rasti, R. Schlögl, J. Meyer and T. Loerting. 2020. J. Phys. Chem. Lett. 11(19): 8268–8274.

Fukazawa, H., A. Hoshikawa, H. Yamauchi, Y. Yamaguchi and Y. Ishii. 2005. Formation and Growth of ice XI: A Powder Neutron Diffraction Study. J. Crystal Growth 282: 251–259.

Fukazawa, H., A. Hoshikawa, Y. Ishii, B. C. Chakoumakos and J. A. Fernandez-Baca. 2006. Existence of Ferroelectric Ice in the Universe. Astrophys. J. 652: L57–L60.

Gasser, T. M., A. V. Thoeny, L. J. Plaga, K. W. Köster, M. Etter, R. Böhmer et al. 2018. Experiments indicating a second hydrogen ordered phase of ice VI. Chem. Sci. 9: 4224–4234.

Gasser, T. M., A. V. Thoeny, A. D. Fortes and T. Loerting. 2021. Structural characterization of ice XIX as the second polymorph related to ice VI. Nat. Comm. 12: 1128.

Goncharov, A. F., V. V. Struzhkin, M. S. Somayazulu, R. J. Hemley and H. K. Mao. 1996. Compression of ice to 210 gigapascals: Infrared evidence for a symmetric hydrogen-bonded phase. Science 273: 218–220.

Goncharov, A. F., V. V. Struhkin, M. Ho-kwang and R. J. Hemley. 1999. Raman spectroscopy of dense H_2O and the transition to symmetric hydrogen bonds. Phys. Rev. Lett. 83: 1998–2001.

Gupta, M. K., R. Mittal, B. Singh, S. K. Mishra, D. T. Adroja, A. D. Fortes et al. 2018. Phonons and anomalous thermal expansion behavior of H_2O and D_2O ice Ih. Phys. Rev. B 98: 104301.

Guthrie, M., R. Boehler, J. J. Molaison, B. Haberl, A. M. dos Santos and C. Tulk. 2019. Structure and disorder in ice VII on the approach to hydrogen-bond symmetrization. Phys. Rev. B 99: 184112.

Hamilton, W. C., B. Kamb, S. J. La Placa and A. Prakash. 1969. *In*: Riehl, N., B. Bullemer and H. Engelhardt (eds.). Physics of Ice. New York: Plenum.

Handa, Y. P., D. D. Klug and E. Whalley. 1986. Difference in energy between cubic and hexagonal ice. J. Chem. Phys. 84: 7009–7010.

Handa, Y. P., D. D. Klug and E. Whalley. 1988. Energies of the phases of ice at low temperature and pressure relative to ice Ih. Can. J. Chem. 66: 919–924.

Hansen, T. C., M. M. Koza and W. F. Kuhs. 2008. Formation and annealing of cubic ice: I. modelling of stacking faults. J. Phys.: Condens. Matter 20: 285104.

Hemley, R. J., L. C. Chen and H. K. Mao. 1989. New transformations between crystalline and amorphous ice. Nature 338: 638–640.

Hernandez, J. A. and R. Caracas. 2018. Proton dynamics and the phase diagram of dense water ice. J. Chem. Phys. 148: 214501.

Hirai, H., H. Kadobayashi, T. Matsuoka, Y. Ohishi and Y. Yamamoto. 2014. High pressure X-ray diffraction and Raman spectroscopic studies of the phase change of D2O ice VII at approximately 11 GPa. High Pressure Res. 34: 289–296.

Hobbs, P. V. 1974. Ice Physics. Oxford: Clarendon Press.

Howe, R. and R. W. Whitworth. 1989. A determination of the crystal structure of ice XI. J. Chem. Phys. 90: 4450–4453.

Huang, Y., C. Zhu, L. Wang, X. Cao, Y. Su, X. Jiang et al. 2016. A new phase diagram of water under negative pressure: The rise of the lowest-density clathrate s-III. Sci. Adv. 2.

Huang, Y., C. Zhu, L. Wang, J. Zhao and X. C. Zeng. 2017. Prediction of a new ice clathrate with record low density: A potential candidate as ice XIX in guest-free form. Chem. Phys. Lett. 671: 186–191.

Jackson, S. M., V. M. Nield, R. W. Whitworth, M. Oguro and C. C. Wilson. 1997. Single-crystal neutron diffraction studies of the structure of ice XI. J. Phys. Chem. B 101: 6142–6145.

Johari, G. P., A. Lavergne and E. Whalley. 1974. Dielectric properties of ice VII and VIII and the phase boundary between ice VI and VII. J. Chem. Phys. 61: 4292–4300.

Johari, G. P. and E. Whalley. 1976. Dielectric properties of ice VI at low temperatures. J. Chem. Phys. 64: 4484–4489.

Johari, G. P. and E. Whalley. 1979. Evidence for a very slow transformation in ice VI at low temperatures. J. Chem. Phys. 70: 2094–2097.

Johari, G. P. 1998. An interpretation for the thermodynamic features of ice Ih<->ice XI transformation. J. Chem. Phys. 109: 9543–9548.

Jorgensen, J. D., R. A. Beyerlein, N. Watanabe and T. G. Worlton. 1984. Structure of D_2O ice VIII from in situ powder neutron diffraction. J. Chem. Phys. 81: 3211–3214.

Jorgensen, J. D. and T. G. Worlton. 1985. Disordered structure of D2O ice VII from *in-situ* neutron powder diffraction. J. Chem. Phys. 83: 329–333.

Journaux, B., J. M. Brown, A. Pakhomova, I. E. Collings, S. Petitgirard, P. Espinoza et al. 2020. Holistic approach for studying planetary hydrospheres: Gibbs representation of ices thermodynamics, elasticity, and the water phase diagram to 2,300 MPa. J. Geophys. Res. 125: e2019JE006176.

Kamb, B. 1964. Ice II: A proton-ordered form of ice. Acta Cryst. 17: 1437–1449.

Kamb, B. 1965. Structure of ice VI. Science 150: 205–209.

Kamb, B., A. Prakash and C. Knobler. 1967. Structure of ice V. Acta Cryst. 22: 706–715.

Kamb, B. and A. Prakash. 1968. Structure of ice III. Acta Cryst. B24: 1317–1327.

Kamb, B., W. C. Hamilton, S. J. La Placa and A. Prakash. 1971. Ordered proton configuration in ice II, from single-crystal neutron diffraction. J. Chem. Phys. 55: 1934–1945.

Kamb, B. 1973. Crystallography of Ice. *In*: Whalley, E., S. J. Jones and L. W. Gold (eds.). Physics and Chemistry of Ice. Royal Society of Canada.

Kamb, B. and S. J. La Placa. 1974. Reversible order-disorder transformation in ice V. Trans. Am. Geophys. Union 56: 1202.

Kamb, W. B. and S. K. Datta. 1960. Crystal structures of the high-pressure forms of ice: Ice III. Nature 187: 140E.141.

Klotz, S., J. M. Besson, G. Hamel, R. J. Nelmes, J. S. Loveday and W. G. Marshall. 1999. Metastable ice VII at low temperature and ambient pressure. Nature 398: 681E.684.

Klotz, S., G. Hamel, J. S. Loveday, R. J. Nelmes and M. Guthrie. 2003. Recrystallisation of HDA ice under pressure by *in-situ* neutron diffraction to 3.9 GPa. Z. Kristallogr. 218: 117–122.

Klotz, S., T. Strässle, C. G. Salzmann, J. Philippe and S. F. Parker. 2005. Incoherent inelastic neutron scattering measurements on ice VII: Are there two kinds of hydrogen bonds in ice? Europhys. Lett. 72: 576–582.

Klotz, S., L. E. Bove, T. Strässle, T. C. Hansen and A. M. Saitta. 2009. The preparation and structure of salty ice VII under pressure. Nat. Mat. 8: 405–409.

Klotz, S., K. Komatsu, F. Pietrucci, H. Kagi, A. A. Ludl, S. Machida et al. 2016. Ice VII from aqueous salt solutions: From a glass to a crystal with broken H-bonds. Sci. Rep. 6: 32040.

Klug, D. D., Y. P. Handa, J. S. Tse and E. Whalley. 1989. Transformation of ice VIII to amorphous ice by "Melting" at low temperature. J. Chem. Phys. 90: 2390–2392.

Knight, C. and S. J. Singer. 2005. Prediction of a phase transition to a hydrogen bond ordered form of ice VI. J. Phys. Chem. B 109: 21040–21046.

Knight, C. and S. J. Singer. 2006. A reexamination of the ice III/IX hydrogen bond ordering phase transition. J. Chem. Phys. 125: 064506.

Knight, C. and S. J. Singer. 2008. Hydrogen bond ordering in ice V and the transition to ice XIII. J. Chem. Phys. 129: 164513.

Knight, C. and S. J. Singer. 2009. Site disorder in Ice VII arising from hydrogen bond fluctuations. J. Phys. Chem. A 113: 12433–12438.

Kohl, I., E. Mayer and A. Hallbrucker. 2001. Ice XII forms on compression of hexagonal ice at 77 K via high-density amorphous water. Phys. Chem. Chem. Phys. 3: 602–605.

Komatsu, K., F. Noritake, S. Machida, A. Sano-Furukawa, T. Hattori, R. Yamane et al. 2016. Partially Ordered State of Ice XV. Sci. Rep. 6: 28920.

Komatsu, K., S. Klotz, S. Machida, A. Sano-Furukawa, T. Hattori and H. Kagi. 2020a. Anomalous hydrogen dynamics of the ice VII–VIII transition revealed by high-pressure neutron diffraction. Proc. Natl. Acad. Sci. USA 117: 6356–6361.

Komatsu, K., S. Machida, F. Noritake, T. Hattori, A. Sano-Furukawa, R. Yamane et al. 2020b. Ice Ic without stacking disorder by evacuating hydrogen from hydrogen hydrate. Nat. Comm. 11: 464.

König, H. 1943. Eine kubische Eismodifikation. Z. Kristallogr. 105: 279–286.

Köster, K. W., V. Fuentes-Landete, A. Raidt, M. Seidl, C. Gainaru, T. Loerting et al. 2015. Dynamics enhanced by HCl doping triggers full Pauling entropy release at the ice XII–XIV transition. Nat. Comm. 6: 7349.

Köster, K. W., A. Raidt, V. Fuentes Landete, C. Gainaru, T. Loerting and R. Böhmer. 2016. Doping-enhanced dipolar dynamics in ice V as a precursor of hydrogen ordering in ice XIII. Phys. Rev. B 94: 184306.

Köster, K. W., V. Fuentes-Landete, A. Raidt, M. Seidl, C. Gainaru, T. Loerting et al. 2018. Author Correction: Dynamics enhanced by HCl doping triggers 60% Pauling entropy release at the ice XII–XIV transition. Nat. Comm. 9: 16189.

Kosyakov, V. I. and V. A. Shestakov. 2001. On the possibility of the existence of a new ice phase under negative pressures. Dokl. Phys. Chem. 376: 49–51.

Koza, M., H. Schober, A. Tölle, F. Fujara and T. Hansen. 1999. Formation of ice XII at different conditions. Nature 397: 660–661.

Koza, M. M., H. Schober, T. Hansen, A. Tölle and F. Fujara. 2000. Ice XII in its second regime of metastability. Phys. Rev. Lett. 84: 4112–4115.

Kuhs, W. F. and M. S. Lehmann. 1983. The structure of ice Ih by neutron diffraction. J. Phys. Chem. 87: 4312–4313.

Kuhs, W. F., J. L. Finney, C. Vettier and D. V. Bliss. 1984. Structure and hydrogen ordering in ices VI, VII, and VIII by neutron powder diffraction. J. Chem. Phys. 81: 3612–3623.

Kuhs, W. F., D. V. Bliss and J. L. Finney. 1987. High-resolution neutron powder diffraction study of ice I$_c$. J. Phys. Colloq., C1 48: 631–636.

Kuhs, W. F., H. Ahsbahs, D. Londono and J. L. Finney. 1989. *In-situ* crystal growth and neutron four-circle diffractometry under high pressure. Physica B 156–157: 684–687.

Kuhs, W. F., C. Lobban and J. L. Finney. 1998. Partial h-ordering in high pressure ices III and V. Rev. High Pressure Sci. Technol. 7: 1141–1143.

Kuhs, W. F., C. Sippel, A. Falentya and T. C. Hansen. 2012. Extent and Relevance of Stacking Disorder in "Ice Ic". Proc. Natl. Acad. Sci. USA 109: 21259–21264.

Kuhs, W. F., T. C. Hansen and A. Falenty. 2018. Filling ices with helium and the formation of helium clathrate hydrate. J. Phys. Chem. Lett. 9: 3194–3198.

Kuo, J.-L. and W. F. Kuhs. 2006. A first principles study on the structure of ice-vi: Static distortion, molecular geometry, and proton ordering. J. Phys. Chem. B 110: 3697–3703.

La Placa, S. J. and W. C. Hamilton. 1973. On a nearly proton-ordered structure for ice IX. J. Chem. Phys. 58: 567–580.

Leadbetter, A. J., R. C. Ward, J. W. Clark, P. A. Tucker, T. Matsuo and H. Suga. 1985. The equilibrium low-temperature structure of ice. J. Chem. Phys. 82: 424–428.

Li, J.-C., J. D. Londono and D. K. Ross. 1991. An inelastic incoherent neutron scattering study of ice II, IX, V, and VI—in the range from 2 to 140 meV. J. Chem. Phys. 94: 6770–6775.

Lin, C., X. Liu, X. Yong, J. S. Tse, J. S. Smith, N. J. English, B. Wang, M. Li, W. Yang and H.-K. Mao. 2020. Proc. Natl. Acad. Sci. U. S. A. 117: 15437.

Lin, C. and J. S. Tse. 2021. J. Phys. Chem. Lett., 12(33): 8024–8038.

Line, C. M. B. and R. W. Whitworth. 1996. A high resolution neutron powder diffraction study of D$_2$O ice XI. J. Chem. Phys. 104: 10008–10013.

Liu, K., M. G. Brown, C. Carter, R. J. Saykally, J. K. Gregory and D. C. Clary. 1996. Characterization of a cage form of the water hexamer. Nature 381: 501–503.

Lobban, C., J. L. Finney and W. F. Kuhs. 1998. The structure of a new phase of ice. Nature 391: 268–270.

Lobban, C., J. L. Finney and W. F. Kuhs. 2000. The structure and ordering of ices III and V. J. Chem. Phys. 112: 7169–7180.

Lobban, C., J. L. Finney and W. F. Kuhs. 2002. The p-T dependency of the ice II crystal structure and the effect of helium inclusion. J. Chem. Phys. 117: 3928–3834.

Loerting, T., I. Kohl, C. Salzmann, E. Mayer and A. Hallbrucker. 2002. The (Meta-)stability domain of ice XII revealed between ca. 158 - 212 K and ca. 0,7 - 1,5 GPa on isobaric heating of high density amorphous water. J. Chem. Phys. 116: 3171–3174.

Londono, D., J. L. Finney and W. F. Kuhs. 1992. Formation, stability, and structure of helium hydrate at high pressure. J. Chem. Phys. 97: 547–552.

Londono, J. D., W. F. Kuhs and J. L. Finney. 1993. Neutron diffraction studies of ices III and IX on under-pressure and recovered samples. J. Chem. Phys. 98: 4878–4888.

Loveday, J. S. and R. J. Nelmes. 2008. High-pressure gas hydrates. Phys. Chem. Chem. Phys. 10: 937–950.

Ludl, A. A., L. E. Bove, D. Corradini, A. M. Saitta, M. Salanne, C. L. Bull et al. 2017. Probing ice VII crystallization from amorphous NaCl–D2O solutions at gigapascal pressures. Phys. Chem. Chem. Phys. 19: 1875–1883.

Ludwig, R. 2001. Water: from clusters to the bulk. Angew. Chem. Int. Ed. 40: 1808–1827.

Malkin, T. L., B. J. Murray, A. V. Brukhno, A. J. and C. G. Salzmann. 2012. Structure of ice crystallized from supercooled water. Proc. Natl. Acad. Sci. USA 109: 1041–1045.

Malkin, T. L., B. J. Murray, C. G. Salzmann, V. Molinero, S. J. Pickering and T. F. Whale. 2015. Stacking Disorder in Ice I. Phys. Chem. Chem. Phys. 17: 60–76.

Martelli, F., N. Giovambattista, S. Torquato and R. Car. 2018. Searching for crystal-ice domains in amorphous ices. Phys. Rev. Mat. 2: 075601.

Matsui, T., M. Hirata, T. Yagasaki, M. Matsumoto and H. Tanaka. 2017. Communication: Hypothetical ultralow-density ice polymorphs. J. Chem. Phys. 147: 091101.

Matsuo, T. and H. Suga. 1987. Calorimetric study of ices Ih doped with alkali hydroxides and other impurities. Journal de Physique C1: 477–483.

Mayer, E. and A. Hallbrucker. 1987. Cubic ice from liquid water. Nature 325: 601–602.

McFarlan, R. L. 1936. The structure of ice III. J. Chem. Phys. 4: 253–259.

Militzer, B. and H. F. Wilson. 2010. New phases of water ice predicted at megabar pressures. Phys. Rev. Lett. 105: 195701.

Millot, M., S. Hamel, J. R. Rygg, P. M. Celliers, G. W. Collins, F. Coppari et al. 2018. Experimental evidence for superionic water ice using shock compression. Nat. Phys. 14: 297–302.

Millot, M., F. Coppari, J. R. Rygg, A. Correa Barrios, S. Hamel, D. C. Swift et al. 2019. Nanosecond X-ray diffraction of shock-compressed superionic water ice. Nature 569: 251–255.

Minceva-Sukarova, B., W. F. Sherman and G. R. Wilkinson. 1984. A high pressure spectroscopic study on the ice III - ice IX, disordered—ordered transition. J. Mol. Struct. 115: 137–140.

Minceva-Sukarova, B., W. F. Sherman and G. R. Wilkinson. 1985. Isolated O-D stretching frequencies in ice II. Spectrochim. Acta 41A: 315–318.

Minceva-Sukarova, B., G. E. Slark and W. F. Sherman. 1986. The raman spectra of ice V and ice VI and evidence of partial proton ordering at low temperature. J. Mol. Struct. 143: 87–90.

Minceva-Sukarova, B., G. Slark and W. F. Sherman. 1988. The raman spectra of the KOH-doped ice polymorphs: V and VI. J. Mol. Struct. 175: 289–293.

Mishima, O., N. Mori and S. Endo. 1979. Thermal expansion anomaly of ice VI related to the order-disorder transition. J. Chem. Phys. 70: 2037–2038.

Moynihan, C. T., P. B. Macedo, C. J. Montrose, P. K. Gupta, M. A. DeBolt, F. J. Dill et al. 1976. Structural relaxation in vitreous materials. Ann. N. Y. Acad. Sci. 279: 15–35.

Murray, B. J., C. G. Salzmann, A. J. Heymsfield, S. Dobbie, R. R. Neely and C. J. Cox. 2015. Trigonal ice crystals in Earth's atmosphere. Bull. Am. Met. Soc. 96: 1519–1531.

Nakamura, T., M. Matsumoto, T. Yagasaki and H. Tanaka. 2016. Thermodynamic stability of ice II and its hydrogen-disordered counterpart: Role of zero-point energy. J. Phys. Chem. B 120: 1843–1848.

Nanda, K. D. and G. J. O. Beran. 2013. What governs the proton ordering in ice XV? J. Phys. Chem. Lett. 4: 3165–3169.

Nelmes, R. J., J. S. Loveday and W. G. Marshall. 1998. Multisite disordered structure of ice VII to 20 GPa. Phys. Rev. Lett. 81: 2719–2722.

Nishibata, K. 1972. Growth of ice IV and equilibrium curves between liquid water, ice IV, ice V and ice VI. Jpn. J. Appl. Phys. 11: 1701–1708.

Nishibata, K. and E. Whalley. 1974. Thermal effects of the transformation ice III-IX. J. Chem. Phys. 60: 3189–3194.

Noguchi, N. and T. Okuchi. 2016. Self-diffusion of protons in H2O ice VII at high pressures: Anomaly around 10 GPa. J. Chem. Phys. 144: 234503.

O'Keeffe, M. 1998. New ice outdoes related nets in smallest-ring size. Nature 392: 879.

Okada, T., T. Iitaka, T. Yagi and K. Aoki. 2014. Electrical conductivity of ice VII. Sci. Rep. 4: 5778.

Parkkinen, P., S. Riikonen and L. Halonen. 2014. Ice XI: Not that ferroelectric. J. Phys. Chem. C 118: 26264–26275.

Pauling, L. 1935. The structure and entropy of ice and other crystals with some randomness of atomic arrangement. J. Am. Chem. Soc. 57: 2680–2684.

Petrenko, V. F. and R. W. Whitworth. 1999. Physics of Ice. Oxford: Oxford University Press.

Pickard, C. J., M. Martinez-Canales and R. J. Needs. 2013. Decomposition and terapascal phases of water ice. Physical Review Letters 110: 245701.

Pistorius, C. W. F. T., E. Rapoport and J. B. Clark. 1968. Phase diagrams of H2O and D2O at high pressures. J. Chem. Phys. 48: 5509–5514.

Polian, A. and M. Grimsditch. 1984. New high-pressure phase of H$_2$O: Ice X. Phys. Rev. Lett. 15: 1312–1314.

Rabideau, S. W., E. D. Finch, G. P. Arnold and A. L. Bowman. 1968. Neutron diffraction study of ice polymorphs. I. Ice IX. J. Chem. Phys. 49: 25142519.

Raza, Z., D. Alfe, C. G. Salzmann, J. Klimes, A. Michaelides and B. Slater. 2011. Proton ordering in cubic ice and hexagonal ice; a potential new ice phase-XIc. Phys. Chem. Chem. Phys. 13: 19788–19795.

Ripmeester, J. A., C. I. Ratcliffe, D. D. Klug and J. S. Tse. 2006. Molecular perspectives on structure and dynamics in clathrate hydrates. Ann. N.Y. Acad. Sci. 715: 161–176.

Ross, R. G., P. Andersson and G. Bäckström. 1978. Effects of H and D order on the thermal conductivity of ice phases J. Chem. Phys. 68: 3967–3972.

Rosu-Finsen, A. and C. G. Salzmann. 2018. Benchmarking acid and base dopants with respect to enabling the ice V to XIII and ice VI to XV hydrogen-ordering phase transitions. J. Chem. Phys. 148: 244507.

Rosu-Finsen, A. and C. G. Salzmann. 2019. Origin of the low-temperature endotherm of acid-doped ice VI: New hydrogen-ordered phase of ice or deep glassy states? Chem. Sci.: 515–523.

Rosu-Finsen, A., A. Amon, J. Armstrong, F. Fernandez-Alonso and C. G. Salzmann. 2020. Deep-glassy ice VI revealed with a combination of neutron spectroscopy and diffraction. J. Phys. Chem. Lett. 11: 1106–1111.

Rosu-Finsen, A. and C. G. Salzmann. 2022. Is pressure the key to hydrogen ordering ice IV? Chem. Phys. Lett. 789: 139325.

Röttger, K., A. Endriss, J. Ihringer, S. Doyle and W. F. Kuhs. 1994. Lattice constants and thermal expansion of H_2O and D_2O ice Ih between 10 and 265 K. Acta Cryst. B50: 644–648.

Salzmann, C., I. Kohl, T. Loerting, E. Mayer and A. Hallbrucker. 2002a. The raman spectrum of ice XII and its relation to that of a new "High-Pressure Phase of H_2O Ice". J. Phys. Chem. B 106: 1–6.

Salzmann, C. G., T. Loerting, I. Kohl, E. Mayer and A. Hallbrucker. 2002b. Pure ice IV from high-density amorphous ice. J. Phys. Chem. B 106: 5587–5590.

Salzmann, C. G., I. Kohl, T. Loerting, E. Mayer and A. Hallbrucker. 2003a. The low-temperature dynamics of recovered ice XII as studied by differential scanning calorimetry: a comparision with ice V. Phys. Chem. Chem. Phys. 5: 3507–3517.

Salzmann, C. G., I. Kohl, T. Loerting, E. Mayer and A. Hallbrucker. 2003b. Raman spectroscopic study on hydrogen bonding in recovered ice IV. J. Phys. Chem. B 107: 2802–2807.

Salzmann, C. G., T. Loerting, I. Kohl, E. Mayer and A. Hallbrucker. 2003c. Pure ices IV and XII from high-density amorphous ice. Can. J. Phys. 81: 25–32.

Salzmann, C. G., E. Mayer and A. Hallbrucker. 2004a. Effect of heating rate and pressure on the crystallization kinetics of high-density amorphous ice on isobaric heating between 0.2 and 1.9 GPa. Phys. Chem. Chem. Phys. 6: 5156–5165.

Salzmann, C. G., E. Mayer and A. Hallbrucker. 2004b. Thermal properties of metastable ices IV and XII: comparison, isotope effects and relative stabilities. Phys. Chem. Chem. Phys. 6: 1269–1276.

Salzmann, C. G., A. Hallbrucker, J. L. Finney and E. Mayer. 2006a. Raman spectroscopic study of hydrogen ordered ice XIII and of its reversible phase transition to disordered ice V. Phys. Chem. Chem. Phys. 8: 3088–3093.

Salzmann, C. G., A. Hallbrucker, J. L. Finney and E. Mayer. 2006b. Raman spectroscopic features of hydrogen-ordering in ice XII. Chem. Phys. Lett. 429: 469–473.

Salzmann, C. G., P. G. Radaelli, A. Hallbrucker, E. Mayer and J. L. Finney. 2006c. The preparation and structures of hydrogen ordered phases of ice. Science 311: 1758–1761.

Salzmann, C. G., P. G. Radaelli, A. Hallbrucker, E. Mayer and J. L. Finney. 2007. New hydrogen ordered phases of ice. *In*: Kuhs, W. F. (ed.). Physics and Chemistry of Ice. Cambridge: The Royal Society of Chemistry.

Salzmann, C. G., P. G. Radaelli, J. L. Finney and E. Mayer. 2008. A calorimetric study on the low temperature dynamics of doped ice v and its reversible phase transition to hydrogen ordered ice XIII. Phys. Chem. Chem. Phys. 10: 6313–6324.

Salzmann, C. G., P. G. Radaelli, E. Mayer and J. L. Finney. 2009. Ice XV: A new thermodynamically stable phase of ice. Phys. Rev. Lett. 103: 105701.

Salzmann, C. G., P. G. Radaelli, B. Slater and J. L. Finney. 2011. The polymorphism of ice: five unresolved questions. Phys. Chem. Chem. Phys. 13: 18468–18480.

Salzmann, C. G., B. J. Murray and J. J. Shephard. 2015. Extent of stacking disorder in diamond. Diam. Relat. Mater. 59: 69–72.

Salzmann, C. G., B. Slater, P. G. Radaelli, J. L. Finney, J. J. Shephard, M. Rosillo-Lopez et al. 2016. Detailed crystallographic analysis of the ice VI to Ice XV hydrogen ordering phase transition. J. Chem. Phys. 145: 204501.

Salzmann, C. G. 2019. Advances in the experimental exploration of water's phase diagram. J. Chem. Phys. 150: 060901.

Salzmann, C. G., Z. Sharif, C. L. Bull, S. T. Bramwell, A. Rosu-Finsen and N. P. Funnell. 2019. Ammonium Fluoride as a hydrogen-disordering agent for ice. J. Phys. Chem. C 123: 16486–16492.

Salzmann, Christoph, G., Alexander Rosu-Finsen, Zainab Sharif, Paolo G. Radaelli and John L. Finney. 2021a. J. Chem. Phys. 154: 134504.

Salzmann, Christoph, John S. Loveday, Alexander Rosu-Finsen and Craig L. Bull. 2021b. Nat. Commun. 12: 3162.

Sharif, Zainab, Ben Slater, Craig L. Bull, Martin Hart and Christoph G. Salzmann. 2021. Effect of ammonium fluoride doping on the ice III to ice IX phase transition. J. Chem. Phys. 154: 114502.

Shephard, J. J. and C. G. Salzmann. 2015. The complex kinetics of the ice VI to Ice XV hydrogen ordering phase transition. Chem. Phys. Lett. 637: 63–66.

Shephard, J. J., S. Klotz and C. G. Salzmann. 2016. A new structural relaxation pathway of low-density amorphous ice. J. Chem. Phys. 144: 204502.

Shephard, J. J. and C. G. Salzmann. 2016. Molecular reorientation dynamics govern the glass transitions of the amorphous ices. J. Phys. Chem. Lett. 7: 2281–2285.

Shephard, J. J., S. Ling, G. C. Sosso, A. Michaelides, B. Slater and C. G. Salzmann. 2017. Is high-density amorphous ice simply a "Derailed" state along the ice I to Ice IV pathway? J. Phys. Chem. Lett. 8: 1645–1650.

Shephard, J. J., B. Slater, P. Harvey, M. Hart, C. L. Bull, S. T. Bramwell et al. 2018. Doping-induced disappearance of ice II from water's phase diagram. Nat. Phys. 14: 569–572.

Stillinger, F. H. and K. S. Schweitzer. 1983. Ice under pressure: Transition to symmetrical hydrogen bonds. J. Phys. Chem. 87: 4281–4288.

Strobel, T. A., M. Somayazulu, S. V. Sinogeikin, P. Dera and R. J. Hemley. 2016. Hydrogen-stuffed, quartz-like water ice. J. Am. Chem. Soc. 138: 13786–13789.

Sun, J., B. K. Clark, S. Torquato and R. Car. 2015. The phase diagram of high-pressure superionic ice. Nat. Comm. 6: 8156.

Tajima, Y., T. Matsuo and H. Suga. 1982. Phase transition in KOH-doped hexagonal ice. Nature 299: 810–812.

Tajima, Y., T. Matsuo and H. Suga. 1984. Calorimetric study of phase transition in hexagonal ice doped with alkali hydroxides. J. Phys. Chem. Solids 45: 1135–1144.

Takii, Y., K. Koga and H. Tanaka. 2008. A plastic phase of water from computer simulation. J. Chem. Phys. 128: 204501.

Tammann, G. 1900. Über die Grenzen des festen Zustandes. Ann. Phys. 2: 1–31.

Tammann, G. 1910. Über das Verhalten des Wassers bei hohen Drucken und tiefern Temperaturen. ZS. Phys. Chem. 72: 609–631.

Teixeira, J. 1998. The double identity of ice X. Nature 392: 232–233.

Thoeny, A. V., T. M. Gasser and T. Loerting. 2019. Distinguishing ice β-XV from deep glassy ice VI: raman spectroscopy. Phys. Chem. Chem. Phys. 21: 15452–15462.

Tielens, A. G. G. M. 2013. The molecular universe. Rev. Mod. Phys. 85: 1021–1081.

Tran, H., A. V. Cunha, J. J. Shephard, A. Shalit, P. Hamm, T. L. C. Jansen et al. 2017. 2D IR spectroscopy of high-pressure phases of ice. J. Chem. Phys. 147: 144501.

Tribello, G. A., B. Slater and C. G. Salzmann. 2006. A blind structure prediction of ice XIV. J. Am. Chem. Soc. 128: 12594–12595.

Tschauner, O., S. Huang, E. Greenberg, V. B. Prakapenka, C. Ma, G. R. Rossman et al. 2018. Ice-VII inclusions in diamonds: Evidence for aqueous fluid in Earth's deep mantle. Science 359: 1136–1139.

Ueda, M., T. Matsuo and H. Suga. 1982. Calorimetric study of proton ordering in hexagonal ice catalysed by hydrogen fluoride. J. Phys. Chem. Solids 12: 1165–1172.

Vonnegut, K. 1963. Cat's Cradle. New York, USA: Holt, Rinehart and Winston.

Walter, M. L. 1990. Science and Cultural Crisis: An Intellectual Biography of Percy Williams Bridgman (1882–1961). Stanford: Stanford University Press.

Whale, T. F., S. J. Clark, J. L. Finney and C. G. Salzmann. 2013. DFT-assisted interpretation of the raman spectra of hydrogen-ordered ice XV. J. Raman Spectrosc 44: 290–298.

Whalley, E. and D. W. Davidson. 1965. Entropy changes at the phase transitions in ice. J. Chem. Phys. 43: 2148–2149.

Whalley, E., D. W. Davidson and J. B. R. Heath. 1966. Dielectric properties of ice VII. Ice VIII: A New Phase of Ice. J. Chem. Phys. 45: 3976–3982.

Whalley, E., J. B. R. Heath and D. W. Davidson. 1968. Ice IX: An antiferroelectric phase related to ice III. J. Chem. Phys. 48: 2362–2370.

Whalley, E. 1983. Cubic ice in nature. J. Phys. Chem. 87: 4174–4179.

Wilson, G. J., R. K. Chan, D. W. Davidson and E. Whalley. 1965. Dielectric properties of ices II, III, V, and VI. J. Chem. Phys. 43: 2384–2391.

Wollan, E. O., W. L. Davidson and C. G. Shull. 1949. Neutron diffraction study of the structure of ice. Phys. Rev. 75: 1348–1352.

Yamane, R., K. Komatsu, J. Gouchi, Y. Uwatoko, S. Machida, T. Hattori, H. Ito and H. Kagi. 2021. Experimental evidence for the existence of a second partially-ordered phase of ice VI. Nat. Comm. 12: 1129.

Yoshimura, Y., S. T. Stewart, H.-k. Mao and R. J. Hemley. 2007. *In situ* Raman spectroscopy of low-temperature/high-pressure transformations of H2O. J. Chem. Phys. 126: 174505.

Yu, X., J. Zhu, S. Du, H. Xu, S. C. Vogel, J. Han et al. 2014. Crystal structure and encapsulation dynamics of ice II-structured neon hydrate. Proc. Natl. Acad. Sci. USA 111: 10456–10461.

Zeng, Q., T. Yan, K. Wang, Y. Gong, Y. Zhou, Y. Huang et al. 2016. Compression icing of room-temperature NaX solutions (X = F, Cl, Br, I). Phys. Chem. Chem. Phys. 18: 14046–14054.

Zeng, Q., C. Yao, K. Wang, C. Q. Sun and B. Zou. 2017. Room-temperature NaI/H2O compression icing: solute–solute interactions. Phys. Chem. Chem. Phys. 19: 26645–26650.

Zha, C.-S., J. S. Tse and W. A. Bassett. 2016. New Raman measurements for H₂O ice VII in the range of 300 cm−1 to 4000 cm−1 at pressures up to 120 GPa. J. Chem. Phys. 145: 124315.

Zhao, J., S. L. Simon and G. B. McKenna. 2013. Using 20-million-Year-old Amber to Test the Super-Arrhenius Behaviour of Glass-forming Systems. Nat. Comm. 4: 1783.

CHAPTER 5

How Many Crystalline Ices are There?

Tobias M. Gasser, Alexander V. Thoeny, Christina Tonauer, Johannes Bachler, Violeta Fuentes-Landete and *Thomas Loerting**

Introduction

Ice is an important part of our lives. Not so much the ice cube in your drink, but much more so the large parts of the surface of Earth that are covered in ice (Hobbs 1974). Currently, roughly 10% of the land of our globe is covered by the ice sheets of Greenland and Antarctica, ice caps and glacial ice (Bartels-Raush et al. 2012). Ice and snow reflect up to 90% of the incoming solar radiation, whereas liquid water reflects only about 10%. Consequently, the current trend of large ice losses due to global warming results in a feedback loop ultimately accelerating surface warming. Also in the troposphere ice plays a crucial role: most of the rainfall that we experience at the surface originates from a high-altitude ice cloud (Bartels-Rausch et al. 2012). Even in the very dry stratosphere ice clouds are of key importance: type 1 polar stratospheric ice clouds are very efficient catalysts converting photostable chlorine reservoir species (such as HCl and $ClONO_2$) into active species (such as Cl_2 and HOCl) that are rapidly photolyzed, thereby initiating ozone depletion in polar springtime (Molina et al. 1987, Molina 1991, Peter 1997, Solomon 1999, Prenni et al. 2001). All of the ice described so far shares the same crystal structure, the one of hexagonal ice (Hobbs 1974, Petrenko et al. 1999). This is, however, not the only type of ice that can be found on Earth. Occasionally some clouds might contain cubic ice. By observing halos, Jesuit priest Scheiner was actually the first to infer that some ice clouds contain crystals that do not have the shapes typical of hexagonal crystals, but the shape of cubic crystals (Hobbs 1974). Cubic ice may exist in cold clouds at temperatures up to 220 K (Murray et al. 2015) (see below). Below 150 K, ice may also be amorphous, for example, in noctilucent clouds that form at the coldest places

Institute of Physical Chemistry, University of Innsbruck, Innrain 52c, A-6020 Innsbruck, Austria.
* Corresponding author: thomas.loerting@uibk.ac.at

and times in our atmosphere in the mesosphere (at altitudes of about 100 km). Once the temperatures exceed 150 K, cubic ice will crystallize (Murray et al. 2005).

For crystalline solids, the concept of polymorphism has been established (Bernstein 2002) and is regarded as one of water's anomalies. It is an important concept in one-component systems, recognized about 200 years ago on the examples of carbonate, phosphate, and arsenate salts (Klaproth 1798, Mitscherlich 1822). In the case of water, an anomalously large variety of crystalline phases has been noted. While cubic and hexagonal ice show distinct space groups ($P6_3/mmc$ vs. $Fd\bar{3}m$), they merely differ in terms of their stacking sequences rather than the local order. Often they are regarded as two polytypes (Guinier et al. 1984) for this reason, but not as two polymorphs. The fact that a single crystal of cubic ice has not been reported so far also speaks against the notion that ice I_c and ice I_h are two polymorphs. By contrast, high-pressure ice crystals are undoubtedly distinct from ice I and need to be regarded as distinct polymorphs. In fact, even in Earth's interior, it is possible to find ice phases that are distinct from hexagonal ice. Some of them, namely ice VI and ice VII, may in fact return to the surface in the form of diamond inclusions (Kagi et al. 2000, Tschauner et al. 2018). Roman numerals are used for ice phases, following the chronological order of the deduction of their crystal structures.

The regions of stability for water vapor, liquid water, and the ice phases are depicted in phase diagrams. The phase boundaries shown in these diagrams indicate p/T-conditions for which two phases may coexist in thermodynamic equilibrium. Some of these phase boundaries are almost parallel to the temperature axis, whereas others are parallel to the pressure axis, so that the former can only be crossed through change of pressure and the latter only through change of temperature. The relation $dp/dT = \Delta S/\Delta V$ describes the slope of the phase boundary. First recognized by Emilé Clapeyron (1834), it implies that the former are density-driven and occur with a finite change of volume ΔV at the phase transition, but no entropy changes ΔS. That is, the density of the ice polymorph at the low-pressure side is lower than the density at the high-pressure side of the phase boundary. One can read from the phase diagram that the density increases from ice I to ice III to ice V to ice VI and to ice VII. Hence, such solid-solid transitions involve rearrangement of the lattice of oxygen atoms and are driven by changes in density, for example, when pressurizing ice I_h to ice II (rhombohedral space group $R\bar{3}$) (Bauer et al. 2008). By contrast, the density barely changes when the polymorphic transition is temperature-induced, that is, $\Delta V \approx 0$. For such transitions, however, entropy changes drive the transition, that is, ΔS is finite. For all the cases seen in the phase diagram in Fig. 1, it is the configurational entropy that changes upon crossing the phase boundary through cooling or heating. No change in volume implies that the oxygen subnetwork in the H_2O crystal remains unchanged, and change in configurational entropy implies that the hydrogen subnetwork changes. At the low-temperature side of the phase boundary, the H-subnetwork is in an ordered configuration, whereas it is disordered on the high-temperature side. In almost all ice polymorphs, each water molecule is tetrahedrally surrounded by four other water molecules, forming the Walrafen-pentamer as the central building block. The Bernal-Fowler rules pose that within each pentamer (i) each oxygen atom has two covalent bonds to hydrogen atoms (obeying the H_2O stoichiometry) and (ii) there must be exactly one H-atom bridging

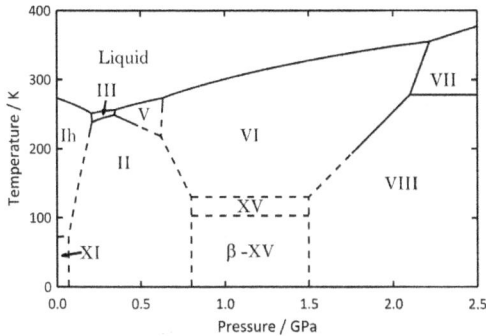

Fig. 1: Phase diagram of water's stable phases in the intermediate pressure regime up to 2.5 GPa and 400 K. Data for the figure are taken from (Gasser et al. 2018). Metastable polymorphs and amorphous ices are not evident in this diagram.

two oxygen atoms (but not two or none) (Bernal et al. 1933). There are exactly six different possibilities for how to arrange hydrogen atoms in accordance with these rules, which differ in terms of the orientation of the water molecules (the direction of the water dipole). In hydrogen-disordered ices, all these configurations are populated with equal probability when averaging over space and time. The reorientation of individual water molecules is caused by concerted hydrogen atom transfer and/or rotation of water molecules. More precisely, mobile point defects such as L-defects (no hydrogen between two oxygen neighbours) migrate through the crystal. While a hydrogen-ordered configuration is more favourable in terms of enthalpy, the hydrogen-disordered configuration is more favorable in terms of entropy.

Currently, 18 ice polymorphs (plus cubic ice) are known. While no new ice was found between 1972 and 1998, the years of the discovery of ice XI (Kawada 1972) and ice XII (Lobban et al. 1998), respectively, the last two decades saw ices XIII through ice XVIII emerge. These new discoveries are in the field of ices that are lower in density than hexagonal ice, which might be thermodynamically stable merely at negative pressures (ices XVI (Falenty et al. 2014) and XVII (del Rosso et al. 2016a,b)), in hydrogen ordered pendants of long known high-pressure ices (ices XIII (Salzmann et al. 2006), XIV (Salzmann et al. 2006), XV (Salzmann et al. 2009), and β-XV (Gasser et al. 2018)) and in superionic ices at ultrahigh pressure (ice XVIII (Millot et al. 2019)).

Despite these recent findings, it is not yet clear how many condensed phases of water can be distinguished. A recent machine learning study on ice phases actually suggests that there are in total 51 ice polymorphs, more than 30 of which have not yet been reported in experiments (Engel et al. 2018). If the stable structures that are predicted in simulations can actually be achieved in experiments, we expect to discover more ice phases in future than have been discovered up to now. Thus, it seems that the search for new ice polymorphs and the scientific quest to understand the hydrogen bond in water and ice will continue for many more decades. Some comprehensive reviews have been published on these topics (Zheligovskaya et al. 2006, Fortes et al. 2009, Fortes et al. 2010). It is the purpose of this contribution to identify what is known about crystalline ices and where the quest might take us in future. In the following we discuss polymorphism of ice in different pressure regimes, starting with the low-pressure regime, in which common hexagonal ice is the most prominent member. We then go on to discuss low-density ices in the vacuum and at negative pressure, move on to the very versatile intermediate pressure range up to 2 GPa before finally discussing the extremes of positive pressure.

The present chapter can be understood as a follow-up to the Fermi summer school proceedings report on amorphous and crystalline ices written by our group (Fuentes-Landete et al. 2015). Even though only five years have passed since its publication, the field of crystalline ices has evolved quite a lot, so that an up-to-date review is warranted. This article can also be understood as complementary to an article about the question on the number of amorphous ices that Loerting et al. (2011) wrote of earlier. Experimental evidence suggests that the answer to the number of distinct amorphous ices that can be equilibrated is three. By contrast, it is clear that ice XVIII is certainly not the last ice polymorph that will be prepared and characterized in experiment.

The Low Pressure Regime (0–0.1 GPa)

Ice I: Cubic Ice and Hexagonal Ice

When cooling liquid water, for example, in the freezer, usually hexagonal ice crystallizes. Hexagonal ice is the dominating phase in the low-pressure regime—see the phase diagram in Fig. 1. Also cubic ice (ice I_c) (Dowell et al. 1960) may crystallize from ambient water under special circumstances (Murray et al. 2005) or may be obtained on heating high-pressure polymorphs or amorphous ice at ambient pressure (Handa et al. 1986b, Handa et al. 1987, Elsaesser et al. 2010). Cubic (ice I_c) and hexagonal ice (ice I_h) are very similar both in terms of interatomic distances and density at ambient pressure (0.92 g/cm^3) (Petrenko et al. 1999). There is particular interest in these two because they represent the low density forms of ice at ambient pressure. Hexagonal ice is ubiquitous and occurs in the form of snowflakes, icicles, ice, hail, etc., whereas cubic ice may appear occasionally in clouds (Murray et al. 2005, Whalley 1983, Mayer et al. 1987). In terms of short-range order, they are in fact indistinguishable—with identical Raman and mid-IR spectra. The difference emerges when comparing the long-range order of the hexagonal rings building the crystal structures, for example, using neutron or X-ray diffraction (Kuhs et al. 1987). Cubic ice shows the stacking sequence ABCABCABC but, hexagonal ice shows the stacking sequence ABABAB. Using the definition of polytypism "for structural modifications, each of which can be regarded as built up by stacking layers of (nearly) identical structure and composition differing only in their stacking sequence"(Guinier et al. 1984), hexagonal ice and cubic ice are regarded as two polytypes, which belong to two different space groups ($P6_3/mmc$ vs. $Fd\bar{3}m$) (Kuhs et al. 1987, Kuhs et al. 1986, Röttger et al. 1994). However, by contrast to hexagonal ice, cubic ice cannot be obtained in the form of a large single crystal, because it always contains hexagonal stacking faults (Kuhs et al. 1987, Kohl et al. 2000, Hansen et al. 2007). Thus, cubic and hexagonal ice can be regarded as two endmembers of a series of mixed states that can be distinguished based on the stacking sequence. Some of them show mainly hexagonal stacking sequences with cubic faults, others mainly cubic sequences with hexagonal faults and others seem to be stacking disordered (Hansen et al. 2008a,

Hansen et al. 2008b). Stacking-disordered ice (ice I$_{sd}$) is what will always crystallize first from supercooled water (Malkin et al. 2012, Malkin et al. 2015). As opposed to the earlier belief that vibrational spectra for ice I are always identical, some subtle differences were noted in Raman spectra and explained based on the lower symmetry of ice I$_{sd}$ compared with ice I$_h$ (Carr et al. 2014). In case of an equal number of hexagonal and cubic stacking sequences, trigonal crystals may in fact grow and be observed in clouds (Murray et al. 2015). This situation corresponds to the mid-member in the continuous series of ice I polytypes. In terms of thermodynamic stability, all these members are very close in enthalpy (within about 50 J/mol) (Kohl et al. 2000, Handa et al. 1986a). However, the transition temperatures to hexagonal ice may vary depending on the extent of cubic stacking faults. Highly hexagonal sequences transform at temperatures near 190 K to hexagonal ice, but highly cubic sequences transform near 220 K. Ice I$_h$ is the thermodynamically most stable variant in all cases, and both ice I$_{sd}$ and ice I$_c$ are metastable compared to it. For this reason the phase diagram of the thermodynamically most stable phases in Fig. 1 does not show stacking-disordered or cubic ice.

Ices XVI and XVII: "Empty" Clathrate Hydrates

Recent developments in the field of ice polymorphism involve low-density phases, which are in fact even less dense than ice I$_h$ and referred to as empty clathrate hydrates. Two such empty clathrates are now known as ices XVI and XVII. A strategy that has proven to be successful to reach ices of even lower densities was pioneered by Falenty and Kuhs (Falenty et al. 2014). While it was previously thought that such ice samples can only be isolated under tensile stress, that is, at negative pressure, they succeeded in preparing them *in vacuo*. As a starting point, they used a clathrate hydrate of cubic structure II that is filled with a particularly small guest, namely, neon. By pumping on the clathrate for five days it was possible to remove the guest atoms, leaving behind the empty clathrate. This had been suggested earlier in studies by Wooldridge et al. (1987) and Jacobson et al. (2009). Some theory work had conjectured that such empty clathrate hydrates would actually collapse to ice I$_h$ at ambient pressure and be thermodynamically stable only if kept under tension. On thermodynamic grounds, one would expect the empty clathrate to collapse, returning to ice I. An interesting aspect here is why the emptied clathrate does not do so, at least at temperatures below 130 K. This may have two reasons: either the clathrate is not fully empty, with a small fraction of cages still occupied with neon, or the empty clathrate is kinetically trapped. While Falenty and Kuhs favor the latter explanation (Falenty et al. 2014), the Rietveld procedure also allows for some residual neon atoms trapped within the ice (as expressed through the standard deviation in the Rietveld procedure). This raises the question how empty the empty clathrates actually are and whether some remaining guest atoms are still trapped within ices XVI/XVII. The crystal structure, thermal expansivity and limit of metastability of the empty hydrate as well as its density of 0.81 g/cm^3, about 12% less than hexagonal ice were determined. The empty hydrate structure exhibits negative thermal expansion below about 55 K, similar to hexagonal ice and is called ice XVI (cubic space group $Fd\overline{3}m$). By refilling the metastable empty clathrate with the same or other guest gases, metastable clathrate hydrates especially with very low occupancies might be incurred in experiments, as suggested in the theory work by Yagasaki (Yagasaki et al. 2016). Upon refilling the host lattice contracts and the negative thermal expansivity disappears again.

A similar strategy was adopted by del Rosso et al. (del Rosso et al. 2016b), producing the novel metastable ice XVII starting from a gas-loaded ice as a precursor. They used even smaller guest atoms, namely molecular hydrogen, which are initially trapped in the filled ice of structure C_0, and then pumped off, where the H_2 release is monitored by means of Raman spectroscopy. Interestingly, the emptied crystal can adsorb and release hydrogen repeatedly, showing a temperature-dependent hysteresis. The diffraction pattern (del Rosso et al. 2016a) is indexed with a hexagonal cell and can be refined within the space group $P6_122$. While the lattice constant a increases with temperature, c decreases from 25 K to 100 K. Above 130 K ice XVII collapses and transforms to stable ice I_h, just like ice XVI.

It seems like the strategy can be exploited also in future, for example, other clathrates such as the one of cubic structure I might be freed from guest molecules. Further possible candidates of metastable ices identified in computer simulations, which might be produced from gas-filled precursors are mentioned below in the next section.

The Negative Pressure Regime: (–0.2 GPa – 0 GPa)

At ambient temperature, freezing of liquid water to hexagonal ice would require considerable negative pressure on the order of –200 MPa. However, liquid water at ambient temperature cannot sustain such tension (Imre et al. 2008). Cavitation, that is, nucleation and growth of a bubble, is observed around –20 MPa (Henderson and Speedy 1987) or –100 MPa (Zheng et al. 1991, Zheng et al. 2002). Ice I crystals stable at T > 0°C have been found in some microscopic inclusions in minerals (Roedder 1967). No phase other than ice I has been observed at negative pressure so far. Natural clathrate hydrates (Davidson 1973) are stabilized by van der Waals-interaction with guest molecules (e.g., methane). The three most common types of clathrate hydrates belong to three different types of structures built from differently sized cages. They are called cubic structure I (sI), cubic structure II (sII), and hexagonal structure (sH). These structures contain differently sized cages built from a differing number of four-, five-, and six-membered rings of water molecules. The empty sII structure is in fact now known as ice XVI. That is, ice XVI has been prepared as a metastable ice *in vacuo*, but never in its stability domain at negative pressure.

The phase diagram at negative pressure is summarized in Fig. 2 for several different calculations. For some water models such as TIP4P/2005, it has been shown that they fare reasonably well in predicting the phase diagram of water (Vega et al. 2008), and so it seems justified to expect that one or the other low-density phase is accessible when doing experiments on stretched ice or stretched water. The TIP4P/2005 model predicts the sII structure to be stable at negative pressure on the order of –400 MPa and the sH structure to be stable at negative pressure on the order of –700 MPa (see Fig. 2, red line). By comparison, the mW-water model predicts the sII structure to be stable even starting at –130 MPa (Jacobson et al. 2009). The sI phase is metastable by only a slight margin in chemical potential in the sII region

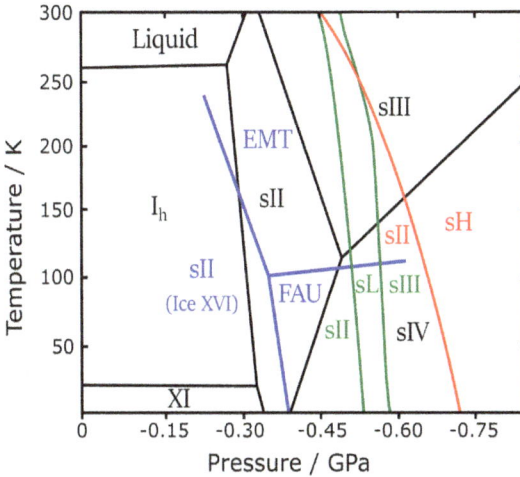

Fig. 2: Predicted phase diagram of water at negative pressure collected from several sources. Conde et al. (2009) predict the sII and sH clathrate structures to be the stable phases at p < –0.30 GPa, based on the TIP4P/2005 model (red lines). Huang et al. (2017) indicate the empty clathrates sIII and sIV to be more stable than sH, also for TIP4P/2005 water (black lines). Liu and Ojamae (2018) identify the stability region for the sL hydrate in the same pressure range, based on non-local dispersion-corrected density functional theory (denoted vdw-DF2 of rPW86-vdW2; green lines). In their most recent work, Liu et al. (2019) identify the FAU and EMT type clathrates to be stable in this pressure domain again based on the same kind of density functional theory calculations. The FAU and EMT ultralow-density ices are named according to their zeolite analogs. The empty clathrate of sII structure is the same as ice XVI.

of stability (Conde et al. 2009). Other crystalline phases, which are necessarily of lower density than hexagonal ice (Kosyakov 2009), have been predicted in other simulations, for example, ice i and ice i' (Fennell et al. 2005). Using the TIP5P model it is predicted that hexagonal ice under tension would transform first to ice i', then at higher tension to sII and finally, at –1 GPa to sH.

Liu and Ojamae (Liu et al. 2018) used density functional theory to search for other possible low-density phases. They identified an ice of ultralow density (0.6 g cm^{-3}) built from nano-cages and named it structure L clathrate (sL). It is predicted to be stable in a small pressure region near –0.50 GPa, where even more negative pressures convert it to the empty sIII hydrate, and less negative pressures to sII (see green lines in Fig. 2). One possible application for such open structures is hydrogen storage, where the sL clathrate shows a storage mass density of 7.7 wt%, which exceeds the 2017 target of the Department of Energy (DOE), namely 5.5 wt%. This would set a new record of hydrogen storage capacity in clathrate hydrates, but structures of even higher capacity are suggested by other groups.

Huang et al. (2016) made predictions employing molecular packing Monte Carlo techniques with dispersion-corrected density functional theory. They predict a crystalline clathrate of cubic structure III (sIII) composed of two large icosihexahedral cavities ($8^6 6^8 4^{12}$) and six small decahedral cavities ($8^2 4^8$) per unit cell. In the framework of the benchmark TIP4P/2005 model, the guest-free sIII clathrate is actually more stable than all other low-density structures mentioned above, provided the pressure is below –0.34 GPa (see black lines in Fig. 2). Specifically, they are also more stable than the sII and sH phases mentioned above. Such pressures can hardly be achieved experimentally, at least when following the most common method of studying liquid water under tension, for example, in mineral inclusions. In such samples, cavitation will occur before the required pressures can be reached, that is, formation of a bubble followed by its sudden collapse and shock-wave propagation. An alternative strategy would be to study solid samples at subzero, say icicles, under tension. Also in this case, rupture might occur, but it has not been studied well under which conditions. Suitably prepared icicles, possibly doped, might be helpful to avoid stresses and might allow reaching tensions of this kind.

Following up on their own work Huang et al. (2017) predict yet another clathrate of cubic structure, which they name sIV. It is composed of even larger cages than the other three cubic clathrate structures, namely eight large icosihexahedral cavities ($12^4 6^4 4^{18}$), eight intermediate dodecahedral cavities ($6^6 4^6$), and sixteen small octahedral cavities ($6^2 4^6$) per unit cell. This results in an extremely low mass density of 0.506 g/cm^3, which makes it a possible candidate for a stable phase below -0.38 GPa (0 K) and -0.73 GPa (200 K), also for the TIP4P/2005 water model (see black lines in Fig. 2).

Matsui et al. (2017) derived more than 300 kinds of porous ice structures from analogous zeolite frameworks and space fullerenes. Based on classical molecular dynamics simulations, they proposed several ices which are less dense and more stable than the sparse ice structures predicted by Huang et al. (2017). Based on the low density of the ice structures and their low density such structures were named "aeroices". As expected thermodynamically, for example, from the Clapeyron equation, structures of high volume and low entropy are supposed to be the most stable near the absolute zero temperature under negative pressure. While the former criterion seems to be fulfilled by the aeroices, the zero-point configurational entropy and the possibility of H-ordering transitions in aeroices have remained untouched in literature so far. Also Liu et al. (2019) exploit the analogy of ice and clathrate structures with SiO$_2$ and zeolites to look into more than 200 hypothetical low-density porous ices that might be stable at negative pressure and produced as metastable ice in vacuum. One of the very interesting candidate structures is their EMT ice, named according to zeolite nomenclature, of extremely low density of 0.5 g/cm^3. First-principles computations and molecular dynamics simulations confirm that the EMT ice is stable under negative pressures and, counter intuitively, exhibits higher thermal stability than other ultralow-density ices. The EMT ice can be viewed as dumbbell-shaped motifs in a hexagonal close-packed structure. Due to its very large cavities, the EMT ice hydrate could, if accessed experimentally, store large amounts of hydrogen of 13 wt% (Liu et al. 2019). Compared with those of ice XI (0.93 g/cm^3), both the bending and stretching vibrational modes of the EMT ice are blue-shifted due to their weaker hydrogen bonds. In the reconstructed temperature-pressure (T-P) phase diagram of water (see blue lines in Fig. 2), the EMT ice is located at deeply negative pressure regions below ice XVI and at higher temperature regions next to the FAU type ice structure. FAU contains two types of dumbbell-shaped $(H_2O)_{48}$ cages, namely $4^6 6^8$ and $4^{18} 6^4 12^4$, arranged in a supramolecular face-centred cubic (fcc) crystal. EMT contains a $(H_2O)_{48}$ cage, again $4^6 6^8$, and a $(H_2O)_{60}$ cage, namely $4^{21} 6^6 12^5$, arranged in a hexagonal close-packing (hcp). It will certainly remain a challenge to confirm or refute such predictions based on experimental work.

The Rich Intermediate Pressure Range (0.1–2 GPa)

High-Temperature Phases: H-disordered

Water's melting line takes quite an interesting path (see Fig. 1). First, it is negatively sloped, that is, the melting point decreases with pressure. This behaviour is most rarely found in other one-component systems. Therefore, the process of pressure-induced melting may take place for ice kept below zero (273 K, 0°C), where pressures of 0.22 GPa suffice to melt the ice between 0°C and 251 K (–22°C). Second, the melting line suddenly changes to a positive slope at 251 K and ~ 0.21 GPa. Now, the opposite process of pressure-induced crystallization may take place. An ice cube kept at 255 K will first melt at ~ 0.21 GPa, but then refreeze upon continued increase of pressure at ~ 0.3 GPa. However, it will not freeze to the same hexagonal ice (ice I_h) it had originated from, but transform into a different polymorph, namely ice III (tetragonal space group $P4_12_12$) (McFarlan 1936, Kamb et al. 1968). An ice cube kept at 263 K (–10°C) will refreeze later above ~ 0.45 GPa, to yet another polymorph, namely ice V (monoclinic space group $C2/c$) (Kamb et al. 1967). Compression of liquid water at ambient temperature to beyond ~ 0.9 GPa will result in the crystallization of ice VI (tetragonal space group $P4_2/nmc$) (Kamb 1965). That is, in a relatively narrow pressure range between 0 and 1 GPa, liquid water freezes to four distinct stable ice polymorphs (see Fig. 1).

Metastable phases are not included in Fig. 1. Such phases may occur because the thermodynamically most stable phase cannot be accessed due to kinetic or geometric constraints. In particular, ice IV (trigonal space group $R\bar{3}c$) (Engelhardt et al. 1972, Engelhardt et al. 1981, Salzmann et al. 2002a, Salzmann et al. 2003a, Salzmann et al. 2004) or ice XII (tetragonal space group $I\bar{4}2d$) (Lobban et al. 2998, Salzmann et al. 2003b, Salzmann et al. 2004, O'Keefee 1998, Hallbrucker 1999, Koza et al. 1999, Koza et al. 2000, Kohl et al. 2000, Kohl et al. 2001, Loerting et al. 2002, Salzmann et al. 2003b) may crystallize from the pressurized liquid or amorphous solid state in the stability fields of ices V and VI. In most experiments especially near 1 GPa, the most stable form does not form directly. Instead, a metastable polymorph usually forms first before transformation to the stable phase occurs—in accordance with the expectations from Ostwald's step rule, which poses that intermediate steps to metastable phases can be taken on the way from an unstable phase to the stable phase.

Ice IV was originally regarded as an intermediate pressure form of ice that often evades preparation (Engelhardt et al. 1972, Engelhardt et al. 1981) and was later described as "a will-o'-the-wisp, a tentative, *ghostly* form of *ice*" (Ball 2001). Organic nucleating agents have increased the chances of ice IV formation from the liquid (Evans 1967), and later it could be demonstrated that it also crystallizes upon slow heating of high-density amorphous ice (Salzmann et al. 2008, Hallbrucker et al. 1999, Koza et al. 1999, Koza et al. 2000, Kohl et al. 2000, Kohl et al. 2001, Loerting et al. 2002, Salzmann et al. 2003a, Kohl et al. 2002, Salzmann et al. 2002a).

Ice II takes a special role. It exists as a stable phase in the intermediate pressure range, but it will never form from liquid water, but always from other ice polymorphs. All of its triple points are a solid-solid-solid triple point. None of the five solid-solid-liquid triple points along water's melting line involve ice II. The six different phases of ice associated with them are ice I_h, ice III, ice V, ice VI, ice VII, and superionic ice (see below). Two high-pressure polymorphs of ice have actually been identified on Earth: the high-pressure forms of ice VI (Kogi et al. 2000) and ice VII (Tschauner et al. 2018) were identified in diamond inclusions, and so these high-pressure forms were actually declared as minerals by the International Mineralogic Association (IMA). At depths of several hundred kilometres, these ices may form in cold subduction slabs, where they might be of relevance to build friction between slabs, resulting in seismic events, which one could term "ice quakes". High-pressure forms of ice are even more abundant in space, where the icy moons are covered in layers of ice up to 1000 km thick so that their own weight suffices to transform the ices to high-pressure polymorphs such as ice III, V, VI, or VII. This requires pressures of a couple of GPa.

Low-Temperature Phases: H-ordered

Ice II is also an exception in terms of its H-subnetwork. The H-atoms are always in an ordered arrangement, that is, the water dipoles are aligned in a typical pattern. For all other H-ordered ice phases, it is possible to disorder the H-sublattice by heating. For ice II, actually the network of O-atoms melts first. It has, therefore, remained elusive to prepare disordered ice II in experiments. The only viable strategy seems to be to prepare disordered ice II at low temperatures as a metastable, kinetically trapped phase. Ice polymorphs in the intermediate pressure range usually come as couples: a high-temperature disordered and a low-temperature disordered phase share the same "house", namely the same O-subnetwork. Currently, the six hydrogen disorder-order pairs I_h-XI (Tajima et al. 1984, Matsuo et al. 1986, Fukazawa et al. 1998, Fukazawa et al. 2002, Kuo et al. 2005, Singer et al. 2005, Knight et al. 2006, Fan et al. 2010), III-IX (Fan et al. 2010, Whalley et al. 1968, LaPlaca et al. 1973, Nishibata et al. 1974, Minceva-Sukarova et al. 1984, Londono et al. 1993, Kuhs et al. 1998, Lobban et al. 2000, Knight et al. 2006), V-XIII (Salzmann et al. 2006, Kuhs et al. 1998, Lobban et al. 2000, Salzmann et al. 2006, Salzmann et al. 2008, Knight et al. 2008), VI-XV (Fan et al. 2010, Kuhs et al. 1984, Knight et al. 2005, Kuo et al. 2006, Salzmann et al. 2009), VII-VIII (Singer et al. 2005, Knight et al. 2006, Fan et al. 2010, Kuhs et al. 1984, Whalley et al. 1966, Jorgensen et al. 1984, Jorgensen et al. 1985, Pruzan et al. 1990, Pruzan et al. 1992, Besson et al. 1994, Besson et al. 1997, Pruzan et al. 2003, Song et al. 2003, Yoshimura et al. 2006, Somayazulu et al. 2008), and XII-XIV (Salzmann et al. 2006, Tribello et al. 2006) are known.

The order-disorder transition temperature varies between 72 K for the ice I_h/XI couple (Tajima et al. 1986, Fukazawa et al. 1998, Fukazawa et al. 2002, Kuo et al. 2005, Singer et al. 2005, Knight et al. 2006, Fan et al. 2010) and 273 K for the ice VII/VIII couple (Singer et al. 2005, Knight et al. 2006, Fan et al. 2010, Kuhs et al. 1984, Whalley et al. 1966, Jorgensen et al. 1984, Jorgensen et al. 1985, Pruzan et al. 1990, Pruzan et al. 1992, Besson et al. 1994, Besson et al. 1997, Pruzan et al. 2004, Song et al. 2003, Yoshimura et al. 2006, Somayazulu et al. 2008). The difference of about 200 K between the hydrogen order-disorder temperature for these two pairs shows that quite different sets of enthalpy and entropy balances as well as activation energies are involved in the process. While the ordering transition is sufficiently fast at 273 K, this is not so true at 72 K. In case of the ordering transition from ice I_h to ice XI (orthorhombic space group $Cmc2_1$) the transition is almost immeasurably slow. Usually, a frustrated crystal of ice I_h is obtained below 72 K, and this frustration manifests itself in a negative thermal expansion coefficient below 65 K—yet another density anomaly of water. In order to avoid the geometric frustration of the H-subnetwork, it is necessary to increase the mobility of the H-sublattice, which is done by introducing point defects externally through doping, for example, with 10^{-4}–10^{-2} M HF, HCl, KOH, or NH_3. These molecules are incorporated as substitutional point defects into the ice lattice (Hobbs 1974). In the presence of these extrinsic defects, the kinetically hindered ordering of hydrogens may be facilitated. Indeed the incorporation of HCl as a dopant was key in the discovery of three ice phases in the last few years, namely ice XIII, XIV, and XV (Salzmann et al. 2006, Salzmann et al. 2009). Dielectric spectroscopy studies have indicated that HCl in fact enhances the reorientation dynamics by up to four orders of magnitude, while other dopants do not show any enhancement (Koster et al. 2015). HF is the second-best alternative as a dopant, enhancing the ice XII reorientation dynamics somewhat. For ice I_h, KOH is effective in facilitating the transition to the H-ordered ice XI (Matsuo et al. 1968). However, even when the ice I_h sample is KOH-doped, it is necessary to be patient and to follow a temperature protocol. One strategy to produce ice XI is to cool KOH-doped ice I_h first to 60 K, nucleate some small domains at of ice XI at this temperature, and then to increase the temperature to 70 K, where the domains grow with time. This temperature protocol is employed because the nucleation rate typically shows a maximum a few percent below the equilibrium temperature of 72 K, whereas the growth rate is highest just below 72 K. Even after weeks of domain growth, only part of the sample has converted to ice XI, though (Tajima et al. 1984, Matsuo et al. 1968). The understanding of the mechanism underlying this increase in mobility is far from complete.

Yet another pair of H-order and disorder is found when cooling ice III (at high-pressure conditions). Below about 170 K, this transforms to ice IX (Fan et al. 2010, Whalley et al. 1968, LaPlaca et al. 1973, Nishibata et al. 1974, Minceva-Sukarova et al. 1984, Londono et al. 1993, Kuhs et al. 1998, Lobban et al. 2000, Knight et al. 2006). In this case the use of external point defects is not necessary. By contrast to ice XI, ice IX cannot be found in the diagram of water's stable polymorphs. This is because ice IX is metastable with respect to ice II (Kamb 1964, Kamb et al. 1971, Finch et al. 1968, Fortes et al. 2003, Fortes et al. 2005). Yet, the activation barrier for O-rearrangement from ice III to ice II is much higher than the barrier for the H-ordering conversion to ice IX, so that the transition to ice II does not take place above 200 K, in spite of ice II being thermodynamically stable.

In case of the ice XII/XIV pair (Salzmann et al. 2006, Tribello et al. 2006), in fact both the H-disordered and the H-ordered polymorph are metastable within the stability domains of ices V and VI. Yet, it is possible to maintain the O-network of ice XII/XIV between 80 K and 200 K under pressure without any oxygen rearrangements taking place to stable ice V. Even at ambient pressure, one can cycle between ice XII (tetragonal space group $I\bar{4}2d$) and XIV (orthorhombic space group $P2_12_12_1$) at 80–150 K without transformation of the O-lattice to the one of ice I_c. The order-disorder transition temperature is ~ 100 K in this case. This makes it an excellent case for study of the influence of dopants on the transition. In fact, in this case HCl seems to be the most suitable dopant, which produces a combination of cationic and Bjerrum L-defects (Fuentes-Landete et al. 2018). By contrast, KOH is not suitable in this case—and it is still unclear why KOH does help to order ice I_h, but not ice XII, and vice versa, why HCl helps to order ice XII. The same holds true for the H-order-disorder transition in ice V/XIII (Koster et al. 2016). In this case KOH accelerates the reorientational dynamics compared to undoped H_2O by one order of magnitude, but HCl by four orders. In a similar manner, ice V (monoclinic space group $C2/c$) can be ordered near 100 K through the use of HCl doping to produce metastable ice XIII (monoclinic space group $P2_1/a$) (Salzmann et al. 2006, Kuhs et al. 1998, Lobban et al. 2000, Salzmann et al. 2008, Knight et al. 2008) in the stability domain of ice VI. Hydrogen-ordered cubic ice I_c (Lokotosh et al. 1993), hydrogen-ordered ices IV, XVI, and XVII as well as hydrogen-disordered ice II have not been prepared in laboratory experiments yet. One specific case that might help to understand the factors governing the kinetics and thermodynamics of H-ordering is the case of ice VI, which is unique among all the ice phases known so far.

Ice VI: A Special Case

One specific case that might help to understand the factors governing the kinetics and thermodynamics of H-ordering is the case of ice VI, which is unique among all the ice phases known so far. It is not straightforward to see that ices have to occur in such H-order-disorder pairs. In fact, there are numerous hydrogen-ordered structures related to each hydrogen-disordered polymorph. Ice VI was discovered more than a hundred years ago by Bridgman when conducting piston-cylinder experiments near 1 GPa (Bridgman 1912). It is a tetragonal ice, where 10 water molecules are arranged in a unit cell of space group $P4_2/nmc$ in a way making two interpenetrating networks ("self-clathrate") (Kamb 1965). Its hydrogen-ordered counterpart ice XV (pseudo-orthorhombic space group $P\bar{1}$), was reported in 2009, where the ordering was found to be antiferroelectric (Salzmann et al. 2009). Ice XV is surrounded in the phase diagram exclusively by other polymorphs of ice (Fan et al. 2010, Kuhs et al. 1984, Knight et al. 2005, Kuo and Kuhs 2006, Salzmann et al. 2009). Also the ferroelectric Cc structure was thought to be a viable candidate (Knight et al. 2005). Komatsu et al. (2016) even list 45 possible different symmetries for ordered water dipoles without violating the ice rules (see also Kuo et al. (2006)). Twelve of them share the same Patterson function so that 33 can possibly be identified in experiments. Eighty years ago, Pauling had regarded all hydrogen-ordered structures related to the same polymorph by simply permuting H atoms as degenerate (Pauling 1935). It is clear now, though, that there are subtle differences in enthalpy, and also in the lattice parameters and density. Thermodynamically, the most stable of these H-ordered variants is not necessarily the one obtained in experiments. Metastable ones may in fact be experimentally more accessible. In other words, it is still unclear for all of the known H-ordered ices whether a more stable arrangement of the H-atoms exists, and whether this might be realized experimentally someday.

Calculations only help to an extent in this context since it is extremely hard to properly account for the dispersion interactions and since all of the H-ordered variants are very close in energy, probably even closer than the error-bar of the best calculations today. That is, there are a number of possibilities in how hydrogens may be ordered, which result in configurations of similar enthalpy. One of these configurations corresponds to the experimentally verified polymorph, and the others might be detected in future laboratory experiments.

Different reaction paths may be experimentally accessed, for example, by using different dopants, by changing temperature protocols or by varying the time spent at specific temperatures. This may lead to H-ordered structures that are different from the known H-ordered structure in the pairs mentioned above.

This has, however, not been achieved experimentally, with the exception being the ice VI structure. Our work in this pressure regime has revealed very interesting phase behaviour that has led to the suggestion of a novel ice polymorph, called ice β-XV. Based on large differences in dielectric activation energies and in Raman spectra, especially related to bands indicating H-order, we have suggested that ice β-XV is an ice polymorph that has an H-order pattern distinct from the one of ice XV. In fact, it might be a ferroelectrically ordered variant of ice VI, for example, the Cc structure mentioned above. The preparation of ice β-XV differs from the preparation of ice XV in that higher pressures used (1.8 ± 0.2 GPa *vs.* 1.0 ± 0.2 GPa) and that slower cooling rates are employed (3 K/min *vs.* 45 K/min), both starting from HCl-doped, disordered ice VI samples. Calorimetry scans differ in that ice XV shows only a single, broad endotherm (middle, dark-grey curve in Fig. 3), whereas ice β-XV shows two endotherms (top curve in Fig. 3). Our interpretation of this observation is that ice β-XV first transforms to ice XV at 103 K, and then ice XV transforms to ice VI at 129 K. The latter is then the previously known disordering transition from ice XV to ice VI which appears at the same temperature as in earlier work (Salzmann et al. 2009, Shephard et al. 2015). The former would then be an order-order transition in the H-atom network—an observation that has not been made for any other ice so far. This transition might also be a transition from ferroelectric to anti-ferroelectric, but this is not fully clear yet and hinges on a structural determination of ice β-XV, which has not been made yet.

Please also note that the first endotherm is followed by a weak exotherm. This is observed both for ice β-XV (top curve in Fig. 3) and "ice XV" (middle, light-grey curve in Fig. 3). This observation can easily be explained when thinking about the mechanism of an order-order transition: any order-order transition has to pass through a disordered transition state that then experiences ordering. This disordered ice VI only appears transiently, cannot be kept stable at these temperatures, and progresses toward the stable ice XV state at 103–129 K. When inspecting marked bands for H-order in the Raman spectrum, as for example, expressed by the Whale index (Whale et al. 2013) or the librational index (Thoney et al. 2019), one notices a transition that runs from highly ordered via disordered to ordered (Thoney et al. 2019). The weak endotherm followed by a weak exotherm in "ice XV" can then be interpreted to indicate that this sample actually contains a small number of ice β-XV domains, next to the majority of domains, which are ice XV (Thoney et al. 2019).

Rosu-Finsen and Salzmann have provided an alternative explanation for the phenomenology by stating that ice β-XV might be a glassy (H-disordered) state that occurs at very low temperatures (Rosu-Finsen et al. 2019). In contrast to our research, Rosu-Finsen and Salzmann do not see this glassy state as a transient state in the course of an order-order transition, but as a kinetically trapped state that experiences a glass transition. The first endotherm in Fig. 3 (top curve) is reinterpreted by them

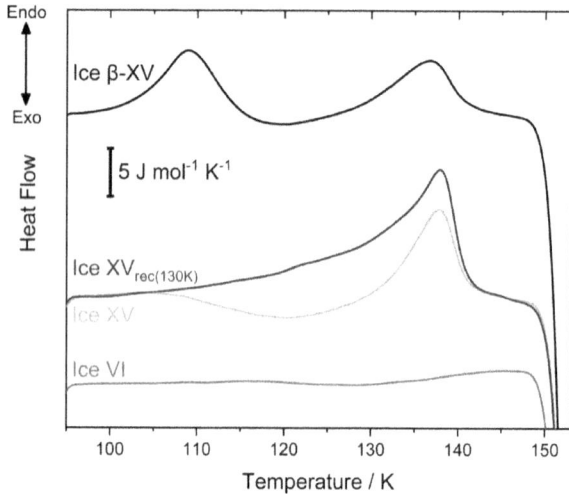

Fig. 3: Calorimetry scans for different types of samples sharing the oxygen network of ice VI, namely ice β-XV (top), ice XV (middle, dark grey), ice XV after recooling from 130 K at ambient pressure (middle, light grey), and ice VI (bottom). All scans are recorded by heating from 80 K at 10 K/min.

as a devitrification, where a huge overshoot makes it look similar to the endotherm signifying the latent heat associated with the order-order transition. For comparison, a devitrification transition associated with the H-subnetwork can be seen in Fig. 3 (bottom). In this case, the H-atoms are immobile below the glass transition at 129 K, but mobile above. One way to distinguish the two scenarios would be to do modulated DSC (differential scanning calorimetry), which is capable of distinguishing reversing (e.g., glass transitions) and non-reversing (e.g., order-order transitions) through modulating the heating ramp with a sinus function. An alternative approach would be to look into the crystal structure using neutron diffraction—this approach is, however, hampered by a large isotope effect. In other words, the protocol that leads to ice β-XV in hydrogenated samples does not do so for deuterated samples. It needs to be figured first how much more time needs to be provided for deuterons to be ordered in the same way the hydrogens do. Currently, we think this is at least 100 times more, so that cooling rates of < 0.03 K/min would be required, which make the experiment hard to realize. Alternatively, long waiting times (in the order of days or weeks) just below the order-order temperature might (or might not) help in reaching D_2O ice β-XV. If our interpretation is correct, the phase diagram contains two stable polymorphs related to ice VI as depicted in Fig. 1 in the pressure range from 0.8–1.5 GPa, whereas ice VIII is more stable at 1.5–2.0 GPa, making ice β-XV and ice XV both metastable in this pressure regime. In terms of density, ice β-XV is actually quite similar to ice χ, which has been predicted from MD simulations to form under the influence of electric fields under similar high-pressure conditions as a ferroelectric ice in an orthorhombic structure (Zhu et al. 2019). This is similar to ice XV, which has a pseudo-orthorhombic crystal structure (Salzmann et al. 2009). While Zhu et al. (2019) state that the ferroelectric phase may only nucleate under very high electric fields, ice β-XV is possibly ferroelectric and nucleates spontaneously under slow cooling at a pressure of 2.0 GPa below 103

K (Gasser et al. 2018). Also Shephard and Salzmann (Shephard et al. 2015) have inferred that ferroelectric intermediates may appear on the ordering transition of ice VI upon cooling. Our experiments actually suggest that these are not intermediates, but rather formed in parallel. A competition between ferroelectric ordering (ice β-XV) and antiferroelectric ordering (ice XV) might actually take place, where ice XV dominates near 1 GPa, but ice β-XV near 2 GPa. This is certainly an issue that deserves future investigation. Similarly, the antiferroelectric ice VIII might have a ferroelectric pendant, which may be stabilized through high electric fields (Hu et al. 2011). So far there is no clear evidence for any ferroelectric ice that can be formed without the help of an external electric field. Also, ice XI, which was regarded for a long time as ferroelectric, is in fact antiferroelectric (Parkkien et al. 2014). However, the growth of ice films at heterointerfaces may be ferroelectric because it is driven by the electrostatics. In these cases, the disordering temperature may be exceptionally high, for example, 175 K on platinum substrates for ferroelectric ice, so that ferroelectric ices may be abundant in cold environments on electrostatic substrates (Sugimoto et al. 2016).

The High- and Ultrahigh-pressure Regime (2–500 GPa)

The Ice VII/VIII Pair

The next dense ice structure, ice VIII (tetragonal space group $I4_1/amd$), needs significantly higher pressures and usually does not form below 2 GPa. It shares a triple point with ices VII (cubic space group $Pn\overline{3}m$) and VI at 2.1 GPa and 273 K. The phase boundary between ices VI and VIII emerging from this triple point recedes to 1.8 GPa at 200 K (see Fig. 1). It may only form under some circumstances in small amounts below this pressure limit. In our own experiments, we have encountered traces of metastable ice VIII even at 2.0 GPa when remaining for about half an hour above 200 K. The corresponding Raman spectra recorded by us at ~ 80 K and 10 m bar are depicted in Fig. 4. We note that the large part of the sample has yielded ice VI spectra (not shown in Fig. 4), but some individual points on the sample were found to show non-ice VI spectra. Thus, we have caught a new phase right at the time of its emerging at 2.0 GPa. We have added on purpose 5% D_2O to this H_2O sample in order to be able to measure both the coupled OH- and the decoupled OD stretching vibrations. For comparison, Wong and Whalley (Wong et al. 1976) have recorded ice VIII Raman spectra independently for both D_2O and H_2O samples. *En gros*, the agreement between the bands in Fig. 4 and the spectra by Wong and Whalley (Wong et al. 1976) is excellent. The deviation of bands is no more than 3 cm^{-1} and also the intensity ratios match. *En détail*, there is, however, a noteworthy difference: the bands in Fig. 4a are sharper and show more features. The band position is in agreement for six of these bands within 3 cm^{-1}. However, the three bands at 92, 111, and 132 cm^{-1} are not observed at all by Wong and Whalley (Wong et al. 1976), even though their measurements allow observation of bands down to 50 cm^{-1}. These features seem to indicate that the phase that emerges in our experiments is ice VIII. However, it seems to differ from the ice VIII prepared by Wong and Whalley (Wong et al. 1976) in that it is either more hydrogen-ordered, or possibly the water dipoles adopt a different kind of hydrogen ordering than in ice VIII. The latter option

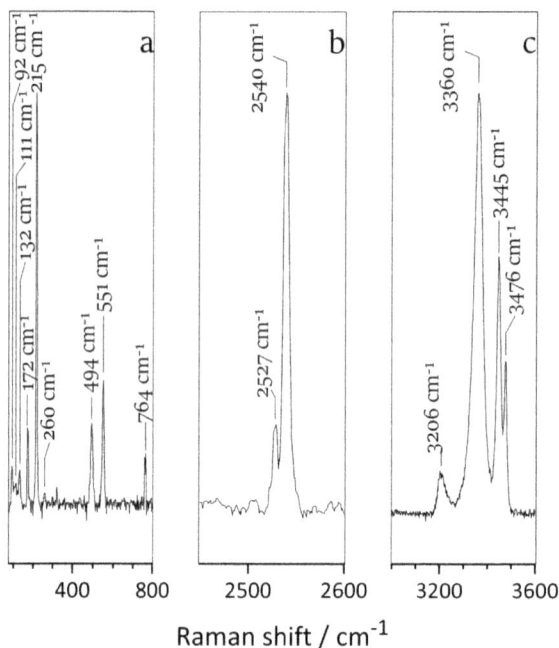

Fig. 4: Raman spectrum for ice VIII recorded at ~ 80 K and 10 mbar. The sample was obtained by keeping a 95 wt% H2O/5 wt% D2O sample above 200 K for 25 minutes at 2.0 GPa and quench-recovering it to 77 K and 1 bar before transfer to the Raman cryostat. The spectrum agrees very well with the spectra recorded by Wong and Whalley in the OH- and OD-stretching regimes at > 2000 cm^{-1} (Wong and Whalley 1976) for pure D_2O-ice VIII and H_2O-ice VIII. The band positions do not deviate by more than 1–3 cm^{-1}. Also in the librational/translational range (at < 500 cm^{-1}), the spectrum presented here agrees very well. Six bands match to better than 3 cm^{-1}. However, the three bands at 132, 111, and 92 cm^{-1} do not appear in the spectra by Wong and Whalley (1976). This could indicate a higher degree of H-order in our sample.

seems more unlikely since the band pattern and intensities looks very similar in both Raman spectra. It might be that the degree of H-ordering is higher when ice VIII is just starting to form at 2.0 GPa in our experiments as opposed to fully transformed at 2.5 GPa in the experiments by Wong and Whalley (Wong et al. 1976).

Ice X and post-ice X-phases

The pressure range, in which ice VII (and its H-ordered counterpart ice VIII) can exist as a stable phase is much broader than for all other ices discussed so far. In the whole pressure range from ~ 2 GPa up to ~ 62 GPa, the oxygen network of ice VII is the only stable arrangement of atoms. Thus, ice VII can be regarded as the first phase that is no longer stable at intermediate pressure conditions, but it is the first stable phase occupying the high pressure regime (see phase diagram in Fig. 5). The phase transition to ice X (cubic space group $Pn\bar{3}m$, just like ice VII) (Schwager et al. 2004) takes place at ~ 50–80 GPa. This phase transition is different from other polymorphic transitions because it involves symmetrisation of the H-bond, but no change in the space group (Hemley et al. 1987). This is identical to saying that it involves the breakdown of the molecular nature of water, transforming it to an atomic crystal of 1:2 O:H ratio (Holzapfel et al. 1972, Teixeira et al. 1998). Water molecules are no longer uniquely defined in ice X, because H atoms occupy the "halfway position" between oxygen atoms, making it a "symmetric ice" (Polian et al. 1984, Poilan et al. 1985). How the phase transition actually takes place is depicted schematically in Fig. 6. At comparatively low pressures of, around 6 GPa, there is a substantial barrier that prevents the H-atoms from hopping between potential wells. In this situation, the ice may either be statically ordered ice VIII (below 273 K) or statically disordered (above 273 K). As the pressure rises, the oxygen atoms come closer to each other, lowering the barrier for the hopping of the H atoms. At a certain pressure point, there is enough thermal energy available for the protons in ice VII to overcome the barrier. At this point, the H-atoms become dynamic and the O-atoms switch from being acceptors to donors and vice versa. Above 1000 K and above 50 GPa, one can regard this dynamic ice as either ice X or as a state, in which the hydrogens in ice VII are dynamically disordered (Goncharov et al. 2005). The time average is a symmetric ice

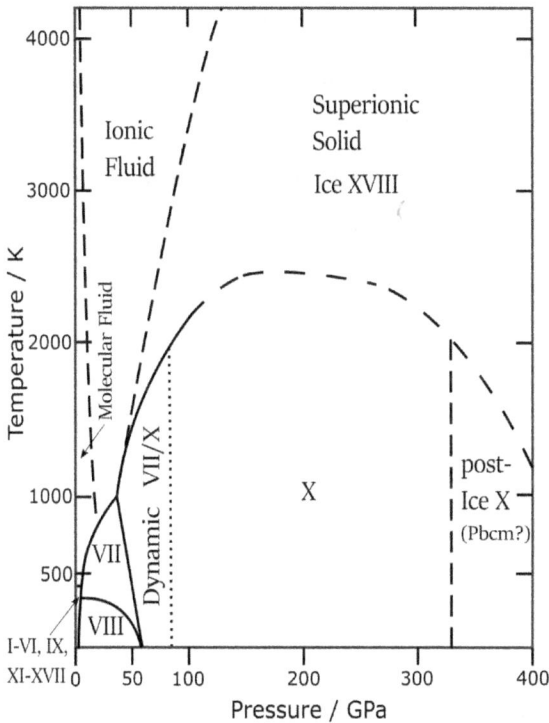

Fig. 5: Phase diagram at extreme conditions, for temperatures up to 4000 K and to ultrahigh-pressures of 400 GPa. The phase boundaries between ices VII, VIII, and X are taken from the experimental data by Pruzan et al. (1997) and Song et al. (2003). Data on the transition from statically disordered ice VII to dynamically disordered ice VII and to ice X are taken from Goncharov et al. (Hirsch and Holzapfel 1984, 1986, Goncharov et al. 1999). Data for the melting line of ice is taken from Schwager et al. (2004) (up to 100 GPa) and Millot et al. (2019). The transition pressure from ice X (cuprite-structure) to post-ice X (anti-luorite structure) is taken from Demontis et al. (1988). The Pbcm ice is named as one of a few predicted post-ice X candidate structures. Other predicted structures extending to the 6 Tpa range are given by Militzer and Wilson (2010). McMahon (2011), Hermann et al. (2012), Pickard et al. (2013). The separation between the molecular and ionic liquid phase is copied from the simulation work by Cavazzoni et al. (1999). The transition line between the atomic solid and the superionic solid as well as ice XVIII follow the experimental work by Millot et al. (2019).

(ice X), but individual snapshots still show the molecular nature (ice VII). Changes in the Raman (Hirsch et al. 1984, Hirsh et al. 1986) and IR (Song et al. 2003, Aoki et al. 1996, Song et al. 1999, Katoh et al. 2000) spectrum of ice VII in the pressure range ~ 35–50 GPa were interpreted to be consistent with H-bond symmetrisation and ice X formation, in accordance with results from Brillouin scattering (Polian et al. 1984, Polian et al. 1985). Pruzan et al. (1997) have argued on the basis of Raman and X-ray diffraction data that the nature of the sample is rather statistically averaged, dynamically disordered ice VII. The main difference between dynamically disordered ice VII and dynamic ice X is that the H-atom distribution in ice VII has two maxima along O...O directions, whereas in ice X the H-atom distribution has only one maximum in the "halfway position" as a result of the quantum effect zero-point motion (Benoit et al. 1999). Consequently, 50–80 GPa marks the transition range between ice VII and X in Fig. 6, where a dynamic situation is encountered. As the pressure increases beyond 100 GPa, the "halfway position" in ice X is no longer a result of statistically averaging H-atoms delocalized in a shallow double-well potential, but a structure with the H-atoms being localized in a single-well potential (Benoit et al. 1998). The ultrahigh pressure has erased the barrier for the proton to hop along a hydrogen bond. While the former is referred to as "dynamic ice X", the latter is referred to as "static ice X". The results by Goncharov et al. are in agreement with a static, ordered ice X (Goncharov et al. 1996, Goncharov et al. 1999) of the cuprite, single-

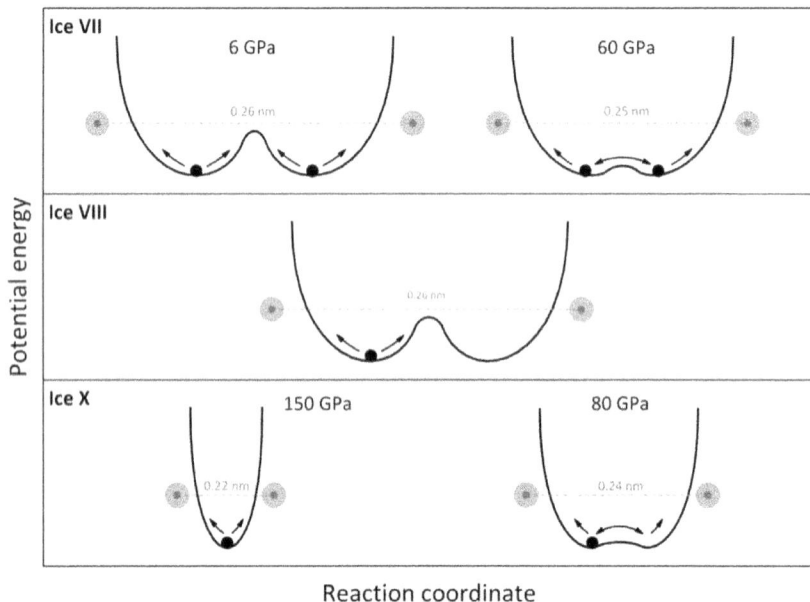

Fig. 6: Schematic illustration of potentials describing the positions of hydrogen atoms in ices VII, VIII, and X. Red spheres connected by dashed lines symbolize oxygen donor and acceptor atoms. The O-O distance is indicated. Both ice VII and ice X may have a double identity, where the hydrogen atoms may either be dynamically switching back and forth between two wells or trapped. The trapping occurs in a single well for ice X, equally in both wells for ice VII, and in one of the two wells for ice VIII. For ice VIII, H-atoms are never switching between wells. Dynamically disordered ice VII and dynamically symmetric ice X only differ in terms of the timescale needed for the switching, that is, there is a continuous transition between the two, as indicated in Fig. 6. A similar figure was also shown by Teixeira (1998).

well type. This ordered ice X is thought to be stable under ultrahigh pressure of at least 330 GPa (Demontis et al. 1988, Demontis et al. 1989). However, in ice X there is still a barrier that prevents the protons to move freely within the O lattice. H-atoms are still localized. This only changes occur at much higher temperatures, where the protons move barrierlessly within the crystal lattice–defining the superionic state as highly mobile H-atoms that are delocalized in a lattice of oxygen atom, where they take a role similar to the role of the electron gas in metals.

In the interior of ice giants such as Neptune or Uranus, water is exposed to much higher pressures, of hundreds of GPa, where they might show some interesting properties such as metal-like behaviour. These ices are called post-ice X phases. The other intriguing possibility in this pressure regime is superionicity. Such a phase has in fact very recently been observed in shockwave experiments, after its original postulation on theoretical grounds three decades ago (Demontis et al. 1988)—and denoted ice XVIII, in spite of a missing crystal structure. Millot et al. (2019) succeeded in recording an X-ray diffraction pattern within a nanosecond after sequences of laser pulses have been fired at a thin layer of water, generating shockwaves that reverberated between diamond plates, thereby heating and compressing the sample to about 3000 K and up to 420 GPa. While the X-ray patterns recorded by them between 100 and 250 GPa indicate the body-centered cubic structure of ice X, above this pressure a transition to a face-centered cubic structure, ice XVIII (cubic space group $Fm\bar{3}m$), is observed. Under such temperature and pressure conditions superionicity was also inferred by Millot et al. (2018). They claim that the transition in the O network would not occur if the hydrogen atoms were not delocalized and highly mobile. Under static conditions the bcc structure would persist under much higher pressures. The superionic fcc structure of ice XVIII does not melt up to 320 GPa and 3800 K. From this study, it might be observed that superionic ice could indeed be stable at depths of about 1/3 of the radius of Uranus and Neptune, possible forming a thick solid icy mantle there.

At lower temperatures, a transition to a post-ice X phase of orthorhombic structure (*Pbcm*) has been predicted (see Fig. 5) (Benoit et al. 1996, Marques et al. 2009, Caracas 2008). Another option was suggested by Demontis et al. (1988) who predict a transition from the body-centered cubic (bcc) cuprite structure in ice X to a face-centered cubic (fcc) anti-fluorite structure. Militzer and Wilson have predicted that at 760 GPa, a transition to *Pbca*, and at 1550 GPa a transition to *Cmcm* takes place (Militzer et al. 2010). The former phase is an insulator and consists of an interpenetrating network, just like ice VII, VIII, and X. The latter phase, however, is metallic and consists of corrugated sheets and the H-atoms no longer occupy tetrahedral sites between nearest O-atoms, but octahedral, midpoint positions between the next-nearest O-atoms. When considering also differences in zero-point energy, the orthorhombic *Pnma* phase (related to the *Cmcm* phase by a slight Peierls lattice distortion) appears as a stable phase in the range 1250–1550 GPa prior to the transformation to the metallic *Cmcm* phase. Such a metallic phase may be accessible only at very high temperature, for example, at T > 7000 K, and a superionic phase is expected below that temperature (Cavazzoni et al. 1999). Other predicted structures extending to the 6 TPa range are given by Militzer and Wilson (Militzer et al. 2010), McMahon (McMahon 2011), Hermann et al. (2012), and Pickard et al. (2013).

To date, there is no experimental evidence for any of these post-ice X phases, but one or the other might well be detectable in the future when new experimental techniques become available.

Summary

The field of ice polymorphism is highly active. Currently, 18 polymorphs (ices I–XVIII) are known. Ice β-XV that is included as a stable phase in Fig. 1 is not included in this count since its crystal structure has not been reported yet. Also ice I is only counted as a single polymorph, even though it has three variants, hexagonal ice (ice I_h), cubic ice (ice I_c), and stacking-disordered ice (ice I_{sd}). This is because ice I constitutes a continuous series of stacking-disordered states with the endmembers hexagonal and cubic ice. For the latter, the observation of a "pure" single crystal of ice I_c is still awaited. Laboratory cubic ice is polycrystalline and always contains hexagonal stacking faults and may even be regarded as hexagonal ice with cubic stacking faults (Hansen et al. 2007, Hansen et al. 2008b).

The crystal structures of these 18 polymorphs are known to a high precision, often also as a function of temperature and/or pressure (Petrenko et al. 1999). One issue, which may be discussed controversially, is the issue of incomplete H-ordering and the question whether H_2O and D_2O ices show identical crystal structures, especially concerning deuterons and hydrogen atoms. Since the ordering involves dynamics of the hydrogen network in an essentially static oxygen network, studies of isotope effects on the ordering transition are essential (Fuentes-Landete et al. 2018). For instance, Fuentes-Landete et al. (2018) examined the isotope effects occurring within the ice XIV-ice XII pair and revealed that deuterated samples are much harder to order than hydrogenated ice XII. This is particularly relevant since the structural information is based on neutron data of D_2O samples. Partial ordering of H-atoms has been observed in ice V (Kuhs et al. 1998, Lobban et al. 2000), but the crystal structure of the completely ordered polymorph ice XIII could be refined only later (Salzmann et al. 2006). In case of the ordered form of ice XII, an incompletely ordered polymorph containing residual disorder was refined to the crystal structure of ice XIV (Salzmann et al. 2006). This has prompted theoreticians to propose that differently ordered variants of ice XIV may exist and be at the origin of the residual disorder (Tribello et al. 2006). Such differently hydrogen-ordered variants related to the same hydrogen-disordered polymorph may be discovered in future experiments. One such candidate is ice β-XV which appears to show a higher degree of order and, foremost, a different arrangement of H_2O dipoles than ice XV based on Raman spectroscopy and our interpretation of the calorimetry data. This is contested based on inelastic neutron scattering measurements and a different interpretation of the calorimetry scans in terms of a deep glassy rather than novel crystalline arrangement of H-atoms (Rosu-Finsen et al. 2019). This also seems possible for ice VIII, where the Raman spectrum shown here in Fig. 4 might indicate a higher degree of order compared to literature data.

In the intermediate pressure range 0.2–2.0 GPa especially disordered ice II and ordered ice IV await their discovery. Other pressure regimes, which have not been explored very much in past experiments pertain to negative pressure ("stretched ice", "aeroice") and extremely high pressure on the order of 1000 GPa. These domains are currently fully in the hands of simulation and theory, with a plethora of predicted ice structures that have remained elusive in experiments, because such conditions can be realized experimentally only with difficulties (McMillan 2003). That is, new ice polymorphs will likely be discovered in the future when new experimental methods become available or when some clever experiments are done using existing technologies. Nanosecond X-ray experiments as used for the discovery of ice XVIII certainly hold promise for more discoveries at ultrahigh pressure and extremes of temperature, but we also want to take up the cudgels for experiments at subzero temperatures in the low- and intermediate pressure regimes.

Acknowledgements

We are indebted to our long-time collaborators on the topic of ice, especially Roland Böhmer (TU Dortmund) and Daniel T. Bowron (Rutherford Appleton Laboratory).

References

Aoki, K., H. Yamawaki, M. Sakashita and H. Fujihisa. 1996. Phys. Rev. B. 54: 15673–15677.
Ball, P. 2001. Life's Matrix: A Biography of Water, University of California Press, Berkeley.
Bartels-Rausch, T., V. Bergeron, J. H. E. Cartwright, R. Escribano, J. L. Finney, H. Grothe, P. J. Gutierrez, J. Haapala, W. F. Kuhs, J. B. C. Pettersson, S. D. Price, C. I. Sainz-Diaz, D. J. Stokes, G. Strazzulla, E. S. Thomson, H. Trinks and N. Uras-Aytemiz. 2012. Rev. Mod. Phys. 84: 885–944.
Bauer, M., M. S. Elsaesser, K. Winkel, E. Mayer and T. Loerting. 2008. Phys. Rev. B. 77: 220105.
Benoit, M., M. Bernasconi, P. Focher and M. Parrinello. 1996. Phys. Rev. Lett. 76: 2934–2936.
Benoit, M., D. Marx and M. Parrinello. 1998. Nature 392: 258–261.
Benoit, M., D. Marx and M. Parrinello. 1999. Solid State Ionics 125: 23–29.
Bernal, J. D. and R. H. Fowler. 1933. J. Chem. Phys. 1: 515–548.
Bernstein, J. 2002. Polymorphism in Molecular Crystals, Oxford University Press, Oxford, UK.
Besson, J. M., P. Pruzan, S. Klotz, G. Hamel, B. Silvi, R. J. Nelmes, J. S. Loveday, R. M. Wilson and S. Hull. 1994. Phys. Rev. B. 49: 12540–12550.
Besson, J. M., M. Kobayashi, T. Nakai, S. Endo and P. Pruzan. 1997. Phys. Rev. B. 55: 11191–11201.
Bridgman, P. W. 1912. Proc. Amer. Acad. Arts Sciences 47: 441–558.
Caracas, R. 2008. Phys. Rev. Lett. 101: 085502.
Carr, T. H. G., J. J. Shephard and C. G. Salzmann. 2014. J. Phys. Chem. Lett. 5: 2469–2473.
Cavazzoni, C., G. L. Chiarotti, S. Scandolo, E. Tosatti, M. Bernasconi and M. Parrinello. 1999. Science 283: 44–46.
Clapeyron, M. C. 1834. Mémoire sur la puissance motrice de la chaleur. Journal de l'École polytechnique. 23: 153–190.
Conde, M. M., C. Vega, G. A. Tribello and B. Slater. 2009. J. Chem. Phys. 131: 034510.
Davidson, D. W. 1973. In Water—A Comprehensive Treatise, Vol. 2 (Ed.: F. Franks), Plenum Press, p. 15.
del Rosso, L., F. Grazzi, M. Celli, D. Colognesi, V. Garcia-Sakai and L. Ulivi. 2016a. J. Phys. Chem. C. 120: 26955–26959.
del Rosso, L., M. Celli and L. Ulivi. 2016b. Nature Comm. 7: 13394.
Demontis, P., R. LeSar and M. L. Klein. 1988. Phys. Rev. Lett. 60: 2284–2287.
Demontis, P., M. L. Klein and R. LeSar. 1989. Phys. Rev. B 40: 2716–2718.
Dowell, L. G. and A. P. Rinfret. 1960. Nature 188: 1144–1148.
Elsaesser, M. S., K. Winkel, E. Mayer and T. Loerting. 2010. Phys. Chem. Chem. Phys. 12: 708–712.
Engel, E. A., A. Anelli, M. Ceriotti, C. J. Pickard and R. J. Needs. 2018. Nature Comm. 9: 2173.
Engelhardt, H. and E. Whalley. 1972. J. Chem. Phys. 56: 2678–2684.
Engelhardt, H. and B. Kamb. 1981. J. Chem. Phys. 75: 5887–5899.
Evans, L. F. J. 1967. Appl. Phys. 38: 4930–4932.
Falenty, A., T. C. Hansen and W. F. Kuhs. 2014. Nature 516: 231.
Fan, X., D. Bing, J. Zhang, Z. Shen and J.-L. Kuo. 2010. Comp. Mater. Sci. 49: S170–S175.

Fennell, C. J., J. D. Gezelter and J. Chem. 2005. Theory Comput. 1: 662–667.

Finch, E. D., S. W. Rabideau, R. G. Wenzel and N. G. Nereson. 1968. J. Chem. Phys. 49: 4361–4365.

Fortes, A. D., I. G. Wood, J. P. Brodholt and L. Vocadlo. 2003. J. Chem. Phys. 119: 4567–4572.

Fortes, A. D., I. G. Wood, M. Alfredsson, L. Vocadlo and K. S. Knight. 2005. J. Appl. Cryst. 38: 612–618.

Fortes, A. D., I. G. Wood, L. Vocadlo, K. S. Knight, W. G. Marshall, M. G. Tucker and F. Fernandez-Alonso. 2009. J. Appl. Crystallogr. 42: 846–866.

Fortes, A. D. and M. Choukroun. 2010. Space Sci. Rev. 153: 185–218.

Fuentes-Landete, V., C. Mitterdorfer, P. H. Handle, G. N. Ruiz, J. Bernard, A. Bogdan, M. Seidl, K. Amann-Winkel, J. Stern, S. Fuhrmann and T. Loerting. 2015. *In*: Debenedetti, P. G., M. A. Ricci and F. Bruni (eds.). Water: Fundamentals as the Basis for Understanding the Environment and Promoting Technology 187: 173–208.

Fuentes-Landete, V., K. W. Koster, R. Bohmer and T. Loerting. 2018. Physical Chemistry Chemical Physics 20: 21607–21616.

Fukazawa, H., S. Ikeda and S. Mae. 1998. Chem. Phys. Lett. 282: 215–218.

Fukazawa, H., S. Ikeda, M. Oguro, T. Fukumura and S. Mae. 2002. J. Phys. Chem. B. 106: 6021–6024.

Gasser, T. M., A. V. Thoeny, L. J. Plaga, K. W. Koster, M. Etter, R. Bohmer and T. Loerting. 2018. Chem. Sci. 9: 4224–4234.

Goncharov, A. F., V. V. Struzhkin, M. S. Somayazulu, R. J. Hemley and H. K. Mao. 1996. Science 273: 218–220.

Goncharov, A. F., V. V. Struzhkin, H.-k. Mao and R. J. Hemley. 1999. Phys. Rev. Lett. 83: 1998–2001.

Goncharov, A. F., N. Goldman, L. E. Fried, J. C. Crowhurst, I. F. W. Kuo, C. J. Mundy and J. M. Zaug. 2005. Phys. Rev. Lett. 94: 125508.

Guinier, A., G. B. Bokii, K. Boll-Dornberger, J. M. Cowley, S. Durovic, H. Jagodzinski, P. Krishna, P. M. De Wolff, B. B. Zvyagin et al. 1984. Acta Crystall. A40: 399–404.

Hallbrucker, A. and E. Mayer. 1987. J. Phys. Chem. 91: 503–505.

Hallbrucker, A. 1999. *In*: Geiger, A. and H.-D. Lüdemann (eds.). Metastable Water, International Bunsen Discussion Meeting, Schloss Nordkirchen, Germany.

Handa, Y. P., D. D. Klug and E. Whalley. 1986a. J. Chem. Phys. 84: 7009–7010.

Handa, Y. P., O. Mishima and E. Whalley. 1986b. J. Chem. Phys. 84: 2766–2770.

Handa, Y. P., D. D. Klug and E. Whalley. 1987. J. Phys. Colloq. 48: 435–440.

Handa, Y. P., D. D. Klug and E. Whalley. 1988. Can. J. Chem. 66: 919–924.

Hansen, T. C., A. Falenty and W. F. Kuhs. 2007. pp. 201–208. *In*: Kuhs, W. F. (ed.). Proc. 11th Intl. Conf. on the Physics and Chemistry of Ice. RSC, Dorchester, UK.

Hansen, T. C., M. M. Koza, P. Lindner and W. F. Kuhs. 2008a. J. Phys.: Condens. Matter 20: 285105.

Hansen, T. C., M. M. Koza and W. F. Kuhs. 2008b. J. Phys.: Condens. Matter 20: 285104.

Hemley, R. J., A. P. Jephcoat, H. K. Mao, C. S. Zha, L. W. Finger and D. E. Cox. 1987. Nature 330: 737–740.

Henderson, S. J. and R. J. Speedy. 1987. J. Phys. Chem. 91: 3069–3072.

Herbert, E., S. Balibar and F. Caupin. 2006. Phys. Rev. E. 74: 041603.

Hermann, A., N. W. Ashcroft and R. Hoffmann. 2012. Proc. Natl. Acad. Sci. USA 109: 745–750.

Hirsch, K. R. and W. B. Holzapfel. 1984. Phys. Lett. A 101A: 142–144.

Hirsch, K. R. and W. B. Holzapfel. 1986. J. Chem. Phys. 84: 2771–2775.

Hobbs, P. V. 1974. Ice Physics, Clarendon Press, Oxford.

Holzapfel, W. B. J. 1972. Chem. Phys. 56: 712–715.

Hu, X., N. Elghobashi-Meinhardt, D. Gembris and J. C. Smith. 2011. Response of water to electric fields at temperatures below the glass-transition: a molecular dynamics analysis. J. Chem. Phys. 135: 134507.

Huang, Y. Y., C. Q. Zhu, L. Wang, X. X. Cao, Y. Su, X. Jiang, S. Meng, J. J. Zhao and X. C. Zeng. 2016. Sci. Adv. 2: e1501010.

Huang, Y. Y., C. Q. Zhu, L. Wang, J. J. Zhao and X. C. Zeng. 2017. Chem. Phys. Lett. 671: 186–191.

Imre, A. R., A. Drozd-Rzoska, A. Horvath, T. Kraska, S. J. Rzoska and J. Non-Cryst. 2008. Solids 354: 4157–4162.

Jacobson, L. C., W. Hujo and V. Molinero. 2009. J. Phys. Chem. B. 113: 10298–10307.

Jorgensen, J. D., R. A. Beyerlein, N. Watanabe and T. G. Worlton. 1984. J. Chem. Phys. 81: 3211–3214.

Jorgensen, J. D. and T. G. Worlton. 1985. J. Chem. Phys. 83: 329–333.

Kagi, H., R. Lu, P. Davidson, A. F. Goncharov, H. K. Mao and R. J. Hemley. 2000. Min. Mag. 64: 1089–1097.

Kamb, B. 1964. Acta Crystallogr. 17: 1437–1449.

Kamb, B. 1965. Science 150: 205–209.

Kamb, B., A. Prakash and C. Knobler. 1967. Acta Cryst. 22: 706–715.

Kamb, B. and A. Prakash. 1968. Acta Crystallogr., Sect. B. 24: 1317–1327.

Kamb, B., W. C. Hamilton, S. J. LaPlaca and A. Prakash. 1971. J. Chem. Phys. 55: 1934–1945.

Katoh, E., M. Song, H. Yamawaki, H. Fujihisa, M. Sakashita and K. Aoki. 2000. Phys. Rev. B. 62: 2976–2979.

Kawada, S. J. 1972. Phys. Soc. Japan 32: 1442.

Klaproth, M. H. 1798. Bergmannische J. I: 294–299.

Knight, C. and S. J. Singer. 2005. J. Phys. Chem. B. 109: 21040–21046.

Knight, C. and S. J. Singer. 2006. J. Chem. Phys. 125: 064506.

Knight, C., S. J. Singer, J.-L. Kuo, T. K. Hirsch, L. Ojamae and M. L. Klein. 2006. Phys. Rev. E. 73: 056113.

Knight, C. and S. J. Singer. 2008. J. Chem. Phys. 129: 164513.

Kohl, I. and E. Mayer. 2000. A. Hallbrucker, J. Phys. Chem. B. 104: 12102–12104.

Kohl, I., E. Mayer and A. Hallbrucker. 2000. Phys. Chem. Chem. Phys. 2: 1579–1586.

Kohl, I., E. Mayer and A. Hallbrucker. 2001. Phys. Chem. Chem. Phys. 3: 602–605.

Kohl, I., T. Loerting, C. Salzmann, E. Mayer and A. Hallbrucker. 2002. NATO Science Series, II: Mathematics, Physics and Chemistry 81: 325–333.

Komatsu, K., F. Noritake, S. Machida, A. Sano-Furukawa, T. Hattori, R. Yamane and H. Kagi. 2016. Sci. Rep. 6: 28920.

Koster, K. W., V. Fuentes-Landete, A. Raidt, M. Seidl, C. Gainaru, T. Loerting and R. Bohmer. 2015. Nature Communications, 6.

Koster, K. W., A. Raidt, V. F. Landete, C. Gainaru, T. Loerting and R. Bohmer. 2016. Physical Review B. 94.

Kosyakov, V. I. J. 2009. Struct. Chem. 50: S60–S65.

Koza, M., H. Schober, A. Tölle, F. Fujara and T. Hansen. 1999. Nature 397: 660–661.

Koza, M. M., H. Schober, T. Hansen, A. Tölle and F. Fujara. 2000. Phys. Rev. Lett. 84: 4112–4115.

Kuhs, W. F., J. L. Finney, C. Vettier and D. V. Bliss. 1984. J. Chem. Phys. 81: 3612–3623.

Kuhs, W. F. and M. S. Lehmann. 1986. pp. 1–65. *In*: Franks, F. (ed.). Water Science Reviews 2, Vol. 2. Cambridge University Press, Cambridge.

Kuhs, W. F., D. V. Bliss and J. L. Finney. 1987. J. Phys. Colloq. C1: 631–636.

Kuhs, W. F., C. Lobban and J. L. Finney. 1998. Rev. High Pressure Sci. Technol. 7: 1141–1143.

Kuo, J.-L., M. L. Klein and W. F. Kuhs. 2005. J. Chem. Phys. 123: 134505.

Kuo, J.-L. and W. F. Kuhs. 2006. J. Phys. Chem. B. 110: 3697–3703.

LaPlaca, S. J., W. C. Hamilton, B. Kamb and A. Prakash. 1973. J. Chem. Phys. 58: 567–580.

Liu, Y. and L. Ojamae. 2018. Phys. Chem. Chem. Phys. 20: 8333–8340.

Liu, Y., Y. Y. Huang, C. Q. Zhu, H. Li, J. J. Zhao, L. Wang, L. Ojamae, J. S. Francisco and X. C. Zeng. 2019. Proc. Natl. Acad. Sci. USA 116: 12684–12691.

Lobban, C., J. L. Finney and W. F. Kuhs. 1998. Nature 391: 268–270.

Lobban, C., J. L. Finney and W. F. Kuhs. 2000. J. Chem. Phys. 112: 7169–7180.

Loerting, T., I. Kohl, C. Salzmann, E. Mayer and A. Hallbrucker. 2002. J. Chem. Phys. 116: 3171–3174.

Loerting, T., K. Winkel, M. Seidl, M. Bauer, C. Mitterdorfer, P. H. Handle, C. G. Salzmann, E. Mayer, J. L. Finney and D. T. Bowron. 2011. Phys. Chem. Chem. Phys. 13: 8783–8794.

Lokotosh, T. V. and N. P. Malomuzh. 1993. Khim. Fiz. 12: 897–907.

Londono, J. D., W. F. Kuhs and J. L. Finney. 1993. J. Chem. Phys. 98: 4878–4888.

Malenkov, G. J. 2009. Phys. Cond. Matt. 21: 283101.

Malkin, T. L., B. J. Murray, A. V. Brukhno, J. Anwar and C. G. Salzmann. 2012. Proc. Natl. Acad. Sci. USA 109: 1041–1045.

Malkin, T. L., B. J. Murray, C. G. Salzmann, V. Molinero, S. J. Pickering and T. F. Whale. 2015. Phys. Chem. Chem. Phys. 17: 60–76.

Marques, M., G. J. Ackland and J. S. Loveday. 2009. High Pressure Res. 29: 208–211.

Matsui, T., M. Hirata, T. Yagasaki, M. Matsumoto and H. Tanaka. 2017. J. Chem. Phys. 147: 091101.

Matsuo, T., Y. Tajima and H. Suga. 1986. J. Phys. Chem. Solids 47: 165–173.

Mayer, E. A. 1987. Hallbrucker, Nature 325: 601–602.

McFarlan, R. L. 1936. J. Chem. Phys. 4: 253–259.

McMahon, J. M. 2011. Phys. Rev. B. 84.

McMillan, P. F. 2003. Chem. Comm. 919–923.

Militzer, B. and H. F. Wilson. 2010. Phys. Rev. Lett. 105: 195701.

Millot, M., S. Hamel, J. R. Rygg, P. M. Celliers, G. W. Collins, F. Coppari, D. E. Fratanduono, R. Jeanloz, D. C. Swift and J. H. Eggert. 2018. Nature Phys. 14: 297.

Millot, M., F. Coppari, J. R. Rygg, A. C. Barrios, S. Hamel, D. C. Swift and J. H. Eggert. 2019. Nature 569: 251–255.

Minceva-Sukarova, W., F. Sherman and G. R. Wilkinson. 1984. J. Mol. Struct. 115: 137–140.

Mitscherlich, E. 1822–1823. Abhl. Akad. Berlin, 43–48.

Molina, M. J. 1991. Atmos. Environ. 25A: 2535–2537.

Molina, M. J., T. L. Tso, L. T. Molina and F. C. Y. Wang. 1987. Science 238: 1253–1257.

Murray, B. J., D. A. Knopf and A. K. Bertram. 2005. Nature 434: 202–205.

Murray, B. J., T. L. Malkin and C. G. Salzmann. 2015a. J. Atmos. Solar-Terrest. Phys. 127: 78–82.

Murray, B. J., C. G. Salzmann, A. J. Heymsfield, S. Dobbie, R. R. Neely and C. J. Cox. 2015b. Bull. Amer. Met. Soc., 96.

Nishibata, K. and E. Whalley. 1974. J. Chem. Phys. 60: 3189–3194.

O'Keeffe, M. 1998. Nature 392: 879.

Parkkinen, P., S. Riikonen and L. Halonen. 2014. J. Phys. Chem. C. 118: 26264–26275.

Pauling, L. J. 1935. Am. Chem. Soc. 57: 2680–2684.

Peter, T. 1997. Ann. Rev. Phys. Chem. 48: 785–822.

Petrenko, V. F. and R. W. Whitworth. 1999. Physics of Ice, Oxford University Press, Oxford.

Pickard, C. J., M. Martinez-Canales and R. J. Needs. 2013. Phys. Rev. Lett. 110.

Polian, A. and M. Grimsditch. 1984. Phys. Rev. Lett. 52: 1312–1314.

Polian, A., J. M. Besson and M. Grimsditch. 1985. Solid State Phys. Pressure: Recent Adv. Anvil Devices, 93–98.

Prenni, A. J. and M. A. Tolbert. 2001. Acc. Chem. Res. 34: 545–553.

Pruzan, P., J. C. Chervin and M. Gauthier. 1990. Europhys. Lett. 13: 81–87.

Pruzan, P., J. C. Chervin and B. Canny. 1992. J. Chem. Phys. 97: 718–721.

Pruzan, P., E. Wolanin, M. Gauthier, J. C. Chervin, B. Canny, D. Haeusermann, M. Hanfland. 1997. J. Phys. Chem. B. 101: 6230–6233.

Pruzan, P., J. C. Chervin, E. Wolanin, B. Canny, M. Gauthier and M. Hanfland. 2003. J. Raman Spectrosc. 34: 591–610.

Roedder, E. 1967. Science 155: 1413–1417.

Rosu-Finsen, A., A. Amon, J. Armstrong, F. Fernandez-Alonso and C. G. Salzmann. 2019. ArXiVe, https://arxiv.org/ftp/arxiv/papers/1911/1911.04368.pdf.

Rosu-Finsen, A. and C. G. Salzmann. 2019. Chem. Sci. 10: 515–523.

Röttger, K., A. Endriss, J. Ihringer, S. Doyle and W. F. Kuhs. 1994. Acta Crystall. B50: 644–648.

Salzmann, C. G., T. Loerting, I. Kohl, E. Mayer and A. Hallbrucker. 2002a. J. Phys. Chem. B 106: 5587–5590.

Salzmann, C., I. Kohl, T. Loerting, E. Mayer and A. Hallbrucker. 2002b. J. Phys. Chem. B. 106: 1–6.

Salzmann, C. G., I. Kohl, T. Loerting, E. Mayer and A. Hallbrucker. 2003a. Can. J. Phys. 81: 25–32.

Salzmann, C. G., I. Kohl, T. Loerting, E. Mayer and A. Hallbrucker. 2003b. Phys. Chem. Chem. Phys. 5: 3507–3517.

Salzmann, C. G., E. Mayer and A. Hallbrucker. 2004. Phys. Chem. Chem. Phys. 6: 1269–1276.

Salzmann, C. G., P. G. Radaelli, A. Hallbrucker, E. Mayer and J. L. Finney. 2006. Science 311: 1758–1761.

Salzmann, C. G., P. G. Radaelli, J. L. Finney and E. Mayer. 2008. Phys. Chem. Chem. Phys. 10: 6313–6324.

Salzmann, C. G., P. G. Radaelli, E. Mayer and J. L. Finney. 2009. Phys. Rev. Lett. 103: 105701.

Schwager, B., L. Chudinovskikh, A. Gavriliuk and R. Boehler. 2004. J. Phys.: Condens. Matter 16: S1177–S1179.

Shephard, J. J. and C. G. Salzmann. 2015. Chem. Phys. Lett. 637: 63–66.

Singer, S. J., J.-L. Kuo, T. K. Hirsch, C. Knight, L. Ojamaee and M. L. Klein. 2005. Phys. Rev. Lett. 94: 135701.

Solomon, S. 1999. Rev. Geophys. 37: 275–316.

Somayazulu, M., J. Shu, C.-s. Zha, A. F. Goncharov, O. Tschauner, H.-k. Mao and R. J. Hemley. 2008. J. Chem. Phys. 128: 064510.

Song, M., H. Yamawaki, H. Fujihisa, M. Sakashita and K. Aoki. 2003. Phys. Rev. B. 68: 014106.

Song, M., H. Yamawaki, H. Fujihisa, M. Sakashita and K. Aoki. 1999. Phys. Rev. B. 60: 12644–12650.

Sugimoto, T., N. Aiga, Y. Otsuki, K. Watanabe and Y. Matsumoto. 2016. Nature Phys. 12: 1063.

Tajima, Y., T. Matsuo and H. Suga. 1984. J. Phys. Chem. Solids 45: 1135–1144.

Tanaka, H. and I. Okabe. 1996. Chem. Phys. Lett. 259: 593–598.

Teixeira, J. 1998. Nature 392: 232–233.

Thoeny, A. V., T. M. Gasser and T. Loerting. 2019. Phys. Chem. Chem. Phys. 21: 15452–15462.

Tribello, G. A., B. Slater and C. G. Salzmann. 2006. J. Am. Chem. Soc. 128: 12594–12595.

Tschauner, O., S. Huang, E. Greenberg, V. B. Prakapenka, C. Ma, G. R. Rossman, A. H. Shen, D. Zhang, M. Newville, A. Lanzirotti and K. Tait. 2018. Science 359: 1136.

Vega, C., E. Sanz, J. L. F. Abascal and E. G. Noya. 2008. J. Phys.: Condens. Matter 20: 153101.

Whale, T. F., S. J. Clark, J. L. Finney and C. G. Salzmann. 2013. J. Raman Spect. 44: 290–298.

Whalley, E. J. 1983. Phys. Chem. 87: 4174–4179.

Whalley, E. and D. W. Davidson. 1966. J. B. R. Heath, J. Chem. Phys. 45: 3976–3982.

Whalley, E., J. B. R. Heath and D. W. Davidson. 1968. J. Chem. Phys. 48: 2362–2370.

Wong, P. T. T. and E. Whalley. 1976. J. Chem. Phys. 64: 2359–2366.

Wooldridge, P. J., H. H. Richardson and J. P. Devlin. 1987. J. Chem. Phys. 87: 4126–4131.

Yagasaki, T., M. Matsumoto and H. Tanaka. 2016. Phys. Rev. B. 93: 054118.

Yoshimura, Y., S. T. Stewart, M. Somayazulu, H.-k. Mao and R. J. Hemley. 2006. J. Chem. Phys. 124: 024502.

Zabrodsky, V. G. and T. V. Lokotosh. 1993. Ukr. Fiz. Zh. 38: 1714–1723.

Zheligovskaya, E. A. and G. G. Malenkov. 2006. Russ. Chem. Rev. 75: 57–76.

Zheng, Q., D. J. Durben, G. H. Wolfe and C. A. Angell. 1991. Science 254: 829–832.

Zheng, Q., J. Green, J. Kieffer, P. H. Poole, J. Shao, G. H. Wolf and C. A. Angell. 2002. NATO Sci. Ser., II 84: 33–46.

Zhu, W. D., Y. Y. Huang, C. Q. Zhu, H. H. Wu, L. Wang, J. Bai, J. L. Yang, J. S. Francisco, J. J. Zhao, L. F. Yuan and X. C. Zeng. 2019. Nature Comm. 10: 1925.

Chapter 6

The Hydrophilic-Hydrophobic Correlations in Water Systems

Francesco Mallamace,[1,*] *Carmelo Corsaro,*[2]
Paola Lanzafame[3]*, Georgia Papanikolaou*[3] *and*
Domenico Mallamace[4]

Introduction

The main aim of the present chapter is two fold: one, we address are view of there search activity in the field, and second, we will high light the importance of new approaches based on molecular properties able to properly define, and evaluate, the HE influences in the system structure and dynamics; hoping that such a work will results in a seminal and training activity for young researchers. It must be taken into account that the HE is a primary physical process border crossing many disciplines and the science life. An introductory idea to HE may be: "A Curious Antipathy for Water, How, does water meet a hydrophobic surface? Like great art, everyone recognizes hydrophobicity but few agree on the details" (Granick and Bae 2008).

A classical approach to the HE is described in terms of the wetting or super-wetting phenomenon. Wetting is commonly defined as the ability of a liquid to maintain contact with a solid surface so that it is ascribed by the intermolecular interactions between the two phases. Research on wetting and wettability started in the begin of the nineteenth century with the Young's equation (Young 1805). Extreme examples of this phenomena range from high tides at the beach to the ion channel since ll membranes. We report in Fig. 1 a suitable example of super-wettable (or

---

[1] Department of Nuclear Science and Engineering, Massachusetts Institute of Technology, Cambridge, MA02139, USA.

[2] Department of MIFT-Section of Physics, University of Messina, V.leF. Stagnod' Alcontres 31, Messina, 98166, Italy

[3] Departments of ChiBio FarAm-Section of Industrial Chemistry, University of Messina, CASPE-INSTM, V.leF. Stagnod' Alcontres 31, Messina, 98166, Italy.

[4] Departments of ChiBio FarAm and MIFT-Section of Industrial Chemistry, University of Messina, CASPE-INSTM, V.leF. Stagnod' Alcontres 31, Messina, 98166, Italy.

* Corresponding author: francesco.mallamace@unime.it

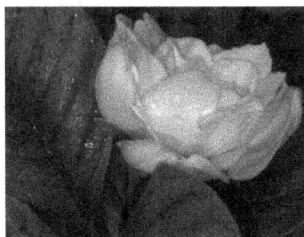

Fig. 1: Millimeter water drops on a gardenia petal.

super-water-repellent materials) that nature provides: the lotus effect where its leaves remain clean despite growing in mud (see for example, Fig. 1). This is characteristic of many plants whose leaves utilize super-hydrophobicity as the basis of a self-cleaning mechanism: water drops completely roll off the leaf, carrying undesirable particulates.

In the case of rough surfaces, wetting may assume either of two regimes: homogeneous wetting (where the liquid completely penetrates the roughness grooves) or heterogeneous wetting (where air or an other fluid is trapped underneath the liquid inside the roughness grooves). The so-called super-hydrophobicity may be related to a sort of transition between these regimes, see Fig. 1. The basic quantities to describe wetting are the contact (θ) and the roll angles (α). A sit is well known, θ represents an experimentally easily accessible parameter ∘ characterizing the surface hydrophobicity and ranges from 180° (for a hypothetical substrate with the same water affinity as vapor) down to 0° for a hydrophilic surface. In the classical Young's law, θ_Y is given in terms of the surface tension (γ) as $cos\ \theta_Y = -(\gamma_{LS} - \gamma_{SV})\ (\gamma_{LV})$, thus, depending on the surface tensions of the liquid/solid (γ_{LS}), solid/vapor (γ_{SV}), and liquid/vapor (γ_{LV}) interfaces.

In terms of a current representation of homogeneous and heterogeneous regimes, the water contact angle θ_W is given as $cos\theta_W = rcos\theta_Y$ in the homogeneous regime (Wenzel 1936), r being the roughness ratio (the ratio of the true area of the solid surface to its projection area). In these terms, it is clearly seen that: if a surface is hydrophobic ($\theta_Y > 90°$) and the roughness ($r > 1$) makes $\theta_W > \theta_Y$. Where as in the heterogeneous regime, the apparent contact angle $\theta\ CB$ can be written as $cos\theta_{CB} = r_f \cdot cos\ \theta_Y + f{-}1$ (Cassie and Baxter 1944). This is the Cassie–Baxter equation and the contact angle increases with an increase in f representing the fraction of the projected area that is wet and r_f is the roughness ratio of the wet area. However, it is important to note that the wetting regime that yields the lowest contact angle is the more stable one from a thermodynamic point of view. The f value at which a hydrophilic surface could turn in to a hydrophobic one is given as $f \geq (r_f\ cos\ \theta_Y)/(r_f\ cos\ \theta_Y + 1)$ (Bhushan and Jung 2006); however, in some conditions hydrophobic surfaces can be achieved above a certain f value as predicted by this equation. The Cassie-Wenzel transition identifies the evolution from a homogeneous to the heterogeneous regime. Figure 2 illustrates such a situation.

Another characteristic of a solid liquid interface is the contact angle hysteresis. If a drops its on a tilted surface, with a roll angles α, the contact angles at the front and back of the droplet correspond to the advancing (θadv) and receding (θrec) contact

Wenzel (homo)   **Cassie-Baxter (hetero)**

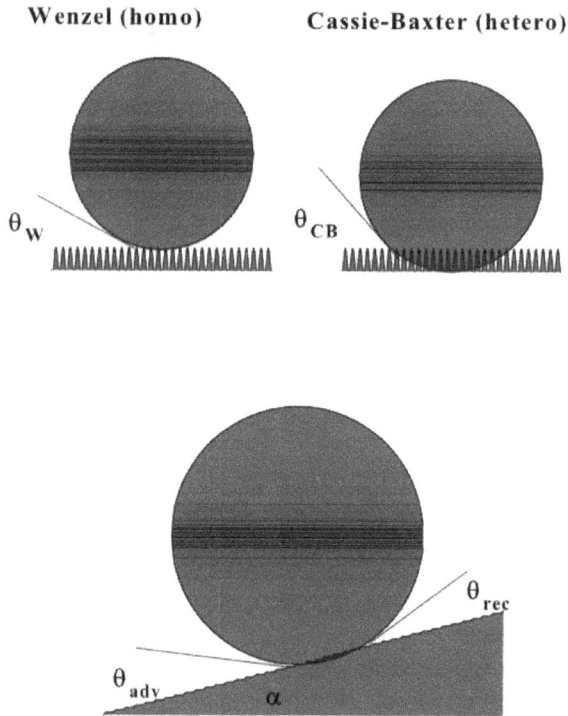

Fig. 2: Schematic of roughness-filling by water according to the Wenzel and Cassie-Baxtermodels.

angle, respectively. The advancing angle is greater than the receding angle, which results in contact angle hysteresis. Contact angle hysteresis occurs due to surface roughness and heterogeneity and is a measure of energy dissipation during the flow of a droplet along a solid surface.

A homogeneous interface is that for which $f = 0$ where as for the heterogeneous one f is a non-zero number. For a homogeneous interface, increasing roughness (high r_f) leads to increasing contact angle hysteresis (high values of $\theta adv - \theta rec$), while for a composite interface, as f approach to unity provides both a high contact angle and a low contact angle hysteresis so that a heterogeneous interface is desirable for super-hydrophobicity and self-cleaning.

Returning to at the lotus (or petal) effect, its low-adhesion super-hydrophobic property was attributed to the microscale papillae incorporated in to hydrophobic epicuticular wax: on the tops of these microscale papillae, there are branch-like nanostructures that created the micro-and nanoscale two-tier roughness responsible for the self-cleaning ability (Feng et al. 2002). Natureal so gives other intriguing cases of an isotropic super-wettability; some common examples arerice and butterfly. In the first case, water droplets roll along the direction parallel to the rice leaf edge rather than perpendicular. This is due to the an isotropic arrangement of microstructures on the leaves; the papillae distribution was in a quasi-one-dimensional order parallel to the leaf edge rather than a uniform arrangement like on the lotus leaf, and thus water droplets would be induced to roll along the natural design. Similarly, the butterfly wings exhibit anisotropic super-hydrophobic states. The water droplets on butterfly wings easily roll away along the radial outward (RO) direction, and their movements are strongly inhibited in the opposite direction.

In recent years, a set of rules has been provided by technology to help arrange super-hydrophilic materials. Two main examples of this are the following: the generation of sufficient roughness on the material surfaces and the tailoring of the material surfaces compositions to be hydrophilic with a water $\theta_w < 65°$, in the opposite case there is a complete wetting (see Su et al. (2016)).

This is a summary of the qualitative description of the hydrophobic effect (HE) and the water dewetting on a hydrophobic surface. At the sametime, HE is also the macroscopic scale on which the manifestation of a local molecular segregation process occurs in water-hydrophobic systems and also depends on the way in which the hydrophobic molecules, individually hydrated, self-assemble in to structures (Ball 2008, Berne and Weeks 2009, Ashbaugh and Pratt 1970).

Nowadays, despite the crucial role played by the hydrophobic hydration in these self-assembly processes overmultiple length scales, from the microscopic noble gas solubility in water to the mesoscopic aggregates (surfactant, polymer, polyelectrolytes, and proteins), its effect is understood only qualitatively (Ball 2008, Garde and Patel 2011, Granick and Bae2008, Hummer 2010, Varilly et al. 2017). Many theoretical models and studies focus on this ubiquitous phenomenon about the molecularly detailed interaction between solute and water, but the extrapolation of molecular scale hypothesis to the large mesoscale is sometime critical and not warranted.

A close examination on the interaction of hydrophobic substances with water reveals a subtle and intriguing situation depending on their sizes, chemistry, and thermodynamics. Simple examples of this are: (i) soluble proteins are globular atone temperature but unfolded upon both heating and cooling; (ii) the water solubility of inert gases is completely different from their dissolution in cyclohexane (typical organic solvent), as the two mediums are opposites in their temperature (Berne and Weeks 2009).

The length-scale dependence of hydrophobicity is, indeed, one of its most fascinating and studied processes in the large area of material sciences and technology, because it shows some aspects of universality. Basically, the difference in the hydration physics of small and large solutes seems to arise from the different manner in which they affect the structure of water (Berne and Weeks 2009, Chandler 2005, Rajamani et al. 2005, Stillinger 1973).

As it is well known, the physics of water is dominated by the hydrogen bond (HB) and its tetrahedral network so that thermal fluctuations of that network can accommodate a small hydrophobic solute (Ball 2008, Hummer et al. 1996) without sacrificing hydrogen bonds. In contrast, a large solute cannot be accommodated in this network and many HBs will be broken, much like at the vapor–liquid interface; hence, interfacial physics (i.e., surface tension) governs the solvation and association of these macroscopic solutes. There is thus the self-assembling effect that creates structures of different geometries (spherical, cylindrical, layers, etc.) in which water will be confined. Such a process is sensitive of the hydrophobic solute and water composition and thermo dynamic variables. For example, depending on the composition micelles of water in the solute or vice versa solute aggregate in water is possible. Water, in one case is the dispersed medium, and in the other case, a continuum system in which solute aggregates are dispersed (Israelachvili 1989, Safran 1994, Tanford 1980).

The accommodation of small solutes however, creates a significant entropic penalty, which comes from the restricted configurations of water molecules in the solute's hydration shell. When small solutes associate, some of the seen tropically restricted water molecules are released, thereby the water entropy increases. In addition, the driving force for molecular assembly increases with temperature (at ambient pressure) reaching a maximum at higher temperatures.

The self-assembling effect of the hydrophobic solute has given researchers an instrument to interpret the largescale HE behavior: when water is confined in a hydrophobic bilayer (Janus interface), the system surface tension becomes the determinant (Huang 2003, Widom 1982). And, if the gap between two hydrophobic surfaces become scritically small, water is ejected, where as water films confined between symmetric hydrophilic surfaces are stable (Zhang 2002). At the same way, water behaves differently if confined in hydrophilic or hydrophobic nanostructures. Also the weakening of hydrophobic interactions upon cooling below room temperature must be considered.

Theoretical considerations and computer simulations propose that for the HE, a key concept may be the size of the hydrophobic object (Ben-Amotz and Widom 2006, Hummer 2010, Janecek and Netz 2007). Water molecules can wrap efficiently around hydrophobic elements with a radius of curvature of 1 nm or less. When water meets hydrophobic surfaces that are flatter than this, it forms a molecularly thin cushion of depleted density around the surface.

Unfortunately, physical insight into the origins of hydrophobicity is not easy to come by. In contrast to many simulations, there are few experiments; thermodynamics maybe one good approach, but interpreting the physical significance is extraordinarily subtle (Garde and Patel 2011). The fly in the ointment is the experiment itself.

One reason for the hurdles in experimentally probing nanoscale hydrophobic cold solutes is that they are essentially water in soluble: the alkane's solubility decreases exponentially with carbon number, falling below an equilibrium mole fraction of about 10–10 for alkanes longer than do decane (Ferguson et al. 2012).Other reasons can be related with the fact that only few techniques have a proper resolution to focus quantitatively on the molecularly detailed interactions between oil and water. Therefore, any analysis on the evolution of hydration phenomena from molecular-scale to larger-length-scale can be only "speculative". This is in some way the same problem encountered in MD simulations studies. Theories, based upon a hypothesis of interpolative evolution from microscopic to macroscopic (Ball 2008, Ashbaugh and Pratt 2006), like the scaled particle theory (Stillinger 1973), can represent a correct approach but sometimes give contradictory findings.

Experiments show the abovementioned universality in the length-scale dependence of hydrophobicity by measuring a surface water depletion layer (the water cushion of depleted density)that, despite the vander Waals attraction, is less than the dimension of even one water molecule. There a son for this small thickness, as proposed by experiments (Zhang 2002) and models (Chandler 2005) is in the large fluctuations of the frustrated water at the interfaces: water meets hydrophobic surfaces softly, because vander Waals attraction out weighs its natural reluctance to do so. However, such a picture deserves further investigation because it could be the connection point between the large and short scale hydrophobicity.

Molecular Interactions and their Effects

As it is well known, at the molecular level, the basic interactions are essentially the electrostatic (V^e), the vander Waals (V^{vW}), and the hydrophobic (V^{HE}). If it involves water, we must consider the Hydrogen bond HB (V^{HB}). A consequence of the molecular self-assembling can be the creation of originated structures in theme so scale: (i) nanoparticles (NP, $2 < D < 20$ nm) and (ii) microsizes colloids (μP, $D > 200$ nm). Having complex macromolecules, other interactions can born (V^N) (e.g., the deplection interaction in grafted colloids V^D, as far as the $\pi-\pi$ typical of peptide self-assembly and thus of biological macromolecules ($V^{\pi\pi}$). The first example in describing dispersed micro-particles is the classical Derjaguin-Landau-Verwey-Overbeek (DLVO) theory with a potential like: $V(r) = V^e(r) + V^{vW}(r) + V^{HE}(r) + V^N(r)$ and due to the coupled structural dynamics of neighboring colloids and surrounding media (Derjaguin and Landau 1941, Verwey et al. 1948). Compared to hydrated ions ($D < 1$ nm) and microsized colloids, NPs are characterized by a structural uniqueness and discreteness. Nowadays, it is well established that the classical colloidal theories developed for microparticles are unable to describe NPs. One reason is in the impossibility of continuum approximations being the sizes of ions, solvent molecules, and NPs with in one order of magnitude of each other. Other reasons are the non-uniformity of the stabilizing layer (usually a surfactant); in the fluctuations of ionic atmospheres and ion-specific effects, enhancing NP an isometry. Thus, multiscale collective effects become essential for accurately accounting the interactions between separate nanoparticles. Hence, the quantitative description of NP forces encounters severe obstacles.

All of this creates the experimentally observed, non additivity of NP interactions, making impossible the decomposition of the potential of mean force (PMF) between two NPs in to separate contributions (Silvera Batista et al. 2015).

A situation due to different factors like: the complexity of interfacial forces at nanometer-scale separations and the discreteness of matter for distances smaller than several tens of nanometers far from the observed interdependence of $V^{vW}(r)$, $V^e(r)$, and $V^{HE}(r)$. Also, it can be stated that the unusual stability of dispersions of small NPs, the extraordinarily sophisticated geometries of numerous NP assemblies characterized by different symmetries, biomimetic behavior of NPs in their complexes with enzymes, and complex dynamics of protein coronas are the experimental manifestations of non-additivity. In principle, this is the real situation of charged disperse systems, but there are many non-ionic systems like the mixtures of water and non-ionic surfactant that as a function of the temperature and composition give to rice sophisticated NPs structures which are spherical, cylindrical, layered, and bilayered. Also, the spherical micelles (with $3 < D < 10$ *nm*) can have a dispersed liquid-like phase (with a critical point and a percolation line) as well as crystalline cubic and hexagonal phases. Additionally, in such a case, the system is characterized by a non-additivity of NP interactions $[V(r) = V^{vw}(r) + V^{HE}(r) + V^{N}(r)]$. Depending on the temperature, there are specific phase regions governed by the V^{vw}, or vice versa by the $V^{HE}(r)$. There are also phase regions in which the droplets are characterized by an attractive interaction and other ones by a repulsive one. The micellar systems can be used to explains correlations effects due to these two main interactions. The HE and its PMF, bringing two nonpolar groups toward eachother, can be also represented as $V(r) = U(r) + \Delta V(r)$, where $U(r)$ and $\Delta V(r)$ are the direct and water mediated contributions to $V(r)$, respectively. Direct interactions are those between the isolated nonpolar groups (in the absence of water), where as $\Delta V(r)$ arises from the additional influence of water in promoting (or suppressing) self-aggregation (Ben-Amotz 2015).

Ion-ion correlations are especially important for charged NPs where $V^{c}(r)$ interactions become much larger than the thermal energy, and the counter ion cloud can be have as a strongly correlated liquid. In nanocolloids, the effect of these correlations may lead to an apparent attractive nanoscale-range interaction that is comparable in strength to $V^{vw}(r)$ interactions and that increases in strength with the NP charge. At the sametime, in non-ionic NPs, the strength of the hydrophilic over the hydrophobic force is dependent on the temperature.

HE has been originally related with the observed alterations of the structure of water around interfaces and solutes. Molecular perturbations, essential for understanding HE, were commonly observed around NPs for different molecular solvents like water, propanol, and ethanol in the hydration layer of large macromolecules. Their $V^{HE}(r)$ characteristic extensions, as proposed by MD simulations (C60 and Au NPs (Kimber et al. 2015)) arrive upto 2 nm; where as fast spectroscopy (Terahertz) able to directly probe solvation dynamics has revealed a similar average width for the dynamic hydration layer around proteins (Ebbinghaus 2007). HE has been related to an entropy reduction of water in solutions with non polar molecules due to the formation of a local water network (like arigidice-like cage) around the solutes. Recent MD simulation (Chandler 2005) and Raman scattering experiments (Davis et al. 2012) have instead proposed that this mechanism holds only for small solutes. If the solute reaches a size of ˜1 nm, this structured hydration shell evolves to a "dry" disordered one (with a lower water density). At the interface of such as hell, there are essentially-OH groups with broken hydrogen bonds (HB). Hence, enthalpy rather than entropy dominates the free energy of solvation, meaning that for large solutes, both HB and dispersion interactions between solutes and solvents make contributions to the balance of solvation energies (Davis et al. 2013).

The interdependence between $V^{c}(r)$ and $V^{HE}(r)$ can be influenced by the non-uniformity charge on a scale comparable to that like the hydrophobic; also, electrostatic interactions can alter the organization of water at the interfaces and HEs with effects on the charge state of the interface due to ions in its vicinity.

The coupling of hydrophobic forces with other interactions was suggested by simulations of proteins with hydrophobic patches (Davis et al. 2013). An atomic force microscopic experiment on co-immobilization of residues (amine or guanidine) has shown a dramatic change in the strength of hydrophobic interaction. Different water shells surrounds these residues; amine protonation doubles the HE strength, whereas guanidinium groups eliminate measurable HE at all investigated pH (Ma et al. 2015).

Another matter of interest is the effects of the NPs multibody on large ensembles of interacting particles (Gray et al. 1999) where entropy can make essential contributions. An example is the simple case of NP hard particles, of different core crystallinity (rods, dumbbells cubes, hexagons, tetrahedrons, etc.), with a meanforce made only of short-range repulsion (no attraction), where the collective maximization of degrees of freedom results in a ordering process into a variety of crystalline and quasi-crystalline super lattices (Haji-Akbari et al. 2009, Tao et al. 2005).

The NPs' collective behavior is reflected in their association process through their orientational preferences and angular an isotropy of the meanforce. An isotropy of interactions between individual NPs manifests particularly well in large NP ensembles; driven by internal or external fields, the NP self-organization can result in many extended assemblies with different structure: for example, chains, super lattices, and dendrites. The collective behavior of NP samplifies the effect of the energetic contributions to the meanforces. Static and dynamic dipolarpolarization in the ground state of NPs, generally neglected, drives the association into several common assembly patterns. Also, the great effects of weak interactions on the geometry of NP assemblies is also evidenced in the formation of helical super structures from chiral and nonchiral NPs (Singh et al. 2014, Yeom et al. 2015).

In summary, the collective NP behavioris the direct consequence of the multiplicity of interdependent molecular processes: grafting layer transitions, stabilizer entropy, faceting, solvent structuring, hydrogen bonding, hydrophobic interactions, and electrostatic repulsion of the charged groups at the solvent-ligand interface. Multiplicity combined with non-additivity are at the core of a possible interpretation of NPs. Thus, these two mechanisms can indicate a new conceptualization and theoretical development of this fieldable to give a new way of thinking about NPs as strongly correlated reconfigurable systems with diverse physical elements and multiscale coupling processes. At the sametime, both these processes must be the bases of new experiments and MD simulations able to give a detailed proof on their dynamic and structural properties in the frame of a unitary model, if possible based on universal concepts like those characterizing polymer solution and systems (de Gennes 1979, Flory 1953).

Thus, researchers must consider both the local configurations evidenced in the discreteness of matter at a scale relevant to NP interactions (molecular and atomic-scale phenomena cannot be average dout) and collective processes. Therefore, from this perspective, nowadays, MD simulation seems more fruitful than experimental techniques. The NPs system in the case of solid inorganic cores (with fewer degrees of freedom) is favorable for atomistic simulations and improvements in the force fields can adequately account for intermolecular interactions, entropic contributions, dispersion interactions between atoms, high polarizability of inorganic materials, and quantum confinement effects (Silvera Batista et al. 2015). Evolving experimental tools that can accurately examine interactions at the nanoscale should help to validate the simulations and stimulate improvement of relevant force fields. In fact, new opportunities for better understanding the electronic origin of classical interactions are likely to develop due to the rapidly improving capabilities in synthesis, simulations, and imaging converge at the scale of NPs.

This is the general situation regarding the NPs' properties. However, the focus of this chapter is to study in a comprehensive way both the properties of the HE and HB and their mutual influences.

Details of the NP Interactions

We start this section by considering that NP self-assembly is a spontaneous thermodynamic-and kinetic-driven process, based on the cooperative add non-additive effect of various intermolecular non-covalent interactions, including HB, the $\pi-\pi$ stacking, electrostatic, hydrophobic, and vander Waals interactions. The proposed collective effect of these interactions determines the NP thermodynamic stability and the state of minimum energy of the ultimately formed nanostructures. Also, due to their relatively weak nature, these interactions can be modulated by means of kinetic parameters (pH, T, number of counter-ions, concentrations, and solvents) so that the assembled structures will result in metastable states. Hence, the competition between the self-assembling kinetic and thermodynamic processes provides the possibility of transformation from one state to another. Such a NP kinetic control is important for understanding the self-assembly mechanisms and gives the opportunity to "create dynamic materials". Thus, non-covalent interactions determine the thermodynamically stable structure, where as for the effect of kinetic parameters can drive the NP system in a metastable phase. Such a picture gives another proof of the relevance of the effects of both the multiplicity and non-additivity in the NP thermodynamics and in their technology. In this respect, it is of pedagogic relevance to have more detail about then on-covalent interactions.

(a) **Hydrogen-bonding:** The HB is a noncovalent attractive interaction between two water molecules, that is, an electropositive hydrogen atom on one molecule and an electronegative oxygen atom on another molecule (the O:H non-covalent vander Waals bond ($\simeq 0.1eV$ binding energy BE)).HB, as it is wellknown, is the basis of the complex water thermodynamics made of many anomalies like the density maximum and the unusual diverging behaviors of the isothermal compressibility (KT) and isobaric specific heat (CP) in the supercooled regime. In water, the HBs are able to arrange a tetrahedral network of water molecules which increases in size on decreasing T. In addition, such a networking originates the system polymorphism for which liquid water is a 'mixture' of two liquid forms with different densities, low-density liquid and high-density liquid, namely, LDL and HDL, respectively. The LDL corresponds to the HB tetrahedral network. The same situation is observable in the water amorphous characterized by LDA and HDA. The HB network originates at $T * \simeq 315K$. For clarity, it must be stressed that the water physics remains again, in science, an open question with more shadow than light, metaphorically speaking. At the same time, HB is the driving force of primary interest for structural organization in biological systems.

Peptides, provide an abundance of HB-formation sites, including a midgroups in the peptide back bone and amino and carboxyl groups in the side chains. In particular, T^* marks the thermodynamic properties of bulk water in the $P-T$

plane, that is, it is the locus of the minimain the water isothermal compressibility $KT (P,T) = -V^{-1}(\partial V/\partial P)T$ for all the pressures and also the crossing point of the thermal expansion $\alpha_p (P,T) = -V^{-1}(\partial S/\partial P) T$. Note that $(\partial \alpha_p/\partial P)T = -(\partial K_T/\partial T) P$ and K_T represents the volume fluctuations $K_T = \langle \delta V^2 \rangle P$, $T/k_B TV$; where as α_p represents the cross correlations $\langle \delta S \delta V \rangle$ and it is $\alpha P = \langle \delta S \delta V \rangle / kBTV$. Thus, by considering this and the water self-diffusion, we see that the T^* temperature marks the onset of tetrahedral water patches (Mallamace et al. 2012, Simpson and Carr 1958). On the otherhand, T_L is the locus of the so-called fragile-to-strong dynamical crossover observed in confined water (Chen et al. 2006) and many other super cooled glass-forming materials (Mallamace et al. 2010a). It is also the locus of the Stokes-Einstein violation (Cerveny et al. 2016, Chen et al. 2006, Mallamace et al. 2014a, Xu et al. 2009).

HB also plays a significant role in the formation and stabilization of the peptide's secondary structure and protein folding. For example, in the process of protein fibril formation, which is thought to be a major element in Alzheimer's disease and other degenerative disorders, water-assisted HB is believed to be the critical factor in self-assembly. Actually, among different non-covalent interactions, hydrogen bonding is probably the most important aspect of peptide self-assembly.

The selectivity and high directionality of hydrogen bonds can induce the conversion of peptides into diverse dimensional ($1 < d < 3$) nanostructures. The HB strengths are mainly $5-10 \, k_B T$ ($\sim 10-40 \, k \, Jmol^{-1}$) per bond at 298K. In contrast to the HB, there is a repulsive intermolecular interaction in water: the Coulomb repulsion between electron lone-pairs on adjacent oxygen atoms. In addition, two H-O covalent bonds originated by the sharing of the electron lone pairs, $\simeq 4.0eV$. Where as the HB dominates water in the stable and supercooled regime, the repulsive lone pairs mainly influence the water physics from above the boiling temperature (T_b) in the subcritical and critical region (the water critical point CP is located at: $T_C = 647.1K$, $P_C = 22.064 \, MPa$).

(b) **π−π Interactions:** π−π stacking can drive the NP self-assembly, especially for π-conjugated peptides, such as the aromatic; these interactions also can induce directional growth, and are very robust in water because of the limited solubility of molecules containing aromatic groups. The π−π stacking is also a more distinct driving force in pure organic solvents, such as toluene.

(c) **Hydrophobic Interactions:** HEs with few directional constraints are the main driving forces in surfactant systems. A Water molecule cannot interact favorably with an apolar one and when amphipathic molecules are put into water, their hydrophobic parts try to aggregate, minimizing their surface are a in contact with the solvent, leaving only the hydrophilic parts exposed to water. The, hydrophobic interactions are thus stabilized by increasing entropy rather than enthalpy (Ben-Amotz 2015). All of this leads to the formation of micellar structures. In a salt-triggered self-assembly process, the role of HE driving forces, due to charge-screening effects, are significantly enhanced. In the case of bio-systems, is must be said that HE forces are important for the rational design of peptide amphiphiles. In particular, both the HE and π−π interaction play important roles in the organization modes of peptide aromatic residues:

in the first case these are disordered, while in the second one, they are well organized and ordered.

(d) **Electrostatic Interactions:** Interaction between charges is a well-known non covalent inter-action of large importance for the self-assembly, processes. Electrostatic bonds, due to Coulombic attractions between opposite charges, create ion-pairs. The strength of an ionic bond is stronger than a HB (~ 500 $kJmol^{-1}$) or ~ 100 $k_B T$).

(e) **vander Waals Interactions:** The vander Waals forces having a strength weaker than that of HB (~ 5 $kJmol^{-1}$) or ~ 1 $k_B T$) and an interaction effective range that can reach 10 nm, are ubiquitous in the assembly of NP systems. They are caused, as a consequence of quantum effects, by correlations in the fluctuating polarizations of nearby particles and represent a distance-dependent interaction.

It must be stressed that thermodynamics influences these interactions. For example, temperature is a powerful factor in varying HB and HE, especially in water solutions. A T increasing weakens and breaks the intermolecular HBs but increases the HE. Also, other solvents properties or factors have a huge influence on the solvation effect of solute molecules: for example, the solution pH affects the competing solvation between the HB donor and acceptor and saltions may capture the solvent molecules by solvation, resulting in a decrease in the HB of solute and solvents. In addition, the salt effects (pH, type of ion, and ionic strength of the solution) have a great impact on electrostatic interactions.

The simplest model with which NP repulsion and attraction is analytically tractable refers to a fluid madeup of hard spheres of HS with an arrow and deep attractive tail. Such an idealized model of adhesive or sticky hard spheres SHS was proposed by Baxter [1968], which allows for an analytical solution within the Percus-Yevick PY approximation and can be used in simulations (Miller and Frenkel 2003). The simplest two-parameter representation of it is: V_B ® = ∞ for $r < r_0$, $-\varepsilon$ for $r_0 < r < r_0 + \Delta r$, and 0 for $r > r_0 + -\varepsilon$, hence it is a simple well with depth $-\varepsilon$ and width$-\varepsilon$. In the literature, some variants have been used to model short-ranged attractions present in real solutions of colloids or reverse micelles by considering vander Waals attractions, hard-sphere-depletion forces, and solvation effects (inparticular hydrophobic bonding and attractions).

A general classification of NP, made in terms of these interactions and of the basic molecules may be the following: amphiphilic-based, polymers and polymeric solutions and gels (hydrogels in particular), and coated non-organic (e.g., Si, Au, CdTe, CoPt$_3$, CdSe, CdS) and organic molecules crystallites giving rice to solid nano-crystallites and biological macromolecules, of which the basic tiles are the peptide aggregates.

The dark red attractive well represents the Baxter model. PMFs calculated according to DLVO (red) and MD simulations reflect the presence of positive ions of different diameters. Blue open circles and squares deal with little and large ions, respectively. PMF for a pair of two amorphous SiO_2 NPs in the presence of $Na+ = 0.00, 0.01, 0.10$, and 1.00 M.

State of the Art

It has been emphasized that: hydrophobicity depends on the eye of the beholder, in the sense that people have different ideas in mind regarding this topic (Granick and Bae 2008); and the hydrophilic-hydrophobic interaction must be treated with special care.

One approach is the macroscopic one which details the wettability of a water droplet on a planar hydrophobic surface in terms of the contact angle θ_γ (independent of droplet size but dependent on the surface roughness) (deGennes 1985, Shirtcliffe et al 2010). One common definition is that in such a situation $\theta_\gamma > 90°$; but given that nothing dramatically changes when θ_γ is minor than 90°, the result is only qualitative and arbitrary, that is, only a convenient definition. Hence, it is elusive (and useless) on the surface contours of proteins when looking, in these terms, at the difference between their polar and hydrophobic patches.

To better understand hydrophobic-hydrophilic interactions, and to over come controversies, the following questions have been proposed: (i) What happens if a surface has the same wettability everywhere, or is "patchy" from spot to spot? (ii) What happens if the system is subject to the change of thermodynamic variables (pressure and temperature in particular)? (iii) Does water truly act unlike other fluids?

Additionally, there are also some other important open questions which remain regarding the dependence of HE on molecular size and temperature, as well as on the balance of direct and water-mediated interactions (Ben-Amotz 2015). The first issue can clarify in a unique way diverse fields like protein folding and surface science. Whereas, in

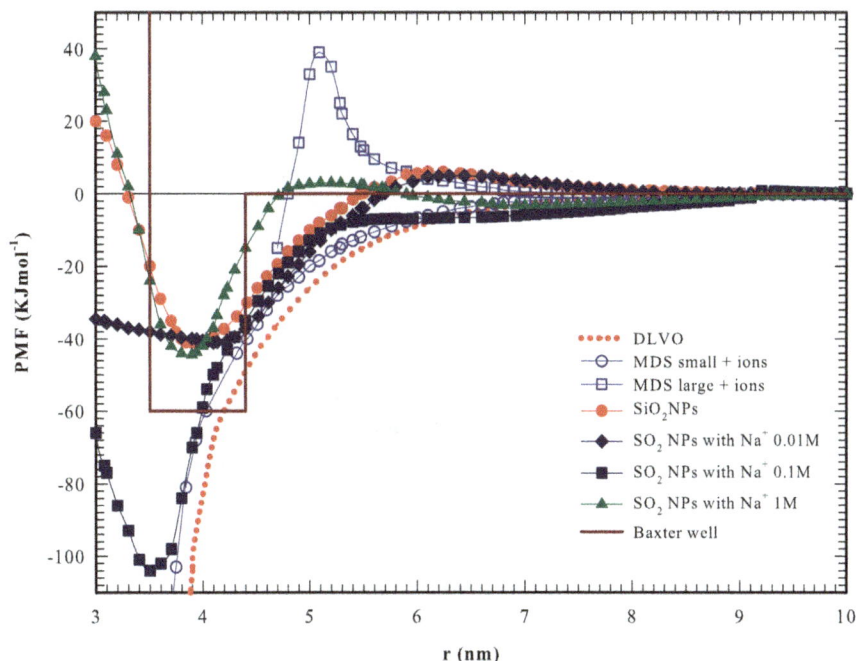

Fig. 3: Examples of PMF for NPs (∼ 4 nm) related to then on additivity.

the second case, it can be quantitatively evaluated how thermodynamics cause proteins to denature and how water systems are sensitive to environmental changes. A great amount of water experimental data, explored on different length and time (frequency) scales, are hard to interpret. At the sametime, many computational models use empirical approaches (point charges and do not explicitly recognize quantum mechanics) so that it may be worth inquiring more critically in to their basic assumptions. However, empiricism also shows that what matters is not just the instantaneous separation between hydrophobic surfaces, but also the time (or frequency) of their contact (Berne and Weeks 2009); the timeline of change also matters.

Finally, in the third case, we have to take in to account the emerging liquid water complexity, made of many anomalies, and also a very recent result for which water in the temperature range above 315 K behaves in the same way as all liquid systems (Mallamace et al. 2012). That is, we have to understand that for the HE, the main challenge is what makes water so special. Nowadays, in this respect, from the large amount of simulations we have only few answers, that however are in agreement with some experiments, and can propose a general theory of hydrophobic surfaces when temperature and pressure change in time and space (Berne and Weeks 2009, Hormoz and Widom 2012, Maibaum and Chandler 2003). Often the theoretical hypotheses are specific to the system understudy, but common responses with experiments can suggest more universality. In these conditions, the HE on short and large scales can also be evaluated and understood on the basis of more general principles of the competition between enthalpy and entropy (Ben-Amotz 2015).

Recent MD Findings & Suggestions

Multiplicity and non-additivity propose, thus, that the nanoscale is unique in its interactions and can constitute a correct pathway to clarifies via a universal approach the emergent physical and biological properties of materials. In such a context, we consider the recent MD simulation and experimental findings together with their suggestions. To simplify, were cognize that NPs deal with a very significant paradigm that shift the basic properties from 'simple identify' of simple replicated objects (molecule) to 'statistical identity" (where every object is some what different). Such a shift can been vis aged in a 'new science' based on concepts typical of statistical physics, which are in turn based essentially on collective properties (and) behaviors and proper energy configurations.

As mentioned earlier, MD simulations are actually more fruit ful than experiments and some NP systems with solid inorganic cores maybe a proper tool for atomistic simulations. In anycase, MD studies can account for the basic and technological properties of these systems. Areas on for this is that further work on the NP interactions deals with NPs as strongly correlated and reconfigurable systems with diverse multiscale coupling processes. A certain advantage in the NP simulation studies comes from their reduced dimensions and in the corresponding reduction in the numbers of atomic degrees of freedom. Taking advantage of this, a correct MD strategy in the understanding of NP interactions can be a direct determination of how to use net inter particle potential force rather than the individual contributions one. On the basis of the potential non-additivity, and related correlations, this appears now as the correct approach to evaluate NP interaction, especially in the complex systems typical of biology where the only key seems to be collectivity. It is, however, clear from the theoretical simulations the special (predominant) role of water molecules (Hummer 2010).

It must be stressed here again that HE operates over multiplelength scales, from the macroscopicp has eseparation to the mesoscopic organization of proteins and surfactant structures (e.g., self-assembling) to origins of inert gas solubility in water or the folding/unfolding processes of polymers and proteins in water at the molecular scale. The physics governing the HE at these length scales appears fundamentally different (Chandler 2005). A large number of theoretical studies focus on the local detailed interactions between solutes and water, but their extrapolation to larger-length-scale hydration phenomena is sometimes not warranted. The original approach to an interpolative view shifting from micro to macro is based on the scaled particle theory, originally proposed by Frank Stillinger (Stillinger 1973), and its further modifications or generalizations (Ashbaugh and Pratt 2006). From the use of such a model, it has been predicted that the crossover from the molecular to the macroscopic regime occurs at a length scale of the order of 1 nm (Chandler 2005).

Such a HE length-scale dependence of hydrophobicity it is not only a very fascinating issue but also a goal in the attempt to explain the HE. Additionally, the scaled approaches support the idea that the difference in the physics of hydration of small and large solutes arises from the different manner in which they affect the structure of water (Chandler 2005, Berne and Weeks 2009, Ashbaugh and Pratt 2006, Huang and Chandler 2000, Lum et al. 1999, Rajamani et al. 2005, Stillinger 1973). Water molecules participating in the local tetrahedral HB network have different behavior when interacting with small or large solutes.

Where as in the first case, because of thermal fluctuations, the HB network can accommodate within them a small hydrophobic solute [Hummer et al . 1996], this is impossible for the large one. The first case thus results in a significant entropic penalty, reflecting restricted configurations of water molecules in the solute's hydration shell; and when these small solutes associate, the network change configuration and some of these entropically restricted watermolecules are released, there by increasing the water entropy. As a result, the driving force for molecular assembly increases with T at ambient conditions be fore reaching a maximum at higher temperatures. The large solute impossibility to be accommodated in the network will result in broken hydrogen bonds, much like at the vapor–liquid interface (Chandler 2005, Stillinger 1973). Hence, interfacial physics (surface tension) governs the solvation and association of macroscopic solutes.

All of this have given a clear idea of how hydration affects molecular-recognition processes, such as in more complex situations of biological macromolecules between proteins and ligands. Now, theoretical simulations provide thermodynamic insight in to cavity–ligand binding, revealing how it is predominantly driven by the behavior of the few surrounding water molecules (Baron et al. 2010). Hydration is also known to affect molecular-recognition processes, such as those between proteins and ligands. Just for accuracy, we have to mention other simulations studies and models that must be considered when looking at water properties and HE. In the firstcase, it might be an important idea that the liquid state contains many defects in the form of bifurcated bonds: a single hydrogen coordinates to two oxygen atoms on different molecules, a situation of central interest in the molecular mobility of the liquid state by lowering the Gibbs energy barrier to diffusion (Sciortino et al. 1991). Another idea deals directly with HE by using lattice Monte Carlo calculations taking into account the aggregation behavior of amphiphilic polymers, showing that the distribution of hydrophobic residues with in the polymer sequence determine show dry/wet interfaces can be created and that such interfaces drive the aggregation process (Varilly et al. 2017).

Another interesting study deals with water in nonpolar confinements (from nanotubes to proteins and novel NP devices) (Rasaiah et al. 2008). Here, the researchers have clarified, by using thermodynamic, structural, and kinetic concepts the unusual properties of water, which are exhibited when its confined to nonpolarpores and cavities of nanoscopic dimensions. The study shows why confined water is strongly cooperative, with the possible coexistence of filled and empty states and sensitivity to small perturbations of the pore polarity and solvent conditions. In these conditions, the liquid water molecules form tight HB wires or clusters. The combination of strong HB interactions between water molecules and the weak attractions to the confining wall allow an exceptionally rapid waterflow (several orders of magnitude larger than macroscopic hydrodynamics expectations).

The proton mobility along the one dimensional water wires also exceeds that in the bulk. It is stressed that proteins appear to exploit these unusual properties of confined water in their biological function (for example, to ensure rapid water flow in aquaporins or to gate proton flow in proton pumps and enzymes).

In the context of the folding and assembly of proteins, the MD studies also give interesting suggestions in clarifying the collapse of a hydrophobic chain in water. It has been demonstrated that such a process is much like that of a first-order phase transition. The evaporation of water in the vicinity of the polymer provides the driving force for collapse, and the rate limiting step is the nucleation of a sufficiently large vaporbubble. The study is made possible through the application of transition path sampling and a coarse-grained treatment of liquid water.

A further improvement occurs by combining solvent coarse-graining and the string method in collective variables, which helps to compute a minimum free-energy pathway (MFEP) by providing an atomistic confirmation for the collapsing chain. In particular, it is shown that length-scale-dependent hydrophobic dewetting is the rate-limiting step in the hydrophobic collapse of the considered chain (Miller et al. 2007, Wolde and Chandler 2002). The Chandler team (2002) has proposed a series of care simulation studies on the role of water fluctuations in the hydrophobic assembly and near the hydrophobic and hydrophilic surfaces. The main findings are that: water near extended hydrophobic surfaces behaves like as at a liquid-vapor interface and its density fluctuations are enhanced compared to those in bulk water. Hence, water near these surfaces is sensitive to the surface changes: conformation, topology, and chemistry, any of which can tip the balance toward or away from the wet state and thus significantly alter biomolecular interactions and function. In particular, the SPC-E model of liquid water and specialized sampling techniques (including umbrella sampling) were used (Patel et al. 2012, Patel et al. 2010, Willard and Chandler 2008).

The *ab initio* molecular dynamics simulations has been used to study hydroxide (OH^-) and hydronium (H^+) ions near a hydrophobic interface, leading to the observation that both ions behave like amphiphilic surfactants that stick to a hydrophobic side of the hydrocarbon surface. Such behavior originates from the asymmetry of the molecular charged is tribution which makes one end of the ions strongly hydrophobic while the other end is even more hydrophilic than water. The effect is more pronounced for the hydroxide ion than for the hydronium ion.

In addition, it has been explained why hydrophobic surfaces in contact with water acquire a net negative charge (Kudin and Car 2008). Together, this latter research a consistent number deep quality MD simulation contributions come from Princeton and in particular from the Debenedetti team. Their research activity covers more than the last decade, ranging from water chemical physics in hydrophobic confinement to a comprehensive understanding of the substrate material features (e.g., geometry, chemistry, and mechanical properties) with special care given to proteins and peptides. In particular, these results highlight the fluctuations role in speeding up the kinetics of numerous phenomena ranging from Cassie–Wenzel transitions on super-hydrophobic surfaces, to hydrophobically driven biomolecular folding and assembly (Altabet et al. 2017a, Altabet et al. 2017, Ferguson et al. 2012, Giovambattista et al. 2007, Giovambattista et al. 2009, Matysiak et al. 2011, Matysiak et al. 2012, Remsinga et al. 2015, Sharma and Debenedetti 2012, Stirnemann et al. 2011).

We close this section by mentioning a very recent and general study that bypasses the hydrophobic or hydrophilic classification of traditional solutes based on chemical experience and a heuristics approach (Pérez-Conesa et al. 2019); however, it is unable to reveal how the local environment modulates the system dynamic and structural properties. This work proposes a local fingerprint—from methane to peptides—inspired by the two-body contribution to entropy, for hydrophobicity and hydrophilicity.

The model represents an "inexpensive, quantitative, and physically meaningful way" of studying HE that only requires as input the water-solute radial distribution functions. In such a manner, the quantitative measure of hydrophilicity is coherent with chemical experience, and moreover, it also shows how the character of an atom can change a sits environment changes. Finally, a very important (universal) issue is that the fingerprint can be used as a collective variable in a funnel meta dynamics simulation of a host-guest system and serves as a desolvation collective variable that enhances transitions between the bound and unbound states (Pérez-Conesa et al. 2019). Such a study is also of relevant interest because it provides evidence of a possible and useful HE description in terms the energy landscape framework (Stillinger 1988)—an approach that can be expected whenever different existing mechanisms compete, and the dominance of one over the other depends on the local environment or driving force.

Recent Experimental Findings & Suggestions

HE and molecular association, from short to large tile molecules, has been studied by using many different experimental techniques ranging from spectroscopy and scattering to calorimetry and imaging (microscopy). Local structure and dynamics

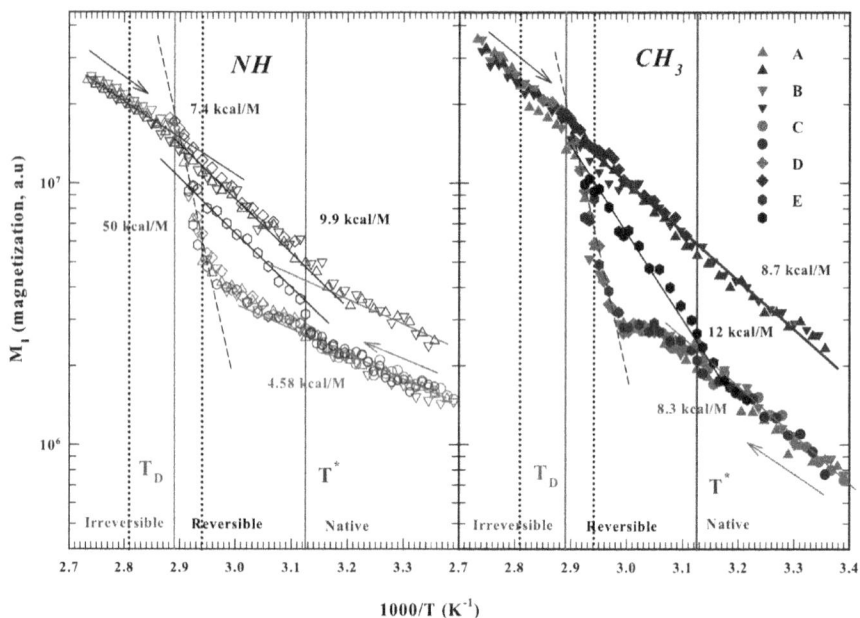

Fig. 4: (a) The Arrhenius representation of the measured magnetization values of the hydrophilic amide groups (NH). Data for all the five different studied thermal cycles (A, B, C, D, and E) are illustrated. The characteristic temperatures T^* and T_D are also reported. Lines represent the Arrhenius behaviors; the corresponding activation energies E_A indicated in kcal/mol.Cycles A, B, and D deal with a complete denaturation; Coperates in the native protein state(N), where as E refers to the native and intermediate states (NRU).(b) The thermal evolution of the magnetization, $M_I(T)$, of the protein methyl (CH_3) groups for all the studied thermal cycles.

have been studied looking to longitudinal, rotational, and vibrational modes by exploring all the available frequency (and temporal) ranges. Despite these many efforts, and the theoretical and MD simulation suggestions, we are unfortunately far from a complete and clear interpretation of the basic phenomenon and processes that occur at the origin of the HE. This maybe also due to the fact that we are still far from a satisfactory understanding of the system of water. Although the experimental literature is not as wide as that of MD simulations, there is however a list of experiments that focus on this process.

It is also important to mention the following works: surface vibrational studies of HB and HE (Du et al. 1994), the aminoacids HE on the structure of water explored by Raman spectroscopy and NMR (Ide et al. 1997), the thermal conductance of HE interfaces (Ge et al. 2006), the heat capacity changes in hydrophobic solvation (Gallagher and Sharp 2003), the di-electric relaxation (Sato et al. 2000), neutron scattering (Russo et al. 2005), and atomic/friction force microscope (AFM/FFM) (Bhushan and Jung 2006, Bhushan et al. 2009, Meyer et al. 2005, Russo et al. 2005).

The nuclear magnetic resonance (NMR) also reveals that at the thermal denaturation of a hydrated protein (lysozyme), the hydrophilic (the amide NH) and hydrophobic (methyl CH3 and methine CH) peptide groups evolve and exhibit different temperature behaviors. This clarifies the role of water and hydrogen bonding in the stabilization of protein configurations (Mallamace et al. 2016c) and also reveals the role of hydrophobic effects on this important protein intramolecular process.

Although the experimental literature is not as wide as that of MD simulations, there is however a list of experiments made on the process of our interest. The following works on the HE structure and dynamic are important: the surface vibrational studies of HB and HE (Du et al. 1994), the exploration of amino acids and the HE influences on the structure of water by means of Raman spectroscopy and NMR (Ide et al. 1997), the thermal conductance of HE interfaces (Le et al. 2006, Shenogina et al. 2009), the dielectric relaxations in different solutions (Cerveny et al. 2016, Sato et al. 2000), the neutron scattering in different solutions and proteins (Russo et al. 2005), atomic/friction force microscope (AFM/FFM)(Bhushan and Jung 2006, Bhushan et al. 2009, Meyer et al. 2005, Russo et al. 2005).

Other example of experimental studies with interesting findings are: (i) a Vibrational Sum-Frequency Generation (VSFG) Spectroscopy study showing that water at the interface is structurally more homogeneous than previously thought (Sovago et al. 2008); (ii) the IR-SFG experiment for which water at the interface of a surfactant coated quartz has a more ordered, ice-like structure (Du et al. 1994), and (iii) the combination of scattering neutron (small angle)and light (static) with viscoelasticity, in a wide composition-temperature range of water-nonionic micellar solutions, is able to determine the microstructure and interaction between micelles and the hydration levels (and effects) just by researchers looking the polymer segmental distribution within a micelle and the intermicellar structure factor (Lobry et al. 1999). The water 'dominance' at a hydrophilic surface reported by these experiments has also been purposed by analogous studies to understand the influence of water on the surface of anionic liquid (Rivera and Baldelli 2004); the resulting indication is that surface hydrophobic non-miscible ionic liquids are more sensitive to the addition of water than miscible ones (hydrophobic ionic liquids respond to the addition of water by reorienting their cations to help solvate water molecules).

It has been also demonstrated that thermal transport measurements can probe the effects of wetting and adhesion on the thermal conductance at water-solid interface (hydrophobic or hydrophilic) quantifying the interfacial bonding strength. This represents, in agreement with MD simulations (Shenogina et al. 2009), a proof that no watervapor layer is present at such interfaces (Le et al. 2006). They find that the Kapitza length (the thermal conductivity of water divided by the thermal conduct an ceper unit area of the interface, which is analogous to the water 'slip length') at hydrophobic interfaces (10–12 nm) is a factor of 2–3 larger than the Kapitza length at hydrophilic interfaces (3–6 nm).

With respect to all these studies, an experiment that must be considered with special care, because it represents a significant breakage with respect to the past, regards the use of the single-molecule force spectroscopy in hydrated hydrophobic polymers of different aromatic side chains (Li and Walker 2011). Its importance is in the comparative exploration of hydrophobicity in the crossover between the microscales (in soluble alkanes) and then a no scales (proteins) water solutions, as a function of temperature. At the sametime, for the obtained findings, it can be considered like a bridge between theory/simulation and experiments. To be precise, the corresponding results show clearly that: (a)the hydration free energy ($\Delta G^{hyd}(T)$)

per monomer strongly depends on T and does not follow interfacial thermodynamics; (b) the T dependence profiles are distinct among the three studied (polystyrene(PS), poly(4-tert-butylstyrene) (PtBS), and poly(4-vinylbiphenyl) (PVBP), hydrophobic polymers as a result of a hydrophobic size effect at the sub-nanometer scale; (c) the $\Delta G^{hyd}(T)$ of a monomer on an a no molecule is different from that of a free monomer and corrections for the reduced hydration free energy due to hydrophobic interaction from neighboring units are required.

This latter work gives the opportunity to get over the previous empiricism because, as confirmed by many recent experiments, it highlights more clearly what happens when we look to the hydrophobic-hydrophilic interaction by the changing thermodynamic variable (T, in particular) by also suggesting new experimental approaches on some important, and never measured HE quantities looking to its thermodynamic (Li and Walker 2011).

Another datum point observation about the effects of temperature on hydrophobicity can be seen in the NMR experiment on water confined in hydrophobic nanotubes (Wang et al. 2008). Such water inside nanotubes switches from hydrophobic at room temperature to hydrophilic when the temperature is lowered to 8°C. A considerables low down in molecular reorientation of such adsorbed water was also detected, demonstrating that the structure of interfacial water could depend sensitively on temperature, which could lead to intriguing T dependences involving interfacial water on hydrophobic surfaces. Originally, these results have been considered in line with predictions that a switch of this kind can happen because the free energy of a full nanotube is very close to that of an emptyone (Hummer et al. 2001). Very recently, a NMR study made in our laboratory, on simple methanol-water solutions, fully confirm this temperature crossover. In addition, the relaxation times process was studied also as a function of the composition (X) on looking directly (and contemporaneously) to the T, X evolution of both the dynamic of hydroxyl groups of water and methanol and the methyl of the methanol (Mallamace et al. 2019). In this work, there was a special consideration for pure bulk water that shows at the crossover a marked decrease in its hydrophilic interaction, in analogy with the confine done (Hummer et al. 2001) becoming "hydrophobic" for $T > 285K$.

We close this section by illustrating some details on how these two NMR experiments (Mallamace et al. 2019, Wang et al. 2008) by confirming the single-molecule forces pectroscopy study (Li and Walker 2011), can open a new and fruitful road in experimental HE exploration. Both these experiments use the Bloembergen, Purcell, and Pound (BPP) model (Bloch 1946, Bloembergen et al. 1948), describing the energy transfer from a radio frequency circuit to a system of nuclear spins immersed in a magnetic field H_0. The system exposure to the radiation perturbation with energy exchange supsets the original equilibrium state by equalizing the populations of the various levels. Such a new equilibrium state is thus a balance between the energy absorption (by the spins from the radiation field), and the transfer to the heat reservoir supplied by all other internal degrees of freedom in the substance nuclei. This is the so-called spin-lattice interaction process described by a characteristic spin-lattice relaxation time T_1 (longitudinal).

Such a time also measures how long one must wait after the application of the constant field H_0 for thermal equilibrium to be established. The competition between resonance absorption and spin-lattice interaction mentioned above also provides away of measuring T_1. In particular, the BPP model proposes that the effect of the nuclear motion on how the dipolar broadening can be treated as a random modulation of the dipolar field caused by the Brownian motion of the atomic nuclei, and if the intermolecular effects are negligible or absent if the system is thus dominated by the dipole-dipole interaction.

In the case of water molecules, it can be studied, by NMR, the way in which a given proton is affected by its nearest neighbor proton (the other H_2O molecule). Thus, the measured $^1HT_1$ is determined by interaction fluctuations induced by molecular motions and is the result of the thermal motion of the magnetic nuclei (affecting the spin-spin interaction) characterized by a correlation time τ_c. This is associated with the local Brownian motion and closely related to the characteristic time that occurs in the Debye theory of polar liquids, for which

$$T_1 = \left\{ \frac{\Delta}{5r^6} \left[\frac{\tau_c}{1 + \omega_0^2 \tau_c^2} + \frac{4\tau_c}{1 + 4\omega_0^2 \tau_c^2} \right] \right\}^{-1} \tag{1}$$

where Δ is a constant determined by the protongy ro magnetic ratio γ and the Planck constant $(2\pi\bar{h}^2)$ as $\Delta = 3/2(\gamma^4\bar{h}^2)$ and r is the interproton distance.

The problem of magnetic resonance absorption was recently reconsidered (Mallamace et al. 2019). The frequency-dependent susceptibility of a magnetic system was found using a quantum-statistical form of the linear theory of irreversible process that uses projection operators (Kubo and Tomita 1954). This model assumes that the perturbation energy changes the distribution of the resonance spectrum, which expands the self-correlation function of the magnetic moment. The system Hamiltonian is thus the sum of a 'secular' term that commutes with the unperturbed terms and a 'non-secular' portion of the perturbation. Secular and non-secular contributions lead to different relaxation functions from which also the spin-spin relaxation time T_2 (transverse) can be evaluated as

$$T_2 = \left\{ \frac{\Delta}{10r^6} \left[3\tau_c + \frac{5\tau_c}{1 + \omega_0^2 \tau_c^2} + \frac{2\tau_c}{1 + 4\omega_0^2 \tau_c^2} \right] \right\}^{-1} \tag{2}$$

In the limit of the rapid decrease of the correlations in which the secular width is strongly narrowed and the non-secular width weakly broadened, $1/\omega_0 \gg \tau_c$. When the nuclear arrangement is isotropic, $T_2\tau_c \sim const$ and $T_2 \simeq T_1$. It must be stressed that in a real system with strong spin interactions and correlations, the experimental behavior varies in a way not fully described in these models. In addition, the spin magnetization when the liquid returns to equilibrium cannot be exponential. Examples include water with its complex thermodynamics inside the super cooled regime and systems characterized by strong correlations or clustering processes. Here T_1 reflects the significant fluctuating interactions between the spin and the rest of the material of which the nucleus is apart.

To understand how the mutual hydrophobicity-hydrophilicity affects solution dynamics, water and methanol (the smallest amphiphilic molecule)solutions have been studied at different concentrations and across a wide temperature range from $330K$ (near the methanol boiling point) to $200K$ (Mallamace et al. 2019). This is because the boiling point of methanol is $TM = 337.8K$ and the melting point $T_M = 175.4K$, where water is in its super cooled regime but methanol is well inside its stable liquid regime. A central point is that the NMR technique allows a simultaneous study of the separate hydrophilic groups of the two molecules (the hydroxyls of water OHW and methanol OH_M and the methanol methylgroups CH_3, the only hydrophobic moiety present in solutions). The water methanol solution has been the subject of numerous previous studies that explain the nonlinear behaviors of transport parameters (see, Mallanace et al. (2016a) and Micali et al. (1996)). According to this, the first step is the use of measured NMR relaxation times, T_1/T_2 data, to obtain the correlation time τ_c. Then, special care is given to the evolution of these relaxation times as a function of τ_c in order to understand the dynamics of the solution molecules and the properties and effects of hydrophilic and hydrophobic interactions.

For comparison, see Fig. 5, the top section, where there is an Arrhenius plot, the correlation times τ_c coming from pure water (Mallamace et al. 2016b)and methanol (OH and CH_3). In the bottom the hydrophilic solution contributions OH_W and OH_M and the hydrophobic CH_3 in the methanol molar fraction range $0.1 < X_M < 0.7$ have been reported. The solution data differ greatly from those of the pure solvent and solute. Note also the behavior of the hydrophilic moieties (of water and methanol) and the hydrophobic solute molecule portion (CH_3). In the hydrophilic moieties, there is approximately the same evolution as in the BPP correlation times corresponding to OH_W and OH_M.

The τ_c data differ for pure water, the solute, and the solutions. τ_c by decreasing T, in pure water and methanol can be observed. Figure 5 (bottom panel)shows that in the high temperature region, the correlation time of both hydroxyl groups are approximately the same, and that at the lowest T the correlationtimes corresponding to the OH_M area factor 2 higher than those in the corresponding OH_W.

The thermal evolution of the methanol hydrophobic correlation time is complex because when the T is decreased there is first a maximum and then a minimum. The τ_c related to the solute methyl group for all X_M increase from high T values toward a maximum located at $\sim 265K$ ($1000/T \simeq 3.78K^{-1}$), after which τ_c decreases toward a minimum near $\sim 225K$. The maximum temperature (T^{HH}) in the solution methyl correlation time corresponds to the flexpoint of the τ_c data for the hydroxyl groups of water and methanol. Thus, the dynamical behavior of the solution exhibits three characteristic temperatures: $T^* \simeq 315K$, $T^{HH} \simeq 265K$, and $T_L \simeq 225K$.

Using these τ_c characteristics and comparing the pure materials (top panel) with the related solutions data (bottom panel), we find that adding methanol to water even at very low concentration ($X_M = 0.1$) changes the local molecular order of water with regard to the effect of the solute on the water and vice versa. In particular, the methanol molecules strongly affect the water HBs and their ability to form the tetrahedral network, and at the sametime, the water influences the solute's properties. Figure 5 also details the two hydrophilic moieties (OH_W and OH_M) at different concentrations, on how hydrophobicity affects the water HB structure. These hydroxyl groups can interact with eachother via HB-originating structures of water and methanol and this networking increases when T decreases despite the hydrophobicity effects due to the methanol methyl groups. In the past, ring-like structures formed by water and methanol molecules have been proposed (Pauling 1960), but such an idea was the subject of some dispute resulting from the findings of X-ray (Guo et al. 2003) and neutron (Dixit et al. 2002) experiments. But Fig. 5's dynamical data indicate that there are hydrophilic-hydrophobic competition effects on the solute and solvent molecules, and that their thermal evolution can provide strong correlations in the intermolecular structures.

Although the local order nature due to hydrophobicity is an open question, experiments fully show that the water's tetrahedral network due to HBs increases instability and size, by decreasing T. Figure 5 show the HB network and the effect of hydrophobicity on it lead to the definition of τ_c and its behavior related to the hydroxyl groups temperature evolution. At the highest temperature ($335K$), pure water has a τ_c value of $\approx 2 \cdot 10^{-9}$sec. Adding methanol in a ratio of one molecule to about 10 of water increases τ_c by approximately one order of magnitude. At this temperature, the water structure is governed by monomers, dimers, and trimers because the HB lifetime is unable to support tetrahedrons.

Thus, the addition of a small amount of methanol imposes an additional molecular mobility (or disorder) and the time required for the local equilibrium to recover is longer than that needed in the pure solvent. This time increase also holds for methanol (see the behavior of the OH_M groups). Because there can be an HB interaction between the OH_W and OH_M moieties, we see that this dramatic change in the system is a HE caused by the solute methyls. By decreasing T, the τ_c of the solution hydroxyl groups adopts an Arrhenius behavior upto $\sim T^*$, after which there is a steep decrease with a flexpoint at T^{HH} followed by a minimum located at approximately T_L, where the correlation time recovers the value $\sim 2.5 \cdot 10^{-10}$ sec measured in pure water in the very deep super cooled regime ($238K$), that is, near the temperature at which the HB network dominates (Mallamace et al. 2007b, Xu et al. 2009).Therefore, in the water super cooled regime (as in bulk water), the HB interaction of the hydroxyl groups and the associated networking exceeds the hydrophobic effects characteristic of the temperatures above the melting point of water. For $T < T_L$ and for all the studied concentrations, we see a moderate τ_c increase, probably due to relaxations in the HB clusters of water and methanol molecules. This is supported by the differences observed at the very low T between the hydroxyl correlation times, with the one of methanol higher than that of water.

The T^{HH} temperature is thus directly connected to the HE and its effect on the solvent and to the hydrophilicity effects on the hydrophobic solute (the CH_3 metabolite). Figure 5 (bottom panel) also shows the methyl groups correlation time τ_c. Note that when T decreases the τ_c increases from an average value of $\sim 8 \cdot 10^{-10}$ sec to $\sim 4 \cdot 10^{-9}$ sec at T^{HH}, then decreases to a minimum at T_L ($5 \cdot 10^{-10}$sec), following which there is an increase analogous to that observed for the hydroxyl groups. It can thus be assumed that the high T increase is caused by the effect of water hydrophilicity on the solute hydrophobic heads; an effect which occurs at all the T, but is affected by the HB interactions that increase with decreasing temperature. In the high T regions, where the HBs are weak and have a short lifetime, the methanol CH_3 are unperturbed and the equilibrium recovery after the NMR perturbation is rapid.

At the sametime, the hydrophilicity determines the water HB structure and the onset of the water network, thus when T decreases, the HB network becomes

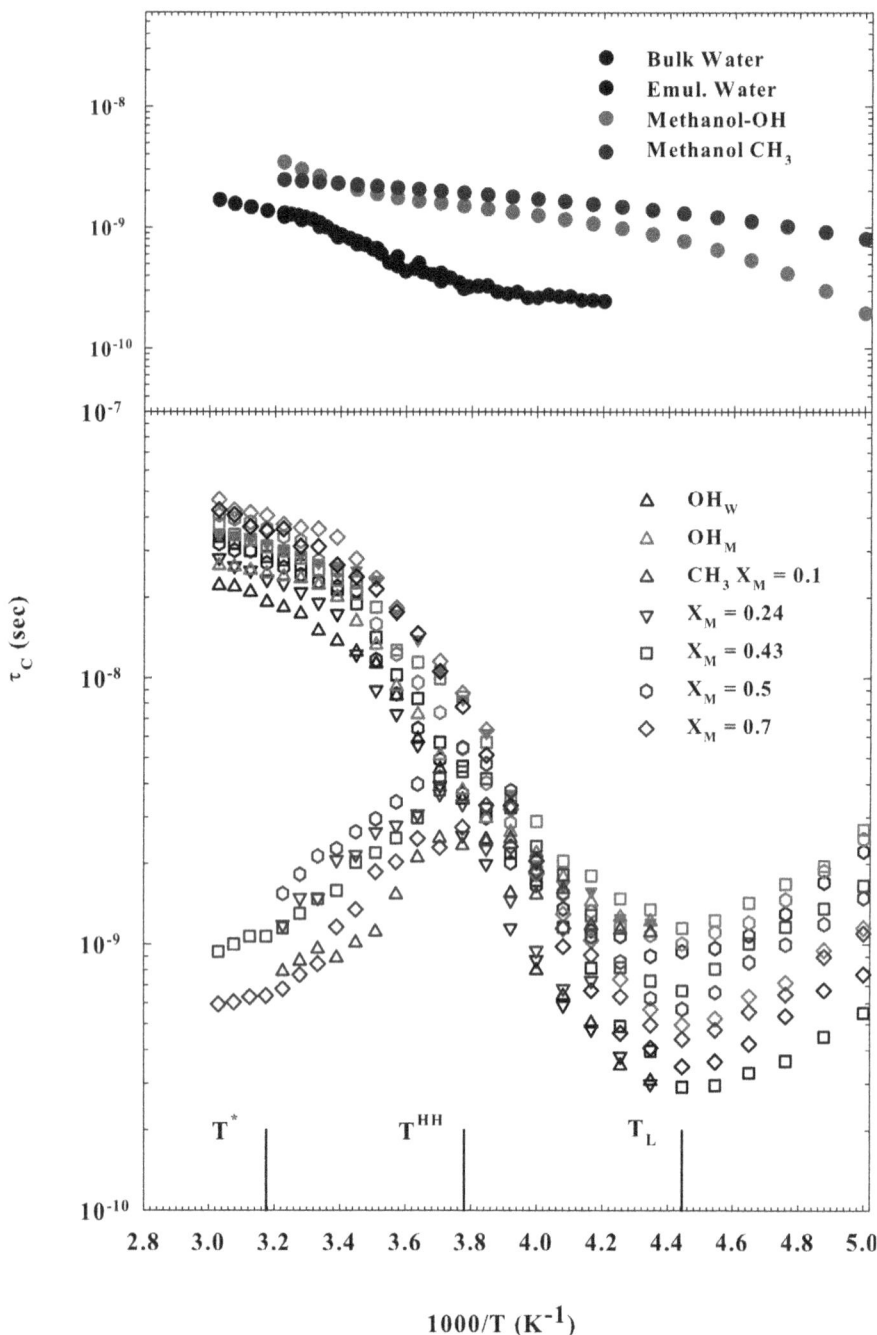

Fig. 5: The Arrhenius plot of the BPP correlation time τ_c obtained by the NMR relaxation times of T_1 and T_2, for the pure solute and solvent; and for some methanol molar fraction in the range $0.1 < XM < 0.7$ In the top panel, the correlation times corresponding to pure water (bulk and emulsioned) and methanol (OH and CH_3) are represented; in the bottom instead there are the hydrophilic solution contributions OH_W and OH_M and the solute hydrophobic part.

increasingly stable and effective, and its influence on the solute hydrophobic metabolites increases and progressively slows the recovery of their equilibrium. A tacertain T, the HB interaction becomes very dominant and controls the system dynamics. Figure 5 (bottom panel) shows that when $T < T^{HH}$, the τ_c measured for all the solution metabolites (hydroxyl and methyl groups) evolves in the same way with approximately the same value. In fact, inspecting these data can be observed that for this latter T regime, all the corresponding data inside the experimental error can be super imposed, that is, the growing tetrahedral structure of water cages all the alcohol molecules at the temperature where the hydrophobic-hydrophilic competition is dominated by the HE.

Thus, all these data indicate that in a solution with hydrophilic and hydrophobic molecules, a crossover temperature exists and defines two different regions, the high T region, which is dominated by HE, and the low T region, which is dominated by hydrophilic HB and by the resulting structures.

Figure 6 shows T_1 (filled symbols) and T_2 (open symbols) versus the corresponding τ_c in bulk water and methanol. For water (blue data), both the longitudinal T_1 and transverse T_2 relaxation times increase as τ_c and temperature increase across the studied range. The region around the water melting temperature (T_M) is a crossover, below which it is $T_2 \simeq T_1$, and above ($T > T_M$, and τ_c at its higher values) T_2, which remains nearly constant where as T_1 evolves with a marked increase. The hyperbolic behavior below T_M is that predicted by the BPP model, that is, $T_2 \tau_c \sim const$ and $T_2 \simeq T_1$ (observable under the experimental condition $1/\omega_0 = 1.429 \times 10^{-9}$ sec $\gg \tau_c$

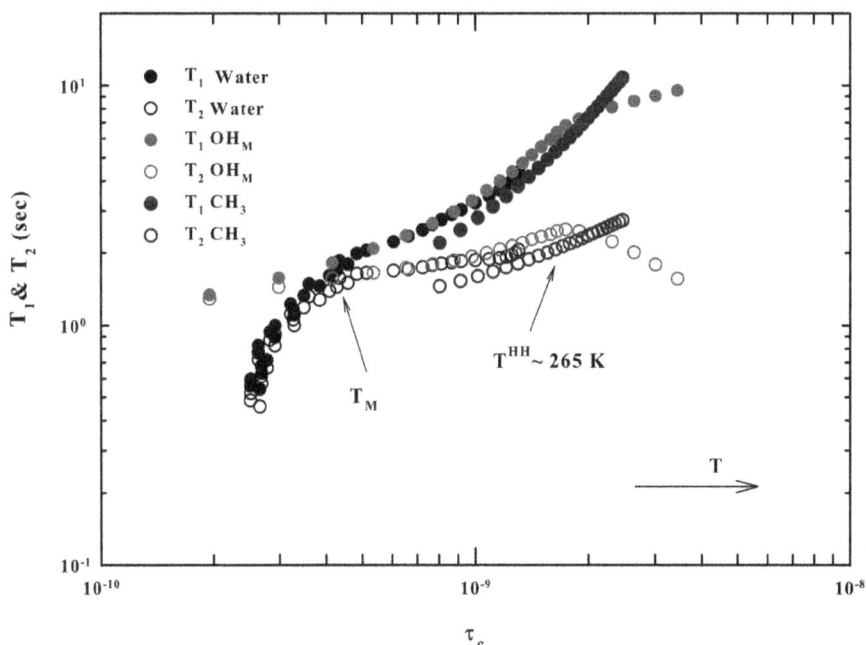

Fig. 6: The T_1 (full symbols) and T_2 (open symbols) behaviors of for bulk water (black) and methanol (grey) as a function of the correlation time τ_c. In the solute case, these times are reported for the hydroxyl (OH$_M$) and methyl (CH$_3$) groups, respectively.

(at ω_0 = 700 MHz)), indicating an increasing stability in the HB network as T decreases inside the water metastable super cooled regime.

The methanol data's hyperbolic behavior holds at the lowest temperature ($T \simeq 205K$).As predicted, T_1 monotonically increases with increasing τ_c and that the T_2 can either slowly grow or decrease. The data indicate that this occurs only for the hydrophobic group CH_3, and in the high-T, regime there are changes in the behavior of the relaxation times corresponding to the hydrophilic OH_M. The methanol spin-spin relaxation time T_2 and that of the OH_M have a maximum at about T^{HH}, where the growth rate of the longitudinal T_1 also changes. This confirms that there is a competition between hydrophobic and hydrophilic heads in determining the bulk alcohol local order. This occurs at a relatively high T for liquid methanol, and the lowered HB lifetime causes the HE to dominate the hydrophilic interactions.

This is consonant with results obtained for water confined in hydrophobic nanotubes that show how hydrophobicity becomes effective in the high T regime, $T > 281K$ (Wang et al. 2008).

Figure 7 shows T_1 and T_2 as a function of τ_c for the solutions at different concentrations: the solvent hydroxyls, the solute hydroxyls, and methyls are in the top, middle and bottom panels, respectively. Note that the τ_c behaviors of the NMR relaxation times for the hydroxyl groups are similar. The only difference is that near T_L, the OH_W data (top panel) are close to the BPP hyperbolic behavior ($T_2\tau_c \sim const$ and $T_2 \simeq T_1$), while in contrast the OH_M data (middle panel)are not (it must be stressed that T_L is for water's temperature inside the metastable super cooled regime where as the methanol lies in the stable liquid phase, are not). For both hydroxyl groups in the high τ_c region, changes are observable in the longitudinal relaxation time T_1 with a cross over between two behaviors at the corresponding T^{HH}, but in the spin-spin relaxation time, this is coincident with the data flexpoints. Also, it can be observed that the τ_c locus of the T_1 crossover increases by increasing X_M, suggesting that a solute amount causes an additional hydrophobic effect that reduces HB networking.

The methanol methyl group relaxation times reported as a function of τ_c (bottom panel) are instead characterized by a cusp-likebehavior at both T^{HH} and T_L, indicating their importance, especially in systems with simultaneous hydrophilic and hydrophobic interactions. The data for the methyl group are mildly dependent on the concentration of the solution, but strongly dependent on T as well as the corresponding transverse NMR relaxation time, and fully reflected in the τ_c correlation time. Both are very sensitive to the water cross over temperature and also to the hydrophobic-hydrophilic interaction thermal balance and thus to T^{HH}. Although low temperature T_L in water indicates the formation of very stable HBs with a tetrahedral network spanning the entire system that causes significant changes in system dynamics, T^{HH} can be proposed to be the temperature above which the hydrophobicity effect dominates in water and water systems. This is again in full agreement with the NMR experiments on water confined inside hydrophobic nanotubes indicating the existence of a hydrophobic-hydrophilic transition around the ambient temperature, demonstrating that the structure of interfacial water on hydrophobic surfaces is T-dependent (Wang et al. 2008). Both these NMR observations in confined water and in water methanol solutions shed new light on the HE and offer additional

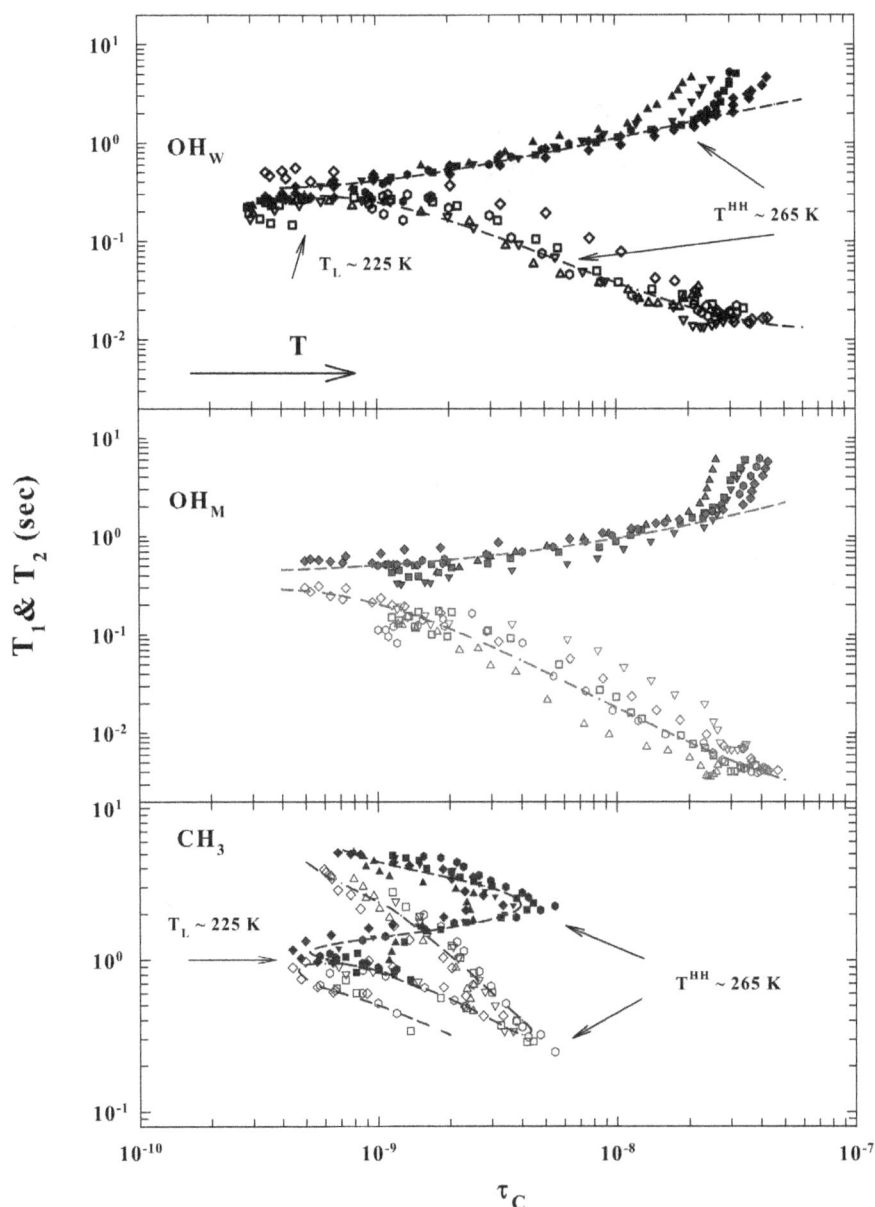

Fig. 7: The figure illustrates the behaviors of T_1 and T_2 as a function of the correlation time τ_c for OH_W (top panel), OH_M (middle panel), and CH_3 (bottom panel). The symbols are the same as those used previously and all the lines are a guide for the eye.

experimental studies to verify its universality. For example, new studies should clarify how both HE and water can influence the biosystems activity (e.g., the folding and unfolding processes), and how the onset of ordered NP structures depends on thermodynamics.

Current Open Questions

In the frame work on whether oil molecules hate or love to be surrounded by water, the hit her to reported experimental findings and theoretical results in this chapter remain open regarding the dependence of HE on temperature, pressure, and solute molecular sizes as well as the balance of direct and water-mediated interactions.

There are, however, many important and hit her to never-experimentally-measured quantities to take into account as well. An example is the pair distribution function $g_{hh}(r)$ between two hydrophobic molecules (this despite the fact that the water $g_{ww}(r)$ is one of the most studied quantities; see Soper (2015)). Having obtained $g_{hh}(r)$, the corresponding force $W(r) = -k_B T \ln g_{hh}(r)$ can be consequently obtained so that hydrophobicity can be treated analytically and all the interesting implications of such forces can thus be evaluated (Hormoz and Widom 2012, Maibaum and Chandler 2003).

Another additional important issue in the NP self-assembling, is that the resulting conformation in the hydration shells about a hydrophobic solute can also affect the structure of the solute itself. This latter process can have great scientific relevance because solute molecules can assume dipolemoments, causing changes in the solution by changing the thermodynamic variables. *In otherwords, the changes in the solute structure when put in solution can affect, not in ordinary way, the hydration thermodynamic functions (entropy and energy) of a hydrophobic solute. Until now, these functions have been ascribed almost exclusively to the effect of the solute on the structure of the neighboring water, ignoring the change in the structure of solute itself.*

This means that on discussing HE, it is important to distinguish two hydrophobicity aspects (related to eachother): hydration and interaction. Hydration pertains to the transfer of an oily molecule into water, where as hydrophobic interaction concerns oily molecules (or nonpolar groups) brought into contact with eachother in water. HE arises from a balance between direct (those involving the isolated solute groups) and water-mediated (a rising from the water's influence in either promoting or suppressing aggregation)intermolecular interactions. While considering the structure of solute molecules (oil and polymer), these interactions must be averaged overall their equilibrium orientations and conformations. Hence, there is, in general, a complex multidimensional free energy surface or energy landscape (depending on the PMF, as well as on the number of solutes in higher-order aggregate structures but also on the relative orientations and conformations of the interacting solute molecules).

In principle, the contact point between the short and largescale hydrophobicity interactions and their possible definitive explanations maybe (oris) in this energy landscape; and in its evolution as a function of the system thermodynamics (P,T) and solute/solvent concentrations. In other words, it is just the competing influence of solute-solute and solute-water interactions, and their thermodynamics that determine the HE properties, and their effects inter and intra the NP. Certainly, such an interpretation that clarify processes like the NP superstructures, the protein folding/unfolding, the polymer self-assembly, and the behaviors of glass forming super cooled liquids (Karplus 2011, Laio and Parrinello 2002, Mallamace et al. 2016c, Okazaki et al. 2006, Yip and Short 2013) is useful for a unique and universal HE interpretation, valid for all its lengthscale. These systems show a dynamical behavior characterized by multi-relaxations rather than the simple Arrhenius from two fixed energy states (Dyre 2006, Stillinger 1988, Yip and Short 2013). This is a situation that can be described by a power-law behavior in the physical response functions (transport quantities, mechanical moduli, relaxation times, etc.) (Dyre 2006, Hecksher et al. 2008, Mallamace et al. 2010a).

The Zhang et al. (2002) shear experiment on a Janus interface is illustrative of the abovementioned situation, that is, the viscoelasticity of water confined between adjoining hydrophobic-hydrophilic surfaces. In this type of confinement, water forms stable films of nanometer thickness whose response to shear deformations are extraordinarily noisy. As quantified in the corresponding power spectrum, the frequency dependence of the complex shear modulus is a power law with a sloping half, indicating a distribution of relaxation processes rather than any dominan tone. The emerging physical picture is the result of an opposition behavior between the two hydrophilic-hydrophobic interactions: where as surface energetics encourages water to dewet the hydrophobic side of the interface, the hydrophilic side constrains the water present.

This is additional evidence of the fluctuating complex behavior of the HE that represents the basis of our attempt to clarify experimentally, by using several complementary techniques, many of the open scientific questions regarding this slushing, intriguing,and interesting phenomenon, which has a large influence on physical, chemical, and biological systems and many possible technological applications. The ground breaking nature of our project lies in our attempt to provide evidence to create a complete study of the HE, as a function of all the thermodynamic variables, in terms of the energy lands cape approach.

Perspectives

With regard to the future, it can be presumed that experiments will focus on HE through the measure of collective and single particle properties in solutions with both hydrophobic solute and water. This also takes advantage of the fact that recently the knowledge on the chemical physics of liquid water have been largely improved just by looking at its thermodynamics int $P-T$ phase diagrams. We have now largely accepted and taken into account water polymorphism (two liquid phases of different densities (Mallamace et al. 2009, Mallamace et al. 2007a)), the Widomline (the locus of the fragile-to-strong (glass forming) dynamical crossover—where the Stokes-Einstein relation is violated (Cerveny et al. 2016, Chen et al. 2006, Xu et al. 2009)), and a line marking the water transition, on increasing T and P, from the well-know nanomalous to simple, normal liquid behavior (Maestro et al. 2016, Mallamace et al. 2012). Also, the dynamical crossover as reported by the cited studies on methanol (Corsaro et al. 2008, Mallamace et al. 2010b) and salt water (Turton et al. 2012) solutions, regarding hydrated proteins (Lagi et al. 2008, Mallamace et al. 2016c, Mallamace et al. 2007c, Mallamace et al. 2014b) seems to be a "universal like" process independent of the system chemistry. In the protein case, it is observed that: (i) the dynamic crossover temperature is unaffected by the hydration level and the first hydration shell remains liquid at all hydrations, even at the lowest temperature; (ii) below T_L, the protein loses its conformational flexibility and its biological activity; as also demonstrated for DNA and RNA, this "dynamic" transition "driven by water" is important for the functioning of biomolecules (Chen et al. 2006, Chu et al. 2008).

While certainly considering that MD simulations will improve our HE knowledge, we choose to conclude our *excursus* looking at the related experimental approaches. All of this as proposed by the response functions theory can be made in terms of the dynamic structure factor or power spectrum $S(k,\omega)$, defined as the Fourier transform (over r and t) of the generalized pair correlation function $G(r,t)$—in other words, considering all the density-density time correlation function of the tagged atom, able to give both the single particle and collective properties. There are many reasons in the HE frame that give special care to the local order (molecular symmetries plus anisotropies) and dynamic changes (from the self-diffusion, to rotational and vibrational spectra, etc.) for its strong sensitivity interactions and thermodynamic.

In this local order frame, experimental techniques must be considered like NMR (Abragam 1961, Bloembergen et al. 1948, Kubo and Tomita 1954), neutron scattering (e.g., elastic and deep-in elastic) (Egelstaff 1971, van Hove 1954), X-ray, Raman, IR, light-fast spectroscopy (Bernie and Pecora 1976, Fabelinskii 1969), strongly sensitive to system atomic and molecular properties and confutations. In these terms, the pair distribution function $g_{hh'}(r)$ of the hydrophobic molecules can be measured, by means of a small angle neutron scattering.

The conformation of the water molecules in the first two (say) solute hydration shells is a serious issue; it will be experimentally studied, by means of NMR and X-ray and Neutron Compton experiments especially for system like Xe with moderate solubility. A related question is that of the effect of the hydration on the structure of the solute itself; in particular, the changes in its structure (including its electronic structure—can take on a dipole moment) in passing from the pure phase into water. The reason of the change in structure of the solute on passing from the pure phase in to water is interesting because until now, the measured entropy and energy of hydration of a hydrophobic solute has been ascribed almost exclusively to the effect of the solute on the structure of the neighboring water, ignoring the change in the structure of the solute itself—may be because so little is known about that.

Many probes deal with collective system responses: examples are sound propagation, viscoelasticity, dielectric analysis, or specific heat techniques. The same holds for quasi-elastic and coherents cattering data where the dynamic structure factor $S(k,\omega)$ gives indication of the evolution of transport quantities and their relaxations (in particular reflecting the onset of some specific structures). This is specifically the case of the self-assembling processes (and the related kinetics) characterizing hydrophobic and hydrophilic systems. In these situations, the observables are directly related to the fluctuations of thermodynamic functions (and variables), like entropy and enthalpy whose competition represents the basis of the HE. For example, the isothermal compressibility K_T reflects the volume fluctuations $\langle \delta V^2 \rangle$ and the adiabatic $K_S = (\rho v^2 - 1)$ those of the pressure $\langle \delta P^2 \rangle$, where as the isobariche at capacity C_p is those of the entropy $\langle \delta S^2 \rangle$ (v is the sound velocity). Here, the thermal expansion coefficient α *P* represents the entropy and volume cross-correlations $\langle \delta S \delta V \rangle$. Such a situation is exactly why we need to explore the HE in the T-P phase diagram.

Just the system fluctuations are a possible key to describe the HE, clarifying its properties by means of the corresponding thermodynamic evolutions of collective measured quantities. For example, if we consider the measured system relaxation times we can obtain according to the Eyring formula (Evans and Polanyi 1935), along with the evaluation of fundamental quantities like the free energy (G), enthalpy (H), and entropy (S), in a separate way for both the hydrophobic solute and water. More precisely, we can evaluate the excess partial molar activation free energy (ΔG_h, ΔG_w), enthalpy (ΔH_h, ΔH_w), and entropy (ΔS_h, ΔS_w), of both the solute (h) and solvent (w), evidencing how these molecules separately contribute to the process understudy.

In conclusion, we are reasonably confident that the certain findings coming from the MD simulation studies on the HE can be supported by the existing and new experimental techniques. The latter is specially designed for a direct simultaneous measure of all the HE forces (or PMF) whose findings can be used to originate a HE theory that should provide a universal quantitative guidance for the study of complex materials and bio-systems; that is, it is able to be arranged under the same rubric short and long range HE.

The physical properties of the hydrophobic-hydrophilic interactions (or hydrophobic effect, HE) are significant in multiple areas of science and technology (in particular in soft materials of special interest in chemical-physics and in medicine and biology), therefore there are in the scientific community many aims to advance its scientific knowledge on this front to new levels.

Infact, the HE interaction can be found in many different research are as like physics, chemistry, biology, and materials science. At the sametime, it is relevant for many technological implications also constituting the bases of the nowadays nanotechnology industries. It is of interest in polymer solution properties in membranes and protein activities. It determines the cellular function, that is, from the protein folding to the macromolecular aggregation. In physics and material science, it can explain the route in the multiscale modelling of materials at theme so scale.

HE involves complex topics, starting from the hydrophobic molecules self-aggregation, typical of soft materials, which takes place over multiplescales. Despite this, the pair distribution function $g_{hh}(r)$ between two hydrophobic molecules has never been measured, despite water $g_{ww}(r)$ being wellknown. This project's first aim was to measure the $g_{hh}(r)$, and its corresponding potential so that HE can be treated analytically. Thus far, many simulations and models, and few experiments fail to provide a reasonable description.

Additional open questions relate to the role of water, the effect of the solute on the solvent (and vice versa) and the overall solute/solvent thermodynamic dependence. Recent experiments highlight the determining role of thermodynamics (temperature, pressure, and concentration), suggesting that aquantitative study on the thermodynamic properties of HE is badly needed and can address a ground breaking issue. All of this suggests that a series of different but synergic experiments (neutron and light scattering, NMR, and viscoelasticity), above and below the self-assembling threshold, detailing the solution's local order and aiding the exploration of the system collective properties will be the future research core of this subject. The basic HE intermolecular interactions, the structure and dynamic properties, and the characteristic relaxation processes (on the multiscales) and its whole thermodynamics can be clarified by detailing the energetic configurations. As a result, useful observables to describe HE undera theoretical scheme based on scaling and universality can be obtained.

We also stress that in order to clarify such unknown interactions, the synergic use of different experimental techniques, including MD simulations, together with the new theoretical ideas are important.

References

Abragam, A. 1961. The Principles of Nuclear Magnetism. Oxford, Oxford, UK.

Ashbaugh, H. S. and L. R. Pratt. 2006. Scaled particle theory and the length scales of hydrophobicity. Rev. Mod. Phys. 78: 159–178.

Altabet, Y. E., A. Haji-Akbari and P. G. Debenedetti. 2017a. Effect of material flexibility on the thermodynamics and kinetics of hydrophobically induced evaporation of water. Proc. Natl. Acad. Sci. USA 114: 2548–2555.

Altabet, Y. E. and P. G. Debenedetti. 2017b. Relationship between local structure and the stability of water in hydrophobic confinement. J. Chem. Phys. 147: 241102.

Gallagher, K. R. and K. A. Sharp. 2003. A new angle on heat capacity changes in hydrophobic solvation. J. Am. Chem. Soc. 125: 9853–9860.

Ball, P. 2008. Water as an active constituent in cell biology. Chem. Rev. 108: 74–108.

Baron, R., P. Setny and J. A. McCammon. 2010. Water in cavity-ligand recognition. J. Am. Chem. Soc. 132: 12092–12097.

Silvera Batista, C. A., R. G. Larson and N. A. Kotov. 2015. Non additivity of nanoparticle interactions. Science 350: 176.

Baxter, R. J. 1968. Percus-Yevick equation for hard spheres with surface adhesion. J. Chem. Phys. 49: 2270.

Ben-Amotz, D. 2015. Hydrophobic ambivalence: teetering on the edge of randomness. J. Phys. Chem. Lett. 6: 1696–1701.

Ben-Amotz, D. and D. B. Widom. 2006. Generalized solvation heat capacities. J. Phys. Chem. B, 110: 19839–19849.

Berne, J. B., J. D. Weeks and R. Zhou. 2009. Dewetting and hydrophobic interaction in physical and biological systems. Ann. Rev. Phys. Chem. 60: 85–103.

Bernie, J. B. and R. Pecora. 1976. Dynamical light scattering. John Wiley, New York.

Bhushan, B. and Jung, Y. C. 2006. Micro- and nanoscale characterization of hydrophobic and hydrophilic leaf surfaces. Nanotechnology 17: 2758.

Bhushan, B., Y. C. Jung and K. Koch. 2009. Micro- nano- and hierarchical structures for super-hydrophobicity, self-cleaning and low adhesion. Phil. Trans. R. Soc. A 367: 1631–1672.

Bloch, F. 1946. Nuclear induction. Phys. Rev. 70: 460–474.

Bloembergen, N., E. M. Purcell and R. V. Pound. 1948. Relaxation effects in nuclear magnetic resonance absorption. Phys. Rev. 73: 679.

Cassie, A. B. D. and S. Baxter. 1944. Wettability of porous surfaces. Trans. Faraday Soc. 95: 546–550.

Cerveny, S., F. Mallamace, J. Swenson, M. Vogel and L. M. Xu. 2016. Confined water as model of super-cooled water. Chem. Rev. 116: 7608–7625.

Chandler, D. 2005. Interfaces and the driving force of hydrophobic assembly. Nature 437: 640–647.

Chen, S. H., L. Liu, Y. Zhang, E. Fratini, P. Baglioni, A. Faraone and E. Mamantov. Experimental evidence of fragile-to-strong dynamic crossover in DNA hydration water. 2006. J. Chem. Phys. 125: 171103.

Chen, S. H., F. Mallamace, C. Y. Mou, M. Broccio, C. Corsaro, A. Faraone and L. Liu. 2006. The violation of the Stokes–Einstein relation in supercooled water. Proc. Natl. Acad. Sci. USA 103: 12974–12978.

Chu, X. Q., E. Fratini, P. Baglioni, A. Faraone and S. H. Chen. 2008. Observation of a dynamic crossover in RNA hydration water which triggers a dynamic transition in the biopolymer. Phys. Rev. E. 77: 011908.

Corsaro, C., J. Spooren, C. Branca, N. Leone, M. Broccio, C. Kim, S.-H. Chen, H. E. Stanley and F. Mallamace. 2008. Clustering dynamics in water/methanol mixtures: A nuclear magnetic resonance study at 205 K < T < 295 K. J. Phys. Chem. B 112: 10449–10454.

Davis, J. G., K. P. Gierszal, P. Wang and D. Ben-Amotz. 2012. Water structural transformation at molecular hydrophobic interfaces. Nature 491: 582.

Davis, J. G., B. M. Rankin and D. Ben-Amotz and K. P. Gierszal. 2013. On the cooperative formation of non-hydrogen-bonded water at molecular hydrophobic interfaces. Nature Chem., 5: 796.

deGennes, P. G. 1979. Scaling concepts in polymer physics. Cornell University Press, Ithaca.

deGennes, P. G. 1985. Wetting—Statistics and dynamics. Rev. Mod. Phys. 57: 827–863.

Derjaguin, B. V. and L. Landau. 1941. Theory of the stability of strongly charged lyophobic sols and of the adhesion of strongly charged particles in solutions of electrolytes. Acta Phys. Chem. URSS, 14: 633.

Dixit, S., J. Crain, W. C. K. Poon, J. L. Finney and A. K. Soper. 2002. Molecular segregation observed in a concentrated alcohol–water solution. Nature 416: 829–832.

Du, Q., E. Freysz and Y. R. Shen. 1994. Surface vibrational spectroscopic studies of hydrogen-bonding and hydrophobicity. Science 264: 826.

Dyre, J. C. 2006. The glass transition and elastic models of glass-forming liquids. Rev. Mod. Phys. 78: 953–972.

Ebbinghaus, S. et al. 2007. An extended dynamical hydration shell around proteins. Proc. Natl. Acad. Sci. U.S.A. 104: 20749.

Egelstaff, P. A. 1971. Thermal neutron scattering. Academic Press, New York.

Evans, M. G. and M. Polanyi. 1935. Trans. Faraday Soc. 31: 875.

Fabelinskii. L. L. 1969. Molecular scattering of light. Plenum Press, New York.

Feng, L., S. H. Li, Y. S. Li, H. J. Li, L. J. Zhang, J. Zha, Y. L. Song, B. Q. Liu, L. Jiang and D. B. Zhu. 2002. Super-hydrophobic surfaces: From natural to artificial. Adv. Mater. 14: 1857–1860.

Ferguson, A. L., N. Giovambattista, P. J. Rossky, A. Z. Panagiotopoulos and P. G. Debenedetti. 2012. A computational investigation of the phase behavior and capillary sublimation of water confined between nanoscale hydrophobic plates. J. Chem. Phys. 137: 8095.

Flory, P. 1953. Principles of polymer chemistry. Cornell University Press, Ithaca.

Garde, S. and A. J. Patel. 2011. Unraveling the hydrophobic effect, one molecule at a time. Proc. Natl. Acad. Sci. USA, 108: 16491–16492.

Ge, Z., D. G. Cahill, and P.V. Braun. 2006. Thermal Conductance of Hydrophilic and Hydrophobic Interfaces. Phys. Rev. Lett., 96: 186101.

Giovambattista, N., P. Rossky and P. G. Debenedetti. 2007. Effect of surface polarity on water contact angle and interfacial hydration structure. J. Phys. Chem. C 11: 1323.

Giovambattista, N., P. Rossky and P. G. Debenedetti. 2009. Effect of temperature on the structure and phase behavior of water confined by hydrophobic, hydrophilic, and heterogeneous surfaces. J. Phys. Chem. B 113: 13723.

Granick, S. and S. C. Bae. 2008. Curious Antipathy for Water. Science 322: 1477–1478.

Gray, J. J., B. Chiang and R. T. Bonnecaze. 1999. Colloidal particles: Origin of anomalous multibody interactions. Nature 402: 750.

Guo, J.-H., Y. Luo, A. Augustsson, S. Kashtanov, J.-E. Rubensson, D. K. Shuh, H. Agren and J. Nord. 2003. Molecular structure of alcohol-water mixtures. Phys. Rev. Lett. 91: 157401–4.

Haji-Akbari, A. and et al. 2009. Disordered, quasicrystalline and crystalline phases of densely packed tetrahedra. Nature 462: 773.

Hecksher, T., A. I. Nielsen, N. BoyeOlsen and J. C. Dyre. 2008. Little evidence for dynamic divergences in ultraviscous molecular liquids. Nat. Phys. 4: 737.

Hormoz, S. and B. Widom. 2012. Volume CLXXVI, Models of the hydrophobic attraction, page 1. IOS Press, Amsterdam. Proc. Int. Sch. of Phys. E. Fermi.

Huang, D. M. and D. Chandler. 2000. Temperature and length scale dependence of hydrophobic effects and their possible implications for protein folding. Proc. Natl. Acad. Sci. USA, 97: 8324.

Huang, X., C. J. Margulis and B. J. Berne. 2003. Hydrophobic hydration and molecular association in methanol–water mixtures studied by microwave dielectric analysis. Proc. Natl. Acad. Sci. USA, 100: 11953.

Hummer, G. 2010. Molecular binding under water's influence. Nature Chem. 2: 906–907.

Hummer, G., S. Garde, A. F. Gracia, A. Pohorille and L. R. Pratt. 1996. An information theory model of hydrophobic interactions. Proc. Natl. Acad. Sci. USA 93: 8951–8955.

Hummer, G., J. C. Rasaiah and J. P. Noworta. 2001. Water conduction through the hydrophobic channel of a carbon nanotube. Nature 414: 188–190.

Ide, M., Y. Maeda and H. Kitano. 1997. Effect of hydrophobicity of aminoacids on the structure of water. J Phys. Chem. B 101: 7022.

Israelachvili, J. N. 1989. Intermolecular and Surphace Forces. Acad. Press, London.

Janecek, J. and R. R. Netz. 2007. Interfacial water at hydrophobic and hydrophilic surfaces: Depletion versus adsorption. Langmuir 12: 8417–8429.

Karplus, K. 2011. Behind the folding funnel diagram. Nat. Chem. Biol. 7: 401.

Kubo, R. and K. Tomita. 1954. A general theory of magnetic resonance absorption. J. of the Phys. Soc. Japan 9(6): 888–919.

Kudin, K. N. and R. Car. 2008. Why are water-hydrophobic interfaces charged? J. Am. Chem. Soc. 130: 3915.

Lagi, M. et al. 2008. The low-temperature dynamic crossover phenomenon in protein hydration water: Simulations vs. experiments. J. Phys. Chem. B 112: 1571–1575.

Laio, A. and M. Parrinello. 2002. Escaping free-energy minima. Proc. Natl. Acad. Sci. U.S.A. 99: 12562–12566.

Le, Z.B. Walker, G.C. and P.V. Braun. 2006. Thermal conductance of hydrophilic and hydrophobic interfaces. Phys. Rev. Lett. 96: 186101.

Li, T. S. and G. C. Walker. 2011. Signature of hydrophobic hydration in a single polymer. Proc. Natl. Acad. Sci. USA 108: 16527–16532.

Lobry, L., N. Micali, F. Mallamace, C. Liao and S. H. Chen. 1999. Interaction and percolation in the l64 triblock copolymer micellar system. Phys. Rev. E 60: 7076.

Lum, K., D. Chandler and J. D. Weeks. 1999. Hydrophobicity at small and large length scales. J. Phys. Chem. B 103: 4570.

Kimber, S. A. J., M. Zobel and R. B. Neder. 2015. Universal solvent restructuring induced by colloidal nanoparticles. Scienc 347: 292.

Ma, C. D., C. Wang, C. Acevedo-Vélez, S. H. Gellman and N. L. Abbott. 2015. Modulation of hydrophobic interactions by proximally immobilized ions. Nature 517: 347.

Maestro, L. M. et al. 2016. On the existence of two states in liquid water: Impact on biological and nanoscopic systems. Int. J. Nanotechnology 13: 667–677.

Maibaum, L. and D. Chandler. 2003. A coarse-grained model of water confined in a hydrophobic tube. J Phys. Chem. B 107: 1189–1193.

Mallamace, D., S.-H. Chen, C. Corsaro, E. Fazio, F. Mallamace and H. E. Stanley. 2019. Hydrophilic and hydrophobic competition in water-methanol solutions. Science China Phys. Mech. & Astron. 62: 107003.

Mallamace, F., M. Broccio, C. Corsaro, A. Faraone, L. Liu, U. Wanderlingh, C. Y. Mou and S. H. Chen. 2006. The fragile-to-strong dynamic crossover transition in confined water: Nuclear magnetic resonance results. J. Chem. Phys. 124: 161102.

Mallamace, F., C. Branca, M. Broccio, C. Corsaro, C. Y. Mou and S. H. Chen. 2007a. The anomalous behavior of the density of water in the range 30 K < t < 373 K. Proc Natl. Acad. Sci. USA, 104: 18387.

Mallamace, F., M. Broccio, C. Corsaro, A. Faraone, D. Majolino, V. Venuti, L. Liu, C. Y. Mou and S. H. Chen. 2007b. Evidence of the existence of the low-density liquid phase in supercooled, confined water. Proc. Nat. Ac. Sci. USA 104: 424–427.

Mallamace, F., S. H. Chen, M. Broccio, C. Corsaro, V. Crupi, D. Majolino, V. Venuti, P: Baglioni, E. Fratini, C. Vannucci and H. E. Stanley. 2007c. Role of the solvent in the dynamical transitions of proteins: The case of the lysozyme-water system. J. Chem. Phys. 127: 045104.

Mallamace, F. 2009. The liquid water polymorphism. Proc. Natl. Acad. Sci. USA, 106: 15097–15098.

Mallamace, F., C. Branca, C. Corsaro, N. Leone, J. Spooren, S. H. Chen and H. E. Stanley. 2010a. Transport properties of glass-forming liquids suggest that dynamic crossover temperature is as important as the glass transition temperature. Proc. Natl. Acad. Sci. USA 107: 22457–22462.

Mallamace, F., C. Branca, C. Corsaro, N. Leone, J. Spooren, H. E. Stanley and S.-H. Chen. 2010b. Dynamical crossover and breakdown of the stokes-Einstein relation in confined water and in methanol-diluted bulkwater. J. Phys. Chem. B114: 1870–1878.

Mallamace, F., C. Corsaro and H. E. Stanley. 2012. A singular thermodynamically consistent temperature at the origin of the anomalous behavior of liquid water. Nat. Sci. Rep. 2: 993.

Mallamace, F., C. Corsaro, D. Mallamace, H. E. Stanley and S. H. Chen. 2013. Water and biological macro-molecules. In: H. E. Stanley (eds.). Liquid polymorphism, number 135 in Adv. Chemical Physics 11: 263–308. Wiley, NewYork.

Mallamace, F., C. Corsaro, N. Leone, V. Villari, N. Micali and S. H. Chen. 2014a. On the ergodicity of supercooled molecular glass-forming liquids at the dynamical arrest: Theo-terphenyl case. Sci.Rep. 4: 3747.

Mallamace, F., P. Baglioni, C. Corsaro, S. H. Chen, D. Mallamace, C. Vasi and H. E. Stanley. 2014b. The influence of water on protein properties. J. Chem. Phys. 141: 165104.

Mallamace, F., C. Corsaro, D. Mallamace, C. Vasi, S. Vasi and H. E. Stanley. 2016a. Dynamical properties of water-methanol solutions. J. Chem. Phys. 144: 064506.

Mallamace, F., C. Corsaro, D. Mallamace, S. Vasi and H. E. Stanley. 2016b. NMR spectroscopy study of local correlations in water. J. Chem. Phys. 145: 214503.

Mallamace, F., C. Corsaro, D. Mallamace, S. Vasi, C. Vasi, P. Baglioni, S. V. Buldyrev, S. H. Chen and H. E. Stanley. 2016c. Energy land scape in protein folding and unfolding. Proc. Natl. Acad. Sci. USA 105: 536.

Matysiak, S., P. G. Debenedetti and P. J. Rossky. 2011. Dissecting the energetics of hydrophobic hydration of polypeptides. J. Phys. Chem. B 115: 14859.

Matysiak, S., P. G. Debenedetti and P. J. Rossky. 2012. Role of hydrophobic hydration in protein stability: A3d water-explicit protein model exhibiting cold and heat denaturation. 2012. J. Phys. Chem. B 116: 144501.

Meyer, E. E., Q. Lin, T. Hassenkam, E. Oroudjev and J. N. Israelachvili. 2005. Origin of the long-range attraction between surfactant-coated surfaces. Proc. Natl. Acad. Sci. USA 102: 6839.

Micali, N., S. Trusso, C. Vasi, D. Blaudez and F. Mallamace. 1996. Dynamical properties in water-methanol solutions studied by rayleigh scattering. Phys. Rev. E. 54: 1720–1724.

Miller, M. A. and D. Frenkel. 2003.Competition of percolation and phase separation in a fluid of adhesive hard spheres. Phys. Rev. Lett. 90: 135702.

Miller, T., E. Vanden-Eijnden and D. Chandler. 2007. Liquid-liquid phase. Proc. Natl. Acad. Sci. USA 104: 14559.

Okazaki, K., N. Koga, S. Takada, J. N. Onuchic and P. G. Wolynes. 2006. Multiple-basin energy landscapes for large-amplitude conformational motions of proteins: Structure-based molecular dynamics simulations. Proc. Natl. Acad. Sci. U.S.A. 103: 11844–11849.

Patel, A. J., P. Varilly, S. N. Jamadagni, M. F. Hagan, D. Chandler and S. Garde. 2012. Sitting at the edge: How biomolecules use hydrophobicity to tune their interactions and function. J. Phys. Chem. B 116: 2498.

Patel, A. J., P. Varilly and D. Chandler. 2010. Fluctuations of water near extended hydrophobic and hydrophilic surfaces. J. Phys. Chem. B 114: 1632.

Pauling, L. 1960. The Nature of the Chemical Bond. Cornell University Press, Ithaca.

Pérez-Conesa, S., P. M. Piaggi and M. Parrinello. 2019. A local fingerprint for hydrophobicity and hydrophilicity: Frommethane to peptides. J. of Chem. Phys. 150: 204103.

Rajamani, S., T. M. Truskett and S. Garde. 2005. Hydrophobic hydration from small to large length scales: Understanding and manipulating the crossover. Proc. Natl. Acad. Sci. USA 102: 9475–9480.

Rasaiah, J. C., S. Garde and G. Hummer. 2008. Water in nonpolar confinement: From nanotubes to proteins and beyond. Annu. Rev. Phys. Chem. 59: 713–740.

Remsinga, R. C., E. Xia, S.Vembanur, S. Sharmac, P. G. Debenedetti, S. Gardeb and A. J. Patelai. 2015. Pathways to dewetting in hydrophobic confinement. Proc. Natl. Acad. Sci. USA 112: 8181.

Rivera-Rubero, S. and S. Baldelli. 2004. Influence of water on the surface of hydrophilic and hydrophobic room-temperature ionic liquids. J. Am. Chem. Soc. 126: 11788–11789.

Russo, D., R. K. Murarka, J. R. D. Copley and T. Head-Gordon. 2005. Molecular view of water dynamics near model peptides. J. Phys. Chem. B 109: 12966.

Safran, S. A. 1994. Statistical thermodynamics of surfaces, interfaces and membranes. Addison-Wesley, Reading.

Sato, T., A. Chiba and Ryusuke Nozaki. 2000. Hydrophobic hydration and molecular association in methanol-water mixtures studied by microwave dielectric analysis. J. Chem. Phys. 112: 2924.2392.

Sciortino, F., A. Geiger and H. E. Stanley. 1991. Effect of defects on Molecular Mobility in liquid water. Nature 354: 218–221.

Sharma, S. and P. G. Debenedetti. 2012. Evaporation rate of water in hydrophobic confinement. Proc. Natl. Acad. Sci. USA 109: 4365.

Shenogina, N., R. Godawat, P. Keblinski and S. Garde. 2009. How wetting and adhesion affect thermal conductance of a range of hydrophobic to hydrophilic aqueous interfaces. Phys. Rev. Lett. 102: 156101.

Shirtcliffe, N. J., G. McHale, S. Atherton and M. J. Newton. 2010. An introduction to super hydrophobicity. Adv. Coll. and Interf. Sci. 161124–138.

Simpson, J. H. and H. Y. Carr. 1958. Diffusion and nuclear spin relaxation in water. Phys. Rev. 111: 1201.

Singh, G. et al. 2014. Self-assembly of magnetite nanocubes into helical superstructures. Science 345: 1149.

Soper, A. K. 2015. Volume CLXXXVII, page 1. IOS Press. Amsterdam, Proc. Int. Sch. of Phys. E. Fermi.

Sovago, M., R. Kramer Campen, G. W. H. Wurpel, M. Muller, H. J. Bakker and 1. Bonn. 2008. Vibrational response of hydrogen-bonded interfacial water is dominated by intramolecular coupling. Phys. Rev. Lett. 100: 173901.

Stillinger, F. H. 1973. Structure in aqueous solutions of nonpolar solutes from the stand point of scaled-particle theory. J. Sol. Chem. 2: 141.

Stillinger, F. H. 1988. Supercooled liquids, glass transition, and the Kauzmann paradox. Phys. 88: 7818–7825.

Stirnemann, G., S. Romero-Vargas Castrillo, J. T. Hynes, P. J. Rossky, P. G. Debenedetti and D. Laage. 2011. Non-monotonic dependence of water reorientation dynamics on surface hydrophilicity: Competing effects of the hydration structure and hydrogen-bond strength. Phys. Chem. Chem. Phys. 13: 19911.

Su, B., Y. Tian and L. Jiang. 2016. Bioinspired Interfaces with Super wettability: From Materials to Chemistry. J. Am. Chem. Soc. 138: 1727–1748.

Tao, A. R., S. Connor P. Yang, J. Huang and F. Kim. 2005. Spontaneous formation of nanoparticle stripe patterns through dewetting. Nature mat. 4: 896.

Tanford, C. 1980. The Hydrophobic Effect. Wiley, New York.

tenWolde, P. R. and D. Chandler. 2002. Drying-induced hydrophobic polymer collapse. Proc. Natl. Acad. Sci. USA 99: 6539–6543.

Turton, D. A., C. Corsaro and D. F. Martin, F. Mallamace and K. Wynne. 2012. The dynamic crossover in water does not require bulk water Phys. Chem. Chem. Phys. 14: 8067–8073.

vanHove, L. 1954. Correlations in space and time and Born approximation scattering in systems of interacting particles. Phys. Rev. 95: 249–262.

Varilly, P., A. P. Willard, J. B. Kirkegaard and T. P. Knowles and D. Chandler. 2017. Intra-chain organisation of hydrophobic residues controls inter-chain aggregation rates of amphiphilic polymers. J. Chem. Phys. 146: 135102.

Verwey, T. J., E. J. W. Overbeek and K. VanNess. 1948. Theory of the stability of lyophobic colloids: The Interactions of Sol Particles Having an Electric Double Layer. Elsevier, New York.

Wang, H. J., X. K. Xi, A. Kleinhammes and Y. Wu. 2008. Temperature-induced hydrophobic-hydrophilic transition observed by water adsorption. Science 322: 80–83.

Wenzel, R. N. 1936. Resistance of solid surfaces to wetting by water. Ind. Eng. Chem. 28: 988–994.

Widom, B. 1982. Potential-distribution theory and the statistical-mechanics of fluids. J. Phys. Chem. 86: 869–872.

Willard, A. P. and D. Chandler. 2008. The role of solvent fluctuations in hydrophobic assembly. J. Phys. Chem. B 112: 6187–6192.

Xu, L., F. Mallamace, Z. Yan, F. W. Starr, S. V. Buldyrev and H. E. Stanley. 2009. Appearance of a fractional Stokes-Einstein relation in water and a structural interpretation of its onset. Nature Physics 5: 565–569.

Yeom, J., B. Yeom, H. Chan, K. W. Smith, S. Dominguez-Medina, J. H. Bahng, G. Zhao, W. S. Chang, S. J. Chang, A. Chuvilin, D. Melnikau, A. L. Rogach. P. Zhang, S. Link, P. Králand N. A. Kotov. 2015. Chiral templating of self-assembling nanostructures by circularly polarized light. Nature Mat. 14: 66–72.

Yip, S. and M. P. Short. 2013. Multiscale materials modelling at the mesoscale. Nat. Mater. 12: 774–777.

Young, T. 1805. Philos. Trans. R. Soc. London, 95: 65.

Zhang, X., Y. Zhu and S. Granick. 2002. Hydrophobicity at a Janus interface. Science 295: 663–666.

CHAPTER 7

Multiscale Dynamics in Diluted Aqueous Solutions Revealed by Extended Depolarized Light Scattering

The Hydration Water Contribution

Lucia Comez,[1,*] *Silvia Corezzi*[2] and *Marco Paolantoni*[3]

Introduction

It is well known that water is essential for life on Earth. It not only stabilizes proteins, nucleic acids, and lipid membranes but also plays a vital role in many cellular processes. Its primary function is that performed in proximity of biomolecules, simply close to or even in the interstitial areas, by the so-called *hydration water*. Although our understanding of hydration water has advanced suggestively in recent years, it remains a challenge to clarify the extent to which solute molecules modify the behavior of nearby water molecules through interactions. A hierarchy of timescales for motions of biomolecules and their hydration environment has been established (Xu and Havenith 2015, Ball 2017, Laage et al. 2017, Schirò and Weik 2019), where fast rearrangements can be seen as precursors of very slow conformational ones. In all this, the relevant role of breaking and reforming water hydrogen bonds with other water and/or solute molecules occurring at the picosecond timescale stands out. The presence of solute, in fact, usually alters the fluctuating hydrogen bond network of water, and the nature of the solute and its amount in the mixture are decisive in determining the effects induced on the aqueous environment. It is possible to find some

[1] IOM-CNR c/o Dipartimento di Fisica e Geologia, Università di Perugia, Via Pascoli 06123 Perugia, Italy.
[2] Dipartimento di Fisica e Geologia, Università di Perugia, Via Pascoli 06123 Perugia, Italy.
[3] Dipartimento di Chimica, Biologia e Biotecnologie, Università di Perugia, Via Elce di Sotto, 06123 Perugia, Italy.
* Corresponding author: comez@iom.cnr.it

small polar solutes filling the space with moderately little solvent rearrangement, as well as small hydrophobic solutes that can be enclosed in pockets around which the surrounding water molecules preserve their hydrogen bond network, and even larger surfaces, such as those of proteins, where strong modifications of the hydrogen bond network certainly occur. Crucial for these events is the *dynamic role* played by water. Hydration is indeed a dynamic phenomenon where water molecules residing in the hydration shell continuously exchange their position with water molecules in the bulk. Literature reports, for example, that hydration water molecules at the solvent-exposed surface of proteins have residence times of several picoseconds before exchanging with the bulk (Ball 2017). The experimental challenge for scientists is therefore to distinguish the hydration from bulk water, and to characterize the hydration water molecules both in terms of their number and dynamics. It has been demonstrated that to obtain realistic estimates, the experimental technique should be fast enough to probe the dynamics of water molecules during their permanence within the hydration shell. From a spectroscopic point of view, this means that the relaxation processes revealed have to be faster than the characteristic time of exchange with the bulk (so-called slow exchange condition); otherwise, the obtained spectral signal results to be an average of the contributions of the two water (hydration and bulk) populations.

Among the multitude of available techniques, the Extended Depolarized Light Scattering (EDLS), that measures the fluctuations of the anisotropy polarizability tensor over the huge frequency range from fractions to several thousands of GHz, has been proven to be a broadband and fast technique, and hence to possess the fundamental requirements needed to reveal in a solution both the solute and solvent contributions and to distinguish the hydration water contribution from the bulk one. To illustrate the EDLS capabilities, a collection of results obtained by this technique on aqueous solutions is here reported. The intent is to contribute to shed light on the interaction between water and biomolecules, starting from simple molecules and then moving to increasingly complex systems. We focus on relatively diluted solutions where it is possible to better discriminate solute and solvent contributions over a broad timescale. At the end, some new perspectives for EDLS will be presented.

Background

The light scattering response measured in EDLS experiments is related to the relaxation of the polarizability anisotropy of the system and can be defined in terms of the time autocorrelation function (TCF) of an off-diagonal component Π_{xz} of the collective polarizability tensor, Π:

$$\Psi(t) = \langle \Pi_{xz}(0)\, \Pi_{xz}(t) \rangle / \Gamma^2 \tag{1}$$

where Γ^2 represents the depolarized light scattering intensity, that is, the mean squared polarizability anisotropy, of a non-interacting system containing the same number and types of molecules as the system of interest. It is given by $\Gamma^2 = N\gamma^2/15$, where $\gamma^2 = (x_S^2 \langle \gamma_S^2 \rangle + x_W^2 \gamma_W^2)$ is the square of the mixture's ideal-gas polarizability anisotropy (S and W indicate solute and water, respectively), with $\gamma_a^2 = [(\alpha_{11,a} - \alpha_{22,a})^2 + (\alpha_{11,a} - \alpha_{33,a})^2 + (\alpha_{22,a} - \alpha_{33,a})^2]/2$ and $\alpha_{jj,a}$ the principal polarizability components of a molecule of type a. In general, the solute molecules are flexible, so γ_S^2 needs to be averaged over molecular conformations.

From a computational point of view, the susceptibility spectra can be obtained from molecular dynamics (MD) simulations after calculating the OKE nuclear response function (Geiger and Ladanyi 1989, Paolantoni and Ladanyi 2002) R(t) defined as:

$$R(t) = -\frac{1}{k_B T} \frac{\partial}{\partial t} \Psi(t) \tag{2}$$

where $1/k_B T$ is the usual Boltzmann factor. The imaginary part of the susceptibility $\chi''(\omega)$ is obtained by performing a Fourier-Laplace transform of R(t):

$$\chi''(\omega) = \Im m[\chi(\omega)] = \int_0^\infty \sin(\omega t) R(t) dt \tag{3}$$

that is related to the susceptibility obtained from the experimental Raman spectrum $I_{HV}(\omega)$ by the relation $\chi''(\omega) = I_{HV}(\omega)/[n_B(\omega) + 1]$, where $n_B(\omega) = 1/[\exp(\hbar\omega/k_B T) - 1]$ is the Bose-Einstein occupation number (Paolantoni et al. 2009). The collective polarizability tensor, Π, for a condensed-phase system composed of N molecules can be written as a sum of molecular Π^M and induced Π^I contributions:

$$\Pi = \Pi^M + \Pi^I \tag{4}$$

where the molecular part is simply given by the sum of the intrinsic molecular polarizabilities α_i

$$\Pi^M = \sum_{i=1}^N \alpha_i. \tag{5}$$

The induced contribution, Π^I, originates from interactions between molecular-induced dipoles. It depends on orientational, conformational, and translational degrees of freedom, and it can be seen as a dynamical modulation of the polarizabilities. Moreover, for a system of two distinct species of molecules, like a solute (S) in water (W), the total molecular polarizability arises from the sum of the polarizability tensors of the two species, and can be expressed by

$$\Pi^M = \Pi^{M,W} + \Pi^{M,S} \tag{6}$$

whereas the interaction-induced polarizability also includes interactions between the two species:

$$\Pi^I = \Pi^{I,W} + \Pi^{I,S} + \Pi^{I,W-S} \tag{7}$$

Based on Eqs. 6 and 7, the TCF of the total collective polarizability, $\Psi_{TOT}(t)$, can be separated into three terms, $\Psi_W(t)$, $\Psi_S(t)$, and $\Psi_{W-S}(t)$, the first two representing the autocorrelations of W and S anisotropic polarizibilities, $\Pi_{xz}^W = \Pi_{xz}^{M,W} + \Pi_{xz}^{I,W}$ and $\Pi_{xz}^S = \Pi_{xz}^{M,S} + \Pi_{xz}^{I,S}$, while the third one is due to cross-correlations between Π_{xz}^W and Π_{xz}^S as well as from terms that include the interspecies interaction-induced anisotropic polarizability, $\Pi_{xz}^{I,W-S}$. Furthermore, based on Eq. 4, $\Psi^W(t)$, $\Psi^S(t)$, and $\Psi^{W-S}(t)$ can each be split into three terms, $\Psi^{MM}(t)$, $\Psi^{II}(t)$, and $\Psi^{MI}(t)$, respectively, arising from the autocorrelation of molecular and induced polarizabilities and the cross-term between them.

For some systems, like water-sugar mixtures, the comparison between MD and EDLS data allowed researchers to verify that in the spectral region of our interest (the water relaxation frequency range) the cross-correlation terms are not dominant (Lupi et al. 2012a,b). We consider this assumption as valid also for the other mixtures discussed hereafter. This is, of course, a crucial point and further computational and theoretical efforts, along with comparisons with experiments, would be desirable for more complex systems.

Experimental

The EDLS spectra are collected over a frequency range from fractions to several thousands of GHz. To cover this broad spectral range, the depolarized scattered light is analyzed by means of two different spectrometers. The low-frequency region, from 0.6 to 90 GHz, is acquired by means of a Sandercock-type (3 + 3)-pass tandem Fabry–Perot interferometer, characterized by a finesse of about 100 and a contrast higher than 10^{10}. In general, three different mirror separations, corresponding to different free spectral ranges, are used to obtain the depolarized spectra over the desired frequency range (Scarponi et al. 2017). The high-frequency region, from 60 to 36000 GHz (2–1200 cm^{-1}), is acquired by using a Jobin-Yvon U1000 double monochromator having 1 m focal length, and equipped with holographic gratings. The detection system is a thermoelectrically cooled Hamamatsu model 943XX photomultiplier. The scattered light is analyzed with a 90° scattering geometry in two different frequency regions: from –10 to 40 cm^{-1} with a resolution of 0.5 cm^{-1} and from 3 to 1200 cm^{-1} with a resolution of 1 cm^{-1}. More details are given in studies by Paolantoni et al. (2009), Perticaroli et al. (2011), and Lupi et al. (2012b). After subtraction of the background contribution, low- and high-frequency spectral signals are merged, exploiting an overlap of about half a decade in frequency. To display the spectra, the susceptibility formalism is typically adopted. To do this, the imaginary part of the dynamic susceptibility $\chi''(\omega)$ is calculated according to the relation $\chi''(\omega) = I_{HV}(\omega)/[n_B(\omega) + 1]$, as previously mentioned. In Fig. 1 a representative EDLS spectrum is reported as a function of frequency ($\nu = \omega/2\pi$) over a broad interval, after reconstruction of the entire profile.

Approach to the Data Analysis

The experimental technique EDLS allows one to investigate disparate dynamical events in diluted bio-solutions. A schematic list of processes occurring in the GHz-THz frequency range and probed by depolarized light scattering is illustrated in Fig. 1. On increasing frequency, phenomena of diffusional rotation of solute molecules, relaxation processes due to the rearrangement of the water H-bond network, and intermolecular vibrational processes take part. The different processes, including solute and solvent molecular rearrangements, cannot be completely explored by using a single spectroscopic technique. EDLS, that couples two instruments, overcomes the limitation of a narrow frequency window and represents a powerful tool for studying the molecular dynamics of aqueous solutions. In the following, we discuss how to analyze and interpret EDLS spectra based on specific examples, by mainly focusing on the water relaxation components.

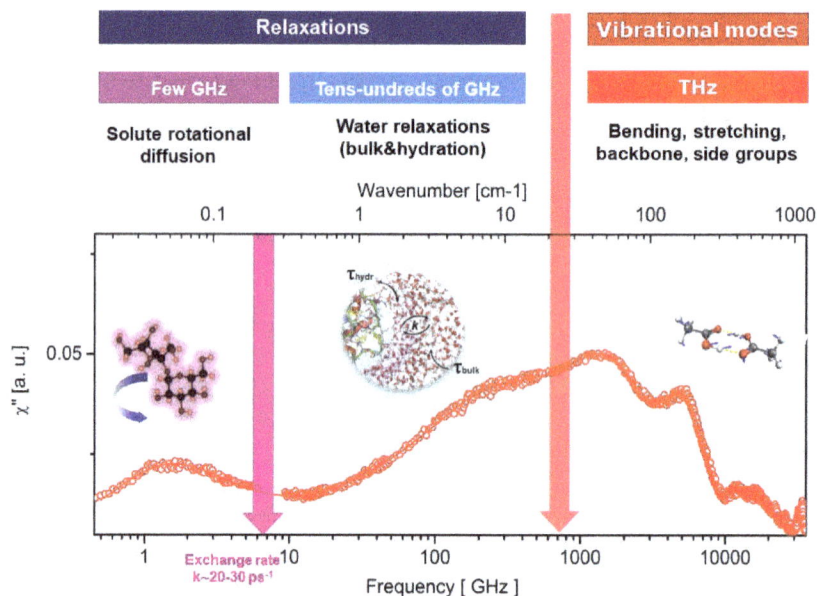

Fig. 1: Imaginary part of the susceptibility of a diluted water solution at ambient temperature, measured by EDLS. The main dynamical processes probed by depolarized light scattering are schematically sketched.

To compare different systems, Fig. 2 reports a set of EDLS susceptibility spectra of pure water and several aqueous solutions collected at ambient temperature. To be as exhaustive as possible, we are including spectra from small hydrophobic and hydrophilic systems, aminoacids, peptides, and proteins (Perticaroli 2010, Lupi 2012a,b, Comez 2013, 2016a), taken at relatively low and comparable concentration, namely 90 mg/ml for *tert*-butyl alcohol (TBA) and for trimethylamine N-oxide (TMAO), 130 mg/ml for lysine, 100 mg/ml for glucose, sucrose, N-acetyl-leucine-amide (NALA), and lysozyme water solutions. All the solutes we have selected are considered model systems to study relevant interactions in more complex materials, but they are also *per se* interesting due to their bioactive role. In particular, TMAO and TBA, having an analogous geometry, with the same hydrophobic part (consisting of three methyl groups) and a different hydrophilic head (a polar N^+O^- group and hydroxyl group, respectively), are among the most widely used model systems for studying the effects of hydrophobic solutes in water, such as hydrophobic hydration and hydrophobic interaction. They have also relevance as cosolutes in biological aqueous solutions: TMAO is a potent osmolyte, known for enhancing the thermodynamic stability of proteins and protecting them against denaturation. TBA is a denaturant, showing a destabilising effect upon the native conformation of proteins. Carbohydrates (like glucose and sucrose) in water can be used as relatively simple hydrophilic systems for investigating the modification imposed by an essentially homogeneous hydrogen bonding solute to the structure and dynamics of water molecules in the hydration layer. It is also known that they are extremely relevant for cryo- and bio-preservation processes. Proline and lysine constitute small parts or subunits of proteins, and NALA is an amphiphilic model peptide

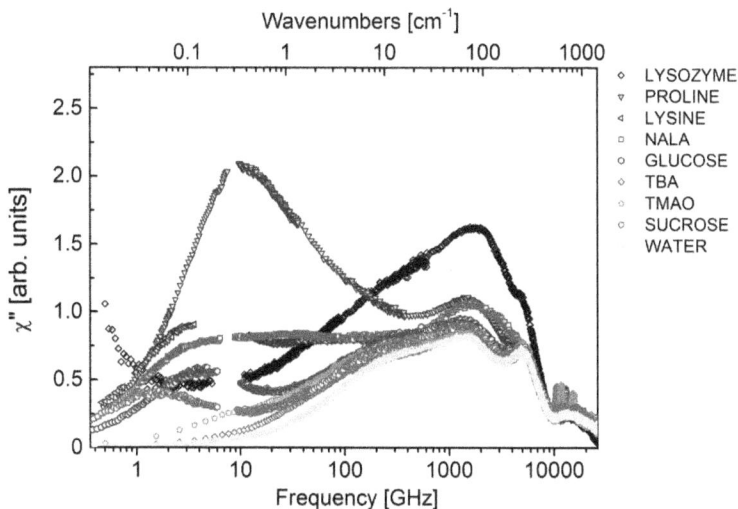

Fig. 2: Imaginary part of the susceptibility of pure water and water solutions at ambient temperature, measured by EDLS. Solutes are indicated in the legend, while the corresponding concentrations are mentioned in the text.

with a hydrophilic backbone and a hydrophobic side chain. These systems offer the possibility to overcome the difficulties arising from the complex surface of a protein, providing the opportunity of analyzing to what extent the perturbation induced by proteins is connected with the exposure of sites of different nature. Finally, lysozyme *is one* of the earliest *characterized* and *most studied enzymes, representing a model for globular proteins*. Looking at Fig. 2, we note that, in comparison with the profile of the solvent,strong differences can be identified in the spectra of the solutions over the whole investigated frequency range. Then, to highlight solute-induced spectral variations, solvent-free (SF) spectra are calculated by subtracting the spectrum of pure water from those of the solutions (Comez 2013). This is an easy practice that allows one to directly detect, without any curve-fitting procedure, the residual features remaining after the removal of the bulk water contribution. To give some examples, Fig. 3 represents the subtraction applied to a selection of systems. TMAO, sucrose, and NALA have been chosen as representatives of small hydrophobes, of hydrophilic systems, and of materials having both hydrophobic and hydrophilic portions, respectively. The left panels of Fig. 3 show, in detail, the total EDLS profiles together with the corresponding solvent contribution, and the calculated SF spectra obtained after a proper normalization of EDLS spectra of the solutions on the librational band of water, which is little perturbed by changes of temperature and solute concentration (Comez et al. 2013). Depending on the system, the SF signal appears to be more or less structured. In general, it includes a high frequency feature above 500 GHz, related to solute vibrational and/or librational modes (Perticaroli et al. 2010, 2011, Comez et al. 2013) and a second broad component, related to solute and solvent relaxation process modes (Perticaroli et al. 2011, Comez et al. 2013), which extend from fractions up to hundreds of GHz.

As regards the high-frequency side, it is evident that the SF spectrum of NALA, as well as of other peptides and amino acids (Comez et al. 2013), presents

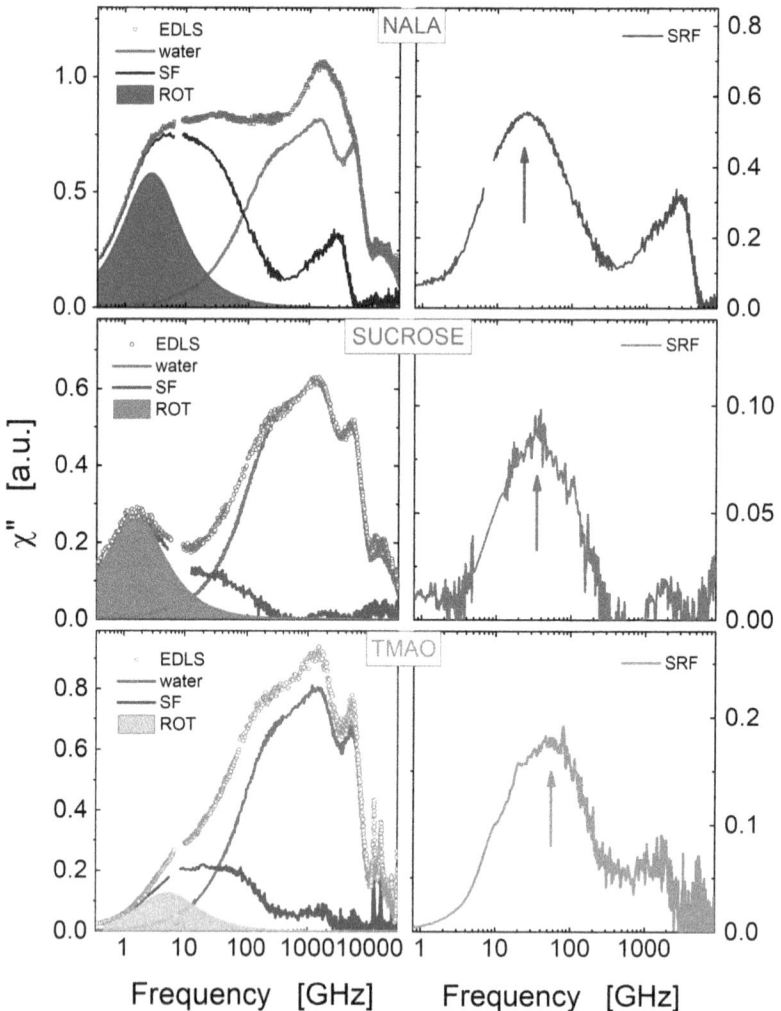

Fig. 3: Left panels: Imaginary part of the susceptibility of pure water and water solutions at ambient temperature, measured by EDLS with the corresponding SF profiles. From top to bottom, results are for NALA, sucrose, and TMAO water solutions. In each case, the rotational diffusion contribution of the solute is represented by a filled colored area. Right panels: The SRF contributions are calculated as described in (Comez 2013).

a THz region strongly modified by the presence of the solute, similarly to the case of proteins (Perticaroli et al. 2010, 2011). This feature arises from librations of methyl groups (Perticaroli et al. 2015) that are thought to play an important role as intrinsic plasticizers favoring protein dynamics, flexibility, and activity (Nickels et al. 2012). The vibrational features are less intense in TMAO (and TBA (Comez 2015, 2016a)), where in addition to the intermolecular vibrational Raman modes of water, there is a little residual contribution due to hindered rotations (librations) of solute molecules within the potential cage formed by the solvent (Comez et al. 2016a). Finally, it is evident that sucrose, and in general sugars, only marginally affect the 1–10 THz region detected by EDLS compared to systems that also have hydrophobic moieties (Comez et al. 2013). However, these residual signals have been analyzed upon changing temperature and/or concentration, thus obtaining precious information on the destructuring effect of carbohydrates on the hydrogen-bonded network of water (Paolantoni et al. 2008, Perticaroli et al. 2008, Rossi et al. 2011).

The complementary experimental window accessed by EDLS (from fraction to hundreds of GHz) is sensitive to the presence of relaxation processes. At very low frequencies, that is, below ~ 10 GHz, the spectrum is dominated by the rotational diffusion of solute molecules, and different intensities and frequencies of the associated peak account for dissimilar optical anisotropy, volume, and concentration of the solute molecules. At higher frequencies (10–100 GHz), the relaxation of perturbed water molecules comes into play, and we observe that while in sucrose it emerges as a distinct feature with respect to the solute rotational diffusion, in NALA and TMAO, the solute and solvent components merge into a single bump and cannot be visually distinguished. For systems that behave in this way, an additional step in the data analysis can be useful to recognize the spectral contribution of hydration water. It consists in the subtraction of the solute rotational contribution (colored area in Fig. 3), according to the method already adopted in Comez et al. (2013). The result of this operation, which we refer to as the solvent-rotation-free (SRF) spectrum (right panels of Fig. 3), is able to enhance the presence of the contribution located at several dozens of GHz, attributed to water molecules whose local translational motions are retarded by a factor of some units with respect to bulk water. A change in the frequency position of the maximum of this component (indicated by an arrow in the right panels of Fig. 3) is evident passing from TMAO to NALA: the more chemically complex the system the lower the peak frequency, indicating a higher hydration-to-bulk retardation factor. In particular, a progressive change from ~ 55 GHz (TMAO) to ~ 38 GHz (sucrose) and finally to ~ 30 GHz (NALA) can be observed, respectively corresponding to a retardation factor with respect to bulk water (relaxation at ~ 230 GHz) of about 4, 6, and 7.5.

Data Analysis

In order to get quantitative information about the dynamical processes probed by EDLS, the imaginary part of the susceptibility, $\chi''(\omega)$, is modeled with a phenomenological function able to reproduce the whole spectral profile. It is in general composed of three parts:

$$\chi''(\omega) = \chi''_{SR}(\omega) + \chi''_{WR}(\omega) + \chi''_{VIB}(\omega) \tag{9}$$

where:

(i) $\chi''_{SR}(\omega) = \Im m \left\{ -\dfrac{\Delta_{ROT}}{[1 + i\omega\tau_{ROT}]} \right\}$ (9a)

(ii) $\chi''_{WR}(\omega) = \Im m \left\{ -\dfrac{\Delta_{hydr}}{[1 + i\omega\tau_{hydr}]^{\beta_{hydr}}} - \dfrac{\Delta_{bulk}}{[1 + i\omega\tau_{bulk}]^{\beta_{bulk}}} \right\}$ (9b)

(iii) $\chi''_{VIB}(\omega) = \Im m \left\{ \dfrac{\Delta_b \omega_b^2}{\omega^2 - \omega_b^2 - i\omega\Gamma_b} + \dfrac{\Delta_s \omega_s^2}{\omega^2 - \omega_s^2 - i\omega\Gamma_s} + \sum_{i=1}^{3} BO_i(\omega) \right\}$ (9c)

(i) The first term, $\chi''_{SR}(\omega)$, related to the rotational diffusion of the solute, consists of a Debye function, where Δ_{ROT} is the amplitude and τ_{ROT} the characteristic time of the diffusion process (Comez et al. 2016a,b, Perticaroli et al. 2018). In the case of proteins, this relaxation is very slow due to the big size of the biomacromolecule, and its contribution falls outside the EDLS frequency window, apart from a tail that is suitably reproduced by a power law (Perticaroli et al. 2010, Corezzi et al. 2019).

(ii) The second term, $\chi''_{WR}(\omega)$, describes the water relaxation processes and is given by the sum of two Cole–Davidson (CD) functions (Paolantoni et al. 2009). The fit parameters $\Delta_{hydr}, \tau_{hydr}, \beta_{hydr}$, and $\Delta_{bulk}, \tau_{bulk}, \beta_{bulk}$ are, respectively, the amplitude, characteristic time, and shape parameter of the hydration and bulk relaxation processe. The use of a CD function to describe the anisotropy fluctuations of bulk water in the frequency domain (corresponding to a stretched exponential in the time domain) is well documented in the literature (Torre et al. 2004). By analogy, the same phenomenological function is employed to describe the relaxation of hydration water. Accordingly, the average relaxation times $\langle \tau_{hydr} \rangle = \beta_{hydr} \tau_{hydr}$ and $\langle \tau_{bulk} \rangle = \beta_{bulk} \tau_{bulk}$ are obtained, respectively, for hydration and bulk water. In order to minimize the number of free parameters in the fit procedure, β_{hydr} and β_{bulk} are fixed to the value 0.6, the same as obtained in pure water (Torre et al. 2004).

(iii) The third term, $\chi''_{VIB}(\omega)$, describes the vibrational part of the EDLS spectrum. In pure water as well as in the simplest solutions of small hydrophilic and hydrophobic molecules (sugars, TBA, and TMAO), it is mainly due to the H-bond intermolecular bending (at about 1.5 THz) stretching (at about 5.1 THz) Raman modes, which are well reproduced by two damped harmonic oscillator (DHO) functions (Walrafen et al. 1986, 1996). On the contrary, in more complex systems other terms must be included in the fitting function to take into account additional vibrational modes of the backbone and side groups. In particular, depending on the system, two or three Brownian oscillator (BO) functions are used; the fit procedure for peptides and proteins is described in detail in Perticaroli et al. (2010) and Corezzi et al. (2019).

The global fit curve with its individual components [Eq. 9a–9c], obtained from the full-spectrum data analysis of NALA, sucrose, and TMAO aqueous solutions, are shown in Fig. 4, demonstrating the good agreement between the model function and experimental data. In this chapter, we will focus our attention on the water relaxation processes, referring elsewhere for a detailed discussion of the other spectral components (Perticaroli et al. 2010, 2015). To highlight the water relaxation contributions within the EDLS spectra, the two CD functions are shown with colored areas. In the next paragraphs, we briefly discuss the main results for the systems presented in Fig. 2.

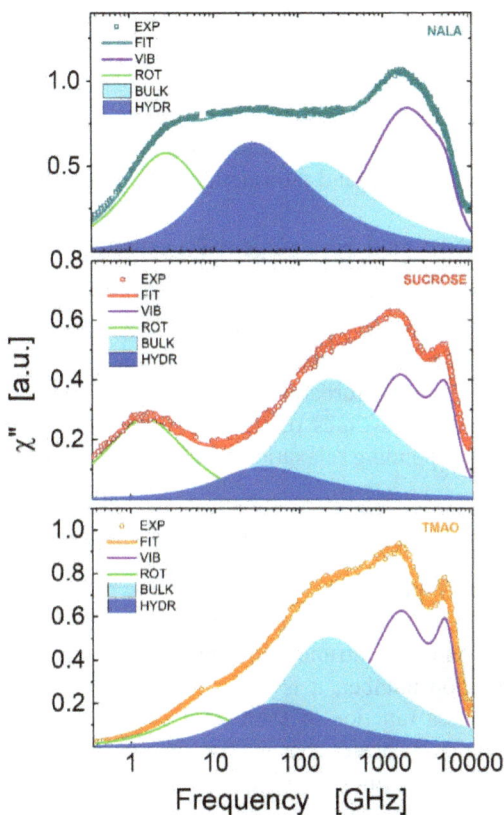

Fig. 4: Susceptibility of NALA, sucrose, and TMAO-water solutions at ambient temperature. Experimental data (symbols) are reported together with the total fitting curve (Eq. 9) and its individual components: rotational diffusion of the solute (green); relaxation of hydration (blue) and bulk (cyan) water; vibrational modes (violet).

Hydration and Bulk Water Relaxations

The ability of EDLS to simultaneously provide the dynamical retardation factor and the number of water molecules perturbed by the solute relies on the assumption that the relaxational part of the spectrum can be described as the sum of separate contributions coming from solute, bulk, and hydration water. This basic assumption has been tested by MD simulations and successfully verified in aqueous solutions of carbohydrates (Lupi et al. 2012a,b), and effectively applied to other small molecules (Corezzi et al. 2014, Comez et al. 2015). We recall that to experimentally discern two separate water contributions the probing technique has to be faster than the time needed for water molecules to exchange between bulk and hydration environments.

Exchange times longer than tens of picoseconds can be inferred for the solutes that we have analyzed, demonstrating that EDLS reasonably fulfills the required "slow exchange" condition (Comez et al. 2015, 2016a,b). Moreover, it should be noted that for the water relaxation processes (10–100 GHz) the most important contribution to the scattering cross-section arises from fluctuations of intermolecular distances triggered by dipole-induced dipole effects (Elola and Ladanyi 2007) and therefore, the obtained two relaxation processes get their strength from local translations of water molecules. The corresponding characteristic times are in the picosecond scale, matching those of the breakage and reformation of hydrogen bonds.

Retardation Factor and Hydration Number

From a quantitative point of view, the fitting results for the water relaxation components obtained from Eq. (9b) provide the two relevant parameters we are interested in: the dynamical retardation factor, ξ, and the hydration number, N_H (providing the extent of the perturbation induced by the solute on water molecules). From the average relaxation times of the two water contributions and from the amplitudes of the corresponding relaxation processes, ξ and N_H can be evaluated as:

(1) $\xi = \langle \tau_{hydr} \rangle / \langle \tau_{bulk} \rangle$

(2) $N_H = f_S^{-1} \, \Delta_{hydr} / (\Delta_{hydr} + \Delta_{bulk})$, with f_s as the solute molar ratio.

Concerning the first point, it is worth mentioning that $\langle \tau_{bulk} \rangle$ for all the systems reported here is found to be very similar to the value of pure water. This allows us to assign the corresponding contribution to free water molecules (i.e., bulk water). Regarding the hydration number, it is important to underline that N_H manifests a decreasing behavior as a function of solute concentration. An explanation for this behavior can be found in the water-sharing phenomenon between hydration shells of close-to-contact solute molecules. To give a quantitative estimation of this effect, a simple numerical model, representing a solute molecule as a sphere (whose radius c is provided by an independent estimate) with an effective hydration shell of constant thickness k, has been applied for different values of the solute molar ratio (Fioretto et al. 2013). The method is based on the generation of random distributions of solute molecules in water, starting from a box only filled with water (27,000 molecules) at the appropriate density, in which some water molecules are replaced by solute molecules randomly placed, at the desired molar ratio. Periodic boundary conditions are used. For each resulting configuration, the number of hydration molecules is evaluated as the total number of water molecules that fall within a distance k from the surface of any solute molecule. This procedure provides a dependence of the hydration number on the solute concentration and allows us to determine its value in the infinite dilution limit, N_0. This information is extremely useful to (i) determine the spatial range over which the dynamical perturbation induced by the solute extends, and to (ii) identify the presence of any aggregation phenomena. Interestingly, ξ and N_H obtained from EDLS by the full-spectrum analysis can be distinguished by a sample's category. Then, we hereafter present a short summary for the selection of systems we made, divided by categories, and emphasizing the generality traits within each of them. A final résumé is reported in Table 1.

Table 1: Resuming table of the hydration properties of solutes in Fig. 2, in terms of temporal and spatial extent of the perturbation induced in the surrounding water molecules.

	MOLECULAR STRUCTURE	MOLECULAR WEIGHT (Da)	RETARDATION FACTOR	SPATIAL EXTENT
TBA		74.12	4-5	1 shell
TMAO		75.11	4-5	1 shell
GLUCOSE		180.16	5-6	< 1 shell
SUCROSE		342.3	5-6	< 1 shell
PROLINE		115.13	7-8	2 shells
LYSINE		146.19	7-8	2 shells
NALA		172.23	7-8	2 shells
LYSOZYME		14307	7-8	3 shells

Small Hydrophobes

Consistently with the previous estimate from Fig. 3, the retardation factor obtained from the fit procedure performed on the total EDLS spectrum for the TMAO-water solution is $\xi = 4.2 \pm 0.6$ (Fig. 5). This value is in agreement with the result obtained for TBA, that is, $\xi = 4.0 \pm 0.8$, and it turns out to be concentration independent within experimental error (Comez et al. 2015, 2016a). Therefore, although the calculated ξ represents an average value over the hydration shell, without being able to distinguish whether the slowdown is uniform at the solute–solvent interface, the common value of ξ for TMAO and TBA suggests that their similar hydrophobic portion could be mainly responsible for the observed dynamical effect on hydration water.

The solvent-sharing model above mentioned helps us to establish how extensive the solute-induced perturbation of the surrounding water molecules is. Following this reasoning, N_H values as obtained from the fitting procedure on TMAO and TBA solutions at different mole fractions are compared with those calculated with the solvent-sharing model; the values of c and k used in the numerical evaluation are given in (Comez et al. 2015, 2016a). The results obtained in diluted conditions indicate that the dynamical perturbation induced by TMAO and TBA falls within

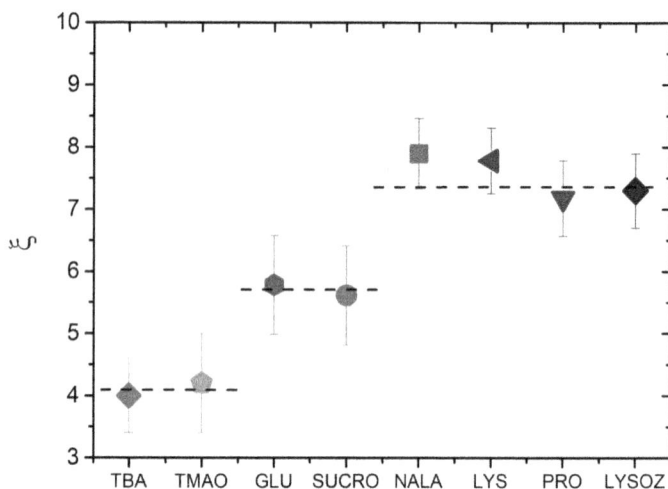

Fig. 5: Retardation factor of hydration water with respect to bulk water for different aqueous solutions. Solutes are TBA, TMAO, glucose, sucrose, NALA, lysine, proline, and lysozyme.

the first hydration layer. Comparable values of hydration number are found by other numerical and experimental studies (Di Michele et al. 2006, Paul and Patey 2006, Bakulin et al. 2011), thus supporting the numerical procedure behind the model. However, whereas in TMAO, the good agreement between numerical and experimental data suggests that the overall decrease of N_H with concentration can be just accounted for by the random close-to-contact events of solute molecules, without the need to invoke aggregation phenomena between them (Comez et al. 2016a), in TBA a definitely steeper behavior of N_H vs f_s has been found by EDLS, which cannot be entirely explained in terms of random overlapping of hydration shells, thus indicating a larger propensity of TBA to self-aggregate (Comez et al. 2015). This difference in tendency to associate is suggested to be at the basis of the different role played by TMAO and TBA as cosolutes in biological aqueous solutions.

Hydrophilic Systems

In this case also, the result of the qualitative estimate of the retardation factor based on Fig. 3 is confirmed from the fit procedure for the sucrose-water solution (Fig. 4, central panel), that gives $\xi = 5.6 \pm 0.9$. Remarkably, this finding is also confirmed for other sugars investigated by EDLS, for which ξ values within 5–6 are obtained (Lupi et al. 2012a,b, 2013). The whole set of results about ξ, for different monosaccharides (glucose, fructose) and disaccharides (sucrose and trehalose) measured at ambient temperature and comparable mixture composition, are shown in Fig. 5. In fact, this finding is even more general since the estimated retardation factor is found to be almost temperature and concentration-independent for all the measured samples.

Concerning the number of water molecules involved in the hydration process as probed by EDLS, we found that a systematic reduction of N_H for increasing the solute concentration was also observed for sugars. In this case, MD simulations performed for glucose and trehalose solutions (at a specific temperature and several concentrations), allowed us to define the geometrical condition according to which the simulated hydration numbers match the experimental ones. The agreement was obtained considering those water molecules whose oxygen (O_w) is within a distance $r = 3.6 \pm 0.1$ Å from the oxygen (O_s) of a solute's hydroxyl group (OH_s). Notably, the same distance criterion was found for both glucose and trehalose. This O_w–O_s distance was close to the value of the first minimum in the O_w–O_s radial distribution function of water-sugar solutions (Lee et al. 2005). Similar concentration behavior of N_H is retrieved if one selects water molecules with O_w distance smaller than 3.1 Å from any solute atom. Since, both in glucose and trehalose, hydroxyl groups are almost uniformly distributed all around the surface of the molecule, it was natural that the two definitions of hydration water provided the same results either way. Joining MD and EDLS data demonstrates that sugar molecules influence the picosecond water mobility at relatively short distances (3–4 Å): a similar short-range dynamical effect,essentially limited to the first hydration layer, was found independently of the sugar size. Furthermore, from the hydration numbers calculated in the infinite dilution limit for trehalose ($N_H = 25$) and glucose ($N_H = 16$), a ratio very close to 8/5 (ratio between the numbers of hydroxyl groups of the two sugar molecules) was obtained, and then an average value of about 3.3 water molecules per OH_s was deduced to be dynamically retarded for both systems in dilute condition. In line with these observations, it was found that by plotting the normalized hydration numbers N_H/n as a function of the OH_s/water mole ratio $f_H = nf_s$, where n was the number of OH_s groups for each sugar molecule, all the data collapsed into a single master plot, suggesting that for sugars, the number of hydroxyl groups is, indeed, the relevant parameter for the hydration process. Finally, by applying the solvent-sharing model, it was found that the calculated N_H decreases with a minor slope as a function of concentration in comparison with both experimental and MD counterparts, indicating that the random overlapping of hydration shells is not sufficient to fully account for the observed effect. Thus, specific intermolecular interactions, which favor solute aggregation at increasing concentrations, also contribute to reducing the effective values of the hydration numbers. The presence of aggregation phenomena in sugars was also reported by other techniques, confirming our results.

Amino Acids and Peptides

Passing from TBA, TMAO, and sugars to diluted aqueous solutions of amino acids and dipeptides, namely proline (PRO), lysine (LYS), and NALA, we found retardation factors of about $\xi \cong 7$–8. Again, this result is general for the whole class of systems, weakly dependent on temperature and concentration. At the selected concentration of about 100 mg/ml (solute/solvent), the perturbation induced by the solute affected ~ 120–200 surrounding water molecules, corresponding to a spatial extension that is twice that of sugars. However, to have more precise information, EDLS data as a function of the solute concentration must be considered. We resume, as a case study, the results for the homologue N-acetyl leucine-methylamide (NALMA) aqueous solutions for which lots of data are available (Perticaroli et al. 2011, Comez et al. 2014). The focus of the discussion regards the behavior of the hydration number as a function of f_s. Experimental data are considered up to very low concentrations and the comparison was made with the results of the solvent-sharing model, as explained for the other systems. At the extremal dilution (a single solute molecule), the model gives $N_0 = 130 \pm 20$, in good agreement with the EDLS fitting procedure, giving $N_0 = 150 \pm 40$ for the most diluted solution we measured (12 mg/ml). The thickness of the water shell perturbed around a single peptide was found to be $k = 6.2$ Å, confirming that the disturbance on water molecules extends beyond the first two shells, and suggesting a longer-range effect compared to small hydrophobic or hydrophilic molecules. It is also evident that this strong effect is not a matter of size of the solute because amino acids and peptides are smaller than disaccharides. On the other hand, since EDLS reveals the collective part of the anisotropic polarizability tensor, it is reasonable to suppose that such an effect cannot simply be due to an addition of the effects related to hydrophilic and hydrophobic moieties of the molecules. A greater chemical complexity seems to be the right *key* to explain the strong influence of amino acids and peptides on the hydration water, and computational simulations reproducing the EDLS susceptibility of this kind of systems could be of enormous help in understanding the experimental results.

Proteins

The representative spectrum of a water-lysozyme solution (100 mg/ml) is reported in Fig. 2. The low frequency rise of the spectrum, due to the protein structural relaxation, is clearly visible. The frequency behavior of this contribution suitably compares with inelastic neutron scattering (INS) data, and is well reproduced by a power law, $\propto v^{-0.3}$ (Roh et al. 2006, Khodadadi et al. 2008, Perticaroli et al. 2010). The signal in the THz region, instead, is rather complex, including both solvent and solute contributions. The part arising from solute vibrations can be reasonably decomposed into three components, modeled with as many Brownian oscillator (BO) functions (see Eq. 9c), associated with librational motions of solvent-exposed side chains and to backbone torsions (Giraud et al. 2003, Perticaroli et al. 2010, Turton et al. 2014).

In particular, the lower frequency component is the boson peak and comes from collective vibrations involving a great number of atoms distributed through the whole protein (Marconi et al. 2008, Perticaroli et al. 2015). In the case of such complex systems, the modeling of the solute contribution with the help of complementary techniques is fundamental to extract the contribution of water dynamics. For a more detailed description of the fitting method, one can refer to Perticaroli et al. (2010). For the purpose of this work, it is important to emphasize that despite the complexity of the water-protein solution, EDLS spectra are compatible with the two water populations (hydration and bulk) treatment, the fast component being surprisingly similar to that of pure water (as in the other solutions), and the retardation factor ξ being about 6–8 as a function of temperature and concentration. Concerning the hydration number, the value in the high dilution limit is found to be around 3000. This means that the perturbation of water dynamics revealed by EDLS extends over a distance of 12 ± 1 Å beyond the surface of the protein. Logically, such a great spatial extent cannot be explained by the mere superposition of the effects induced by the individual molecular groups constituting the protein (Comez et al. 2013). However, EDLS results contribute, together with those obtained by different techniques sensitive to the collective dynamics of water such as THz, Optical Kerr effect, dielectric, and neutron spectroscopies (Orecchini et al. 2009, Hunt et al. 2012, Mazur et al. 2012, Comez et al. 2016b) to unveil complex aspects of the dynamics of water interacting with biomolecules, which are still demanding for a complete theoretical explanation. Unfortunately, in the case of peptide and protein solutions, simulations are still unable to reproduce all the spectroscopic signatures as they appear in EDLS spectra (Martin et al. 2014). There is, therefore, huge computational work required on this front in order to entirely understand the result found experimentally.

Finally, very recently, EDLS has been employed to study ternary solutions relevant for biopreservation (Corezzi et al. 2019, Corezzi et al. 2021). The measurements have been carried out on a lysozyme-trehalose aqueous solution at the same solute composition as that previously studied by computer simulations (Corradini et al. 2013). The sugar concentration (> 1 M) is in the range in which trehalose is found to be particularly effective in protecting biomolecules. The ternary mixture has been investigated on cooling, providing experimental evidence that the sugar strongly modifies the solvation properties of the protein. The comparison of aqueous solutions of lysozyme with and without trehalose showed that the combined action of sugar and protein produces an exceptional dynamic slowdown of a fraction of water molecules around the protein. These waters are dynamically retarded more than twice with respect to those in the binary solution, and their dynamics gets slower and slower upon cooling. Overall, three timescales have been observed, similarly to what found in computer simulations: the smallest one refers to the relaxation of bulk water (revealed also in the water-lysozyme (WL) and water-trehalose (WT) solutions, and in pure water), the intermediate one is associated with the hydration water detected in binary solutions (WT and WL), and the longest timescale is related to the ultraslow water in the ternary mixture. Remarkably, these results validate the picture derived from numerical simulations and experimentally support the view in which trehalose induces an enormous modification of the hydration of lysozyme by forming a shell that cages and strongly slows down some water molecules close to the protein surface. This has allowed us to suppose that such ultraslow water is likely to be involved in the mechanism of bioprotection. Through this research EDLS has proved to be also a powerful tool to investigate molecular mechanisms responsible for the effectiveness of sugars in preserving proteins, either by reducing temperature (cryopreservation) or by dehydrating through freeze-drying procedures.

Conclusions and Perspectives

In this chapter, we have reported on the results of EDLS experiments on water solutions of small hydrophobics, sugars, amino acids, peptides, and proteins. The wealth of results testifies to the ability of EDLS to probe the hydration dynamics, by providing a simultaneous estimate of the retardation factor between hydration and bulk water and of the average number of water molecules whose dynamics is affected by the solute (hydration number), and also by revealing the possible presence of aggregation processes. The light-scattering response measured in EDLS experiments indicates that the perturbation induced by a biomacromolecule on the surrounding water molecules cannot be traced back to the relaxation of polarizability anisotropy of its small parts. In fact, on increasing the complexity of the solute molecule, new properties of the hydration shell develop, which are not explained by a simple superposition of the effects induced by its constituent moieties. Both the retardation factor and the hydration number are found to heavily depend on the solute chemistry. These experimental findings require further theoretical studies to get a deeper understanding of the mutual dependence of the properties of water and biomolecule mixed together in a solution.

As a future perspective for EDLS, we would like to mention the new concept of fully scanning multimodal micro-spectroscopy for simultaneous detection of Brillouin and Raman light scattering (Scarponi et al. 2017). Recently, a μ-Brillouin-Raman set up, operating over an exceptionally wide spectral range with an unprecedented 150-dB contrast, and spatial resolution on a subcellular scale, has been realized and the first EDLS susceptibility spectrum of a cellular line of murine fibroblast, measured in the central region of the nucleus, has been obtained. This is the first step to assess the molecular properties of water inclusions in cells. We also believe that the upgraded technique, by virtue of the high contrast, which is especially important for the analysis of opaque or turbid media such as biomedical samples, may open the route for *in situ* analysis of different phenomena in material sciences and potentially for *in vivo* diagnoses of pathologies involving mechanical changes as well as altered structure and composition.

Acknowledgments

M.E. Gallina, D. Fioretto, L. Lupi, D. Matyushov, A. Morresi, A. Paciaroni, S. Perticaroli, P. Sassi, and F. Scarponi, are gratefully acknowledged for a number of discussions on the issues presented here. M.P. acknowledges funding by Centro Nazionale Trapianti (Project: "Indagini spettroscopiche di sistemi liposominali modello di membrana biologica, di soluzioni di proteine e di crioconservanti, per uno studio molecolare dei processi di crioconservazione").

References

Ball, P. 2017. Water is an active matrix of life for cell and molecular biology. Proc. Natl. Acad. Sci. USA 114: 13327–13335.

Bakulin. A. A., M. S. Pshenichnikov, H. J. Bakker and C. Petersen. 2011. Hydrophobic molecules slow down the hydrogen-bond dynamics of water. J. Phys. Chem. A 115: 1821–1829.

Comez, L., L. Lupi, A. Morresi, M. Paolantoni, P. Sassi and D. Fioretto. 2013. More is different: Experimental results on the effect of biomolecules on the dynamics of hydration water. J. Phys. Chem. Lett. 4: 1188–1192.

Comez, L., S. Perticaroli, M. Paolantoni, P. Sassi, S. Corezzi, A. Morresi and D. Fioretto. 2014. Concentration dependence of hydration water in a model peptide. Phys. Chem. Chem. Phys. 16: 12433–12440.

Comez, L., M. Paolantoni, L. Lupi, P. Sassi, S. Corezzi, A. Morresi and D. Fioretto. 2015. Hydrophobic hydration in water-tert-butyl alcohol solutions by extended depolarized light scattering. J. Phys. Chem. B. 119: 9236–9240.

Comez, L., M. Paolantoni, S. Corezzi, L. Lupi, P. Sassi, A. Morresi and D. Fioretto. 2016a. Aqueous solvation of amphiphilic molecules by extended depolarized light scattering: The case of trimethylamine-N-oxide. Phys. Chem. Chem. Phys. 18: 8881–8889.

Comez, L., M. Paolantoni, P. Sassi, S. Corezzi, A. Morresi and D. Fioretto. 2016b. Molecular properties of aqueous solutions: A focus on the structural dynamics of hydration water. Soft Matter 12: 5501–5514.

Corezzi, S., P. Sassi, M. Paolantoni, L. Comez, A. Morresi and D. Fioretto. 2014. Hydration and rotational diffusion of levoglucos an in aqueous solutions. J. Chem. Phys. 140: 184505–184514.

Corezzi, S., M. Paolantoni, P. Sassi, A. Morresi, D. Fioretto and L. Comez. 2019. Trehalose-induced slowdown of lysozyme hydration dynamics probed by EDLS spectroscopy. J. Chem. Phys. 151: 015101–015108.

Corezzi, S., B. Bracco, P. Sassi, Paolantoni, M. and L. Comez. 2021. Protein hydration in a bioprotecting mixture. Life., 11: 995.

Corradini, D., E. D. Strekalova, E. Stanley and P. Gallo. 2013. Microscopic mechanism of protein cryopreservation in an aqueous solution with trehalose. Sci. Rep. 3: 1218.

Di Michele, A., M. Freda, G. Onori, M. Paolantoni, A. Santucci and P. Sassi. 2006. Modulation of hydrophobic effect by cosolutes. J. Phys. Chem. B 110: 21077–21085.

Elola, M. D. and B. M. Ladanyi. 2007. Intermolecular polarizability dynamics of aqueous formamide liquid mixtures studied by molecular dynamics simulations. J. Chem. Phys. 126: 084504.

Fioretto, D., L. Comez, S. Corezzi, M. Paolantoni, P. Sassi and A. Morresi. 2013. Solvent sharing models for non-interacting solute molecules: The case of glucose and trehalose. Food Biophysics 8: 177–182.

Geiger, L. C. and B. M. Ladanyi. 1989. Molecular dynamics simulation study of nonlinear optical response of fluids. Chem. Phys. Lett. 159: 413–420.

Giraud, G., J. Karolin and K. Wynne. 2003. Low-frequency modes of peptides and globular proteins in solution observed by ultrafast OHD-RIKES Spectroscopy. Biophys. J. 85: 1903–1913.

Hunt, N. T., L. Kattner, R. P. Shanks and K. Wynne. 2012. The dynamics of water–protein interaction studied by ultrafast optical kerr-effect spectroscopy. J. Am. Chem. Soc. 129: 3168–3172.

Khodadadi, S., S. Pawlus and A. P. Sokolov. 2008. Influence of hydration on protein dynamics: Combining dielectric and neutron scattering spectroscopy data. J. Phys. Chem. B. 112: 14273–14280.

Laage, D., T. Elsaesser and J. T. Hynes. 2017. Water dynamics in the hydration shells of biomolecules. Chem. Rev. 117: 10694–10725.

Lee, S. L., P. G. Debenedetti and J. R. Errington. 2005. A computational study of hydration, solution structure, and dynamics in dilute carbohydrate solutions. J. Chem. Phys. 122: 204511.

Lupi, L., L. Comez, M. Paolantoni, D. Fioretto and B. M. Ladanyi. 2012a. Dynamics of biological water: Insights from molecular modeling of light scattering in aqueous trehalose solutions. J. Phys. Chem. B. 116: 7499–7508.

Lupi, L., L. Comez, M. Paolantoni, S. Perticaroli, P. Sassi, A. Morresi, B. M. Ladanyi and D. Fioretto. 2012b. Hydration and Aggregation in Mono- and Disaccharide Aqueous Solutions by Gigahertz-to-Terahertz Light Scattering and Molecular Dynamics Simulations. J. Phys. Chem. B. 116: 14760–14767.

Lupi, L., L. Comez, M. Paolantoni, P. Sassi, A. Morresi and D. Fioretto. 2013. Influence of sucrose on surrounding water by extended frequency range depolarized light scattering. pp. 47–60. *In*: Magazù, S. (ed.). Sucrose: Properties, Biosynthesis, and Health Implications. Nova Science Publishers, Inc.

Marconi, M., E. Cornicchi, G. Onori and A. Paciaroni. 2008. Comparative study of protein dynamics in hydrated powders and in solutions: A neutron scattering investigation. Chem. Phys. 345: 224–229.

Martin, D. R., D. Fioretto and D. V. Matyushov. 2014. Depolarized light scattering and dielectric response of a peptide dissolved in water. J. Chem. Phys. 104: 035101.

Mazur, K., I. A. Mazur and S. R. Meech. 2012. Water dynamics at protein interfaces: Ultrafast optical kerr effect study. J. Phys. Chem. A. 116: 2678–2685.

Nickels, J. D., J. E. Curtis, H. O'Neill and A. P. Sokolov. 2012. Role of methyl groups in dynamics and evolution of biomolecules. J. Biol. Phys. 38: 497–505.

Orecchini, A., A. Paciaroni, A. De Francesco, C. Petrillo and F. Sacchetti. 2009. Collective dynamics of protein hydration water by brillouin neutron spectroscopy. J. Am. Chem. Soc. 131: 4664–4669.

Paolantoni, M. and B. M. Ladanyi. 2002. Polarizability anisotropy relaxation in liquid ethanol: A molecular dynamics study. J. Chem. Phys. 117: 3856–3873.

Paolantoni, M., L. Comez, D. Fioretto, M. E. Gallina, A. Morresi, P. Sassi and F. Scarponi. 2008. Structural and dynamical properties of glucose aqueous solutions by depolarized Rayleigh scattering. J. Raman Spectrosc. 39: 238–243.

Paolantoni, M., L. Comez, M. E. Gallina, P. Sassi, F. Scarponi, D. Fioretto and A. Morresi. 2009. Light scattering spectra of water in trehalose aqueous solutions: Evidence for two different solvent relaxation processes. J. Phys. Chem. B 113: 7874–7878.

Paul, S. and G. N. Patey. 2006. Why tert-Butyl alcohol associates in aqueous solution but trimethylamine-n-oxide does not. J. Phys. Chem. B 110: 10514–10518.

Perticaroli, S., P. Sassi, A. Morresi and M. Paolantoni. 2008. Low-wavenumber Raman scattering from aqueous solutions of carbohydrates. J. Raman Spectrosc. 39: 227–232.

Perticaroli, S., L. Comez, M. Paolantoni, P. Sassi, L. Lupi, D. Fioretto, A. Paciaroni and A. Morresi. 2010. Broadband depolarized light scattering study of diluted protein aqueous solutions. J. Phys. Chem. B 114: 8262–8269.

Perticaroli, S., L. Comez, M. Paolantoni, P. Sassi, A. Morresi and D. Fioretto. 2011. Extended frequency range depolarized light scattering study of NALMA-water solutions. J. Am. Chem. Soc. 133: 12063–12068.

Perticaroli, S., D. Russo, M. Paolantoni, M. A. Gonzalez, P. Sassi, J. D. Nickles, L. Comez, E. Pellegrini, D. Fioretto and A. Morresi. 2015. Painting biological low-frequency vibrational modes from small peptides to proteins. Phys. Chem. Chem. Phys. 17: 11423–11431.

Perticaroli, S., L. Comez, P. Sassi, A. Morresi, D. Fioretto and M. Paolantoni. 2018. Water-like behavior of formamide: Jump reorientation probed byextended depolarized light scattering. J. Phys. Chem. Lett. 9: 120–125.

Roh, J. H., J. E. Curtis, S. Azzam, V. N. Novikov, I. Peral, Z. Chowdhuri, B. Gregory and A. P. Sokolov. 2006. Influence of hydration on the dynamics of lysozyme. Biophysical Journal 91: 2573–2588.

Rossi. B., L. Comez, L. Lupi, S. Caponi and F. Rossi. 2011. Vibrational properties of cyclodextrin–water solutions investigated by low-frequency raman scattering: Temperature and concentration effects. Food Biophysics 6: 227–232.

Scarponi, F., S. Mattana, S. Corezzi, S. Caponi, L. Comez, P. Sassi, A. Morresi, M. Paolantoni, L. Urbanelli, C. Emiliani, F. Palombo, J. Sandercock and D. Fioretto. 2017. High-performance versatile setup for simultaneous brillouin-raman microspectroscopy. Phys. Rev. X 7: 031015–031026.

Schirò, G. and M. Weik. 2019. Role of hydration water in the onset of protein structural dynamics. J. Phys. Condens. Matter 31: 463002.

Torre, R., P. Bartolini and R. Righini. 2004. Structural relaxation in supercooled water by time-resolved spectroscopy. Nature 428: 296–299.

Turton, D. A., H. M. Senn, T. Harwood, A. J. Lapthorn, E. M. Ellis and K. Wynne. 2014. Terahertz underdamped vibrational motion governs protein-ligand binding in solution. Nature Communications 5: 3999.

Walrafen, G. E., M. R. Fisher, M. S. Hokmabadi and W.-H. Yang. 1986. Temperature dependence of the low- and high-frequency Raman scattering from liquid water. J. Chem. Phys. 85: 6970.

Walrafen, G. E., Y. C. Chu and G. J. Piermarini. 1996. Low-frequency raman scattering from water at high pressures and high temperatures. J. Phys. Chem. 100: 10363–10372.

Xu, Y. and M. Havenith. 2015. Perspective: Watching low-frequency vibrations of water in biomolecular recognition by THz spectroscopy. J. Chem. Phys. 143: 170901–170908.

CHAPTER 8

Neutron Diffraction Studies of Water and Aqueous Solutions

Maria Antonietta Ricci

Introduction

It's river, sea, lake, pond, ice and more...
It's sweet, salty, brackish....
It's a place where you stay and travel. (Ηράκλειτος, 535 b.C. – 475 b.C)

Heraclitus' delicate words about water are suggestive of its fascination and ubiquity on Earth. Since water is ubiquitous, it has become the iconic liquid in the collective image, and we cannot imagine life without water. While the latter is probably true (Ball and Ben-Jacob 2014), and traces of water on other planets (Orosei et al. 2018) and in their atmosphere (Tsiaras et al. 2019) can be considered as hints of life, or life-friendly ambience; the idea that water is the iconic liquid contrasts with the thermodynamic properties of what the physicists call liquid. Several water behaviours are indeed anomalous, when compared with those of the physicists' iconic liquid, that is, Argon (Hansen and McDonald 2013). Nevertheless, we are so familiar with water properties and behaviours that its extravagance shows up to our eyes only when we look at it from a scientific perspective. At that point, we discover exactly the anomalies that make water so relevant to our life and ambient: it is well known, for instance, that the crystalline form of water at ambient pressure, namely ice Ih, is less dense than water. Ice allows life to continue beneath it, because it floats on the surface of lakes. The density anomaly is only the most evident anomaly of water; about 70 of them can be itemized. Why is water so anomalous? The easy answer is: because water molecules form intermolecular Hydrogen bonds (HB), resulting at ambient conditions in a continuous percolating network of molecules with a slightly distorted tetrahedral symmetry. Distorted tetrahedral bonds allow water molecules in the liquid phase to approach closer to one another, leading to a higher density with respect to ice. Water density increases as the temperature decreases up to a maximum

Dipartimento di Scienze, Universita' Roma Tre via della Vasca Navale 84, 00146 Roma.
Email: mariaantonietta.ricci@uniroma3.it

at T ≈ 280 K, then starts decreasing while tetrahedrality is recovered and eventually ice nucleates. However, this explanation of the anomalous density of water is not completely satisfactory, as it is a description of facts, not a model based on a solid understanding of the interactions causing water to behave as it does. As a matter of fact, although water has been and is the most studied liquid and we have plenty of experimental data and computer simulations of its properties, yet our understanding of this substance is inadequate and a matter of great debate. We mention here that the last quarter of the twentieth century has seen huge and unsuccessful research activity aimed at finding solid experimental evidence in favor of a clever water model proposed by Stanley and coworkers (Poole et al. 1992). Stanley's model proposes the existence of a second critical point of water (below T ≈ 220 K) to explain water anomalies, and in particular the divergence of its response functions. This point should be the extreme of the line of coexistence of two forms of liquid, the Low (LDL) and High (HDL) Density Liquid, respectively, originating from two amorphous polymorphes, the Low Density Amorphous (LDA) and High Density Amorphous (HDA), found at much lower temperatures (≈ 100 K) (Mishima et al. 1985). Notably, polymorphism is an interesting characteristic of water, particularly of its crystalline phase. Unfortunately, so far this model has not been experimentally verified, due to the difficulty of supercooling water well below 273 K, approaching the homogeneous nucleation barrier, which conceals the second critical point.

Given the central role of HBs in determining the properties of water, this chapter is dedicated to review what is known from neutron diffraction experiments about H-bonding in water and aqueous solutions. The structure of water confined within small volumes or in contact with solid surfaces is not discussed here, as this issue has been recently addressed in a similar review by A. K. Soper (Fernandez-Alonso and Price 2017).

Neutron Diffraction Experiments

Neutrons are a particularly suitable probe to investigate the structure of water and aqueous solutions, because, at odds with X-rays, they are strongly scattered by hydrogen and deuterium atoms and are able to distinguish between the two isotopes: they don't see only the water oxygens, and consequently are sensible to the relative positions and orientations of the water molecules.

The measured quantity in a neutron diffraction experiment is the Differential Cross-Section (DCS):

$$\frac{d\sigma}{d\Omega}(k,2\theta) = \sum c_\alpha \langle b_\alpha^2 \rangle S_\alpha^{self}(k,2\theta) + \sum c_\alpha c_\beta (2-\delta_{\alpha\beta})\langle b_\alpha \rangle \langle b_\alpha \rangle H_{\alpha\beta}(Q) \quad (1)$$

where $\delta_{\alpha\beta}$ is the Kroneker delta, k and Q are the incident and exchanged neutron wave vector, in the elastic approximation, 2θ is the scattering angle, and c_α, b_α the atomic fraction and neutron scattering length (Sears 1992) of the species α. S_α^{self} is the integral of the self contribution to the van Hove dynamic structure factor and does not bring information on the structure of the material, being a flat (in the ideal case) or smoothly decreasing profile. The structural information is embedded in the second addendum of Eq. (1), where $H_{\alpha\beta}(Q)$ is the integral of the distinct dynamical structure factor (Lovesey 1986, Fernandez-Alonso and Price 2017). Indeed, the $H_{\alpha\beta}(Q)$ are related by

Fourier transform to the site-site pair distribution functions (PDF), which are the basis of the statistical mechanical theory of liquids (Hansen and McDonald 2013):

$$H_{\alpha\beta}(Q) = 4\pi\rho \int r^2 \left[g_{\alpha\beta}(r) - 1 \right] \frac{\sin Qr}{Qr} \quad (2)$$

The $g_{\alpha\beta}(r)$ gives the density of probability of finding an atom β at distance r from the atom α at the origin of the reference frame. Thus, this is the primary quantity that we want to extract from a diffraction experiment. In the case of water, the DCS is a combination of three PDFs, which cannot be extracted from a single diffraction experiment. Here the neutron's ability of distinguishing hydrogen from deuterium comes through: since the scattering lengths of H and D are respectively $\langle b_H \rangle = -3.74$ fm, and $\langle b_D \rangle = +6.67$ fm, experiments performed on different isotopic mixtures of H_2O and D_2O will give markedly different DCS. In the case of pure liquid water, three experiments, usually performed on pure H_2O, pure D_2O, and a 50% mixture of the two, will be sufficient to extract the three $H_{\alpha\beta}(Q)$ and the radial distribution functions, namely OO, OH, and HH (Soper and Silver 1982), within the assumption that isotopic substitution does not sensibly affect the structure of the liquid. When solutes are present in the sample, the number of atomic pairs increases, and so does the number of independent experiments in theory required. Since in practice there are limits to the number of available isotopically substituted samples, direct Fourier transform of the data is not helpful. In these cases, the adopted strategy is to interpret the experiments via computer simulation techniques constrained by the diffraction data, as Reverse Monte Carlo (McGreevy and Pustzai 1988) or the Empirical Potential Structure Refinement (EPSR) (Soper 1996). For a discussion about differences and advantages of the two methods, the reader must refer to chapter 3 of Fernandez-Alonso and Price (2017). These simulation codes give access to a collection of molecular configurations compatible with the experimental data, which can be asked for calculating several structural properties of the sample, in addition to the PDFs. In particular, in this chapter we will show the distribution functions of the angles defined by selected atomic sites, $P(\theta)$, the distribution functions of molecular clusters, $P(n)$, and the spatial density functions (SDF) (Svishchev and Kusalik 1993). The latter represent the region of space around a given atomic site or group, where the probability of finding another atom, group, or molecule exceedes a threshold value and provide a nice and immediate three-dimensional image of the correlations between neighbouring molecules.

All the results discussed in the following have been obtained by using the EPSR code. Briefly, this method starts by equilibrating an ensemble of molecules representative of the real sample and interacting according to a site-site potential model (called reference potential), which catches the distinctive characteristics of the sample, as for instance intramolecular structure, intermolecular HBs and so on. From the molecular configurations, the code calculates the interference differential scattering cross-section for each i sample:

$$F_i(Q) = \sum c_\alpha c_\beta (2 - \delta_{\alpha\beta}) \langle b_\alpha \rangle \langle b_\alpha \rangle H_{\alpha\beta}(Q_e) \quad (3)$$

to be compared with the corresponding experimental data. The difference between the two is used as feedback correction to the potential model and the cycle is iterated until the best achievable fit is obtained, along with a correction to the reference potential, called empirical potential. This is a small correction at very short atomic distances, if the initial guess of the reference potential is sensible.

Pure Water at Ambient Conditions

The PDFs of pure water at ambient conditions are shown in Fig. 1A. Observation of the $O_w O_w$ function clearly evidences the presence of orientational correlation between water molecules: in fact, the O_w sites are almost coincident with the molecule centres of mass. Thus, if the molecules were randomly oriented, the second peak should be centered at twice the position of the first one, as in the case of simple liquids (Hansen and McDonald 2013). On the contrary, in water, the ratio between the positions of the second and first peak is $\approx \sqrt{8/3}$: this is the signature of a tetrahedral

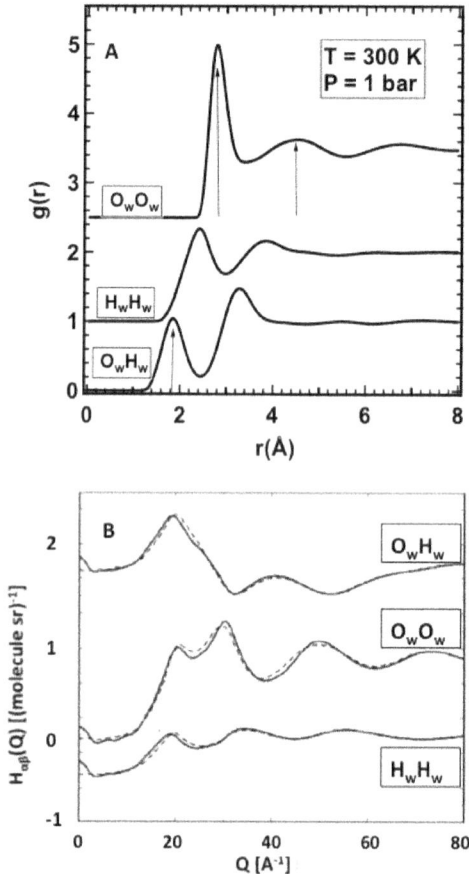

Fig. 1: A—The PDFs of water at ambient conditions (data made available by courtesy of A. K. Soper). The $O_w O_w$ and $H_w H_w$ functions have been offset for clarity. The arrows evidence the most relevant peaks of these functions, namely the HB peak of the $O_w H_w$ PDF and the two first peaks of the $O_w O_w$ function. B—The $H_{\alpha\beta}(Q)$ of supercooled water (solid lines), compared with those of ambient water (dashed lines), data from Botti et al. (2002). Data have been offset for clarity.

coordination. This is due to the presence of HB between first neighbouring molecules, evidenced by the first peak of the O_wH_w function at ≈ 1.8 Å, which is usually called the HB peak.

The number of β atoms within a given distance from an α atom can be easily evaluated from the $g_{\alpha\beta}(r)$ functions, as:

$$n_{\alpha\beta}(r_{min},r_{max}) = 4\pi\rho c_\beta \int_{r_{max}}^{r_{min}} r^2 g_{\alpha\beta}(r)dr \qquad (4)$$

where ρ is the sample density. If water molecules in the liquid state were tetrahedrally coordinated (as in ice I_h), Eq. 4 would give $n_{O_wO_w} = 4$. In liquid ambient water, we find instead a number close to 4.5, suggesting that there are interstitial molecules within the tetrahedra, which confer to water a higher density compared to ice, a result that is shared with many computer simulations of liquid water. The number of HBs, as evaluated by integrating the O_wH_w function under the first peak, is found of the order of 1.8, instead of 2, implying that each water molecule is on average H-bonded to 3.6 other molecules, not 4: thus about 10% of the HBs are broken in the liquid state. These inferences are confirmed by the distribution function of the angle formed by three neighboring oxygen atoms. In a perfect tetrahedal system, this should be peaked at $\approx 109°$; in ambient water instead this distribution is very broad and covers the range from 60° to 180°, with a maximum at the expected tetrahedral angle, plus a sharp peak at $\approx 60°$, corresponding to the interstitial molecules (Bernabei et al. 2008a,b, Soper 2013 and in Fernandez-Alonso and Price 2017, Chapter 3).

This picture is compatible with the existence of a percolating network of H bonds, covering all the available volume in the sample. According to a model proposed by Stanley (Stanley 1979) and corroborated by experimental and simulation tests (Stanley and Teixeira 1980, Geiger and Stanley 1982, Blumberg and Stanley 1984), the anomalies of water at low temperatures can be explained by the extension of this network. This makes water similar to a "transient gel", which continuously reconstructs itself on the picosecond time scale of the H bonds. Incidentally, the network allows reorientation of molecules and supports charge transfer via the tunneling of hydrogens along the HBs.

Pure Water At Low Temperatures

The most recent determination of the PSFs of pure supercooled water at ambient pressure is reported in (Botti et al. 2002). The level of supercooling is not exceptional (T = 267 K), due to the experimental requirements, nevertheless it has been possible to highlight clear differences of the three $H_{\alpha\beta}(Q)$ with those of ambient water (Fig. 1B). In particular, we notice that all peaks are down shifted in the supercooled state: this translates after Fourier transform into a sharpening of the first intermolecular peaks of the three PDFs (data not shown), implying a better definition of the first neighbours shell, without substantial changes of the orientational order. As a consequence, the thermal effect on the structure of water seems limited to the gradual freezing of the dynamical degrees of freedom.

Changes of pressure are instead much more effective in distorting the H-bonding network of water, as demonstrated by measurements performed at pressures varying from 260 bar to 4000 bar (Soper and Ricci 2000) at T = 268 K. This is the lowest temperature in the stable phase of water which allows a wide range of pressures to be examined and, interestingly, the thermodynamic range where dynamical anomalies of water are most evident: we mention as an example the behaviour of the coefficient of diffusion (Harris and Newitt 1997). At short distances, that is, within the first neighbouring shell, we observe only minor changes: the position of the first peaks of the three functions do not appreciably change while the peaks weakly broaden. Important changes are instead visible at the level of the second and third peaks. This suggests that the orientational correlations of water at the level of the first shell are robust, while pressure induces strong modifications of the second neighbouring shell. These are particularly evident in the O_wO_w PDF (Fig. 2A), where the second peak moves to shorter distances as pressure is applied and eventually merges with the first peak, as a shoulder at its right side.

Modifications of the relative orientations of water molecules with increasing pressure are also visible in Fig. 2B, where the maximum of the distribution function

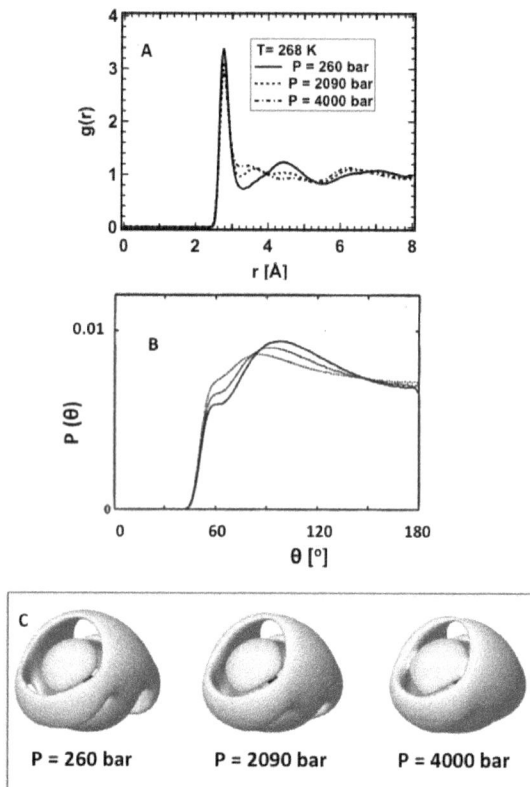

Fig. 2: A—The PDF of water oxygens at T = 268 K and pressure increasing from 260 bar to 4000 bar. B—Distribution function, $P(\theta)$, of the angle formed by three first neighbour oxygen sites at T = 268 K and three pressure states, namely: 260 bar (solid line), 2090 bar (dashed line), and 4000 bar (dotted line). C—The SDF at the same thermodinamic states (the pressure increases going from left to right): these show the regions where the probability of finding a water molecule in the first or second shell of the molecule at the origin of the reference frame is larger than ≈ 30% (Soper and Ricci 2000). We notice the progressive collapse of the second shell on the first one.

of the O_w-O_w-O_w moves to smaller values and the intensity of the secondary peak at 60° increases with increasing pressure. Panel C, in the same figure, reports the SDF of water molecules within the first and second shell of neighbours of a central molecule: these show that changes of the $O_w O_w$ PDF and of the $P(\theta)$ are compatible with a collapse of the second neighbouring shell on the first one. On the other hand, the first peak of the $O_w H_w$ PDF does not sensibly change upon application of pressure. Consequently, the average number of HB is stable, while the percolating HB network is distorted, but not broken by pressure.

Given the evolution of the PDFs with pressure, it is tempting to extrapolate the PDFs and SDFs of the two water polymorphs, LDL and HDL, if they exist (Soper and Ricci 2000). In doing so, we obtain the PDFs shown in Fig. 3A. The corresponding SDF (Fig. 3B and C) highlight the collapse of the second shell on the first neighbour one. We want to stress here that these results do not demonstrate the existence of two polymorphs of liquid water (LDW and HDW) and give only an insight of what these could be, if they exist, as the experiment has been done in the stable phase of water, at a temperature higher than that of the proposed second critical point.

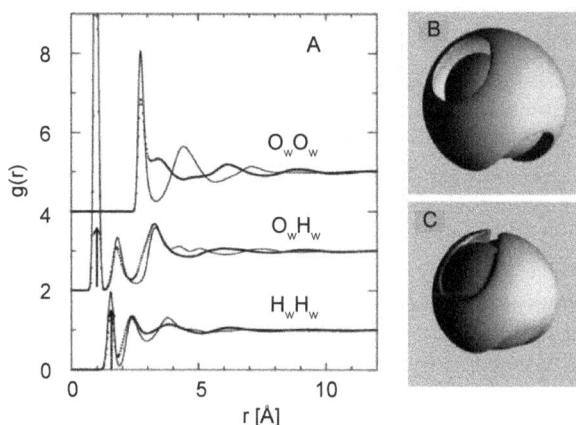

Fig. 3: A—Extrapolated PDFs of LDW (thin line) and HDW (thick line) (Soper and Ricci 2000). The arrows indicate the intramolecular peaks. B—SDF of LDW around a central water molecule. C—SDF of HDW around a central molecule.

Pure Water At Supercritical Temperatures

Supercritical water and aqueous solutions are important media for greenhouse chemistry and geological research (Franck 2000). The peculiar properties of supercritical water as a solvent, and in particular the tunable solubility of minerals and aromatic complexes at these states have several industrial applications and are of fundamental interest for research in geology. Pressure changes at constant temperature may indeed enable selective precipitation of solutes, opening perspectives for waste treatment, and possibly elucidating the origin and formation of rocks under the Earth's mantle. On the other hand, curiosity-driven research on water is based on the observation that while the percolating HB network of water at low temperatures is very robust, as demonstrated in previous sections, its characteristics may change at supercritical conditions. Indeed, since the breaking of HBs is an energy activated process (Conde and Teixeira 1984), one could expect that the network can be more easily broken by increasing the temperature. Is it true? At what temperature, may we expect that the network is substantially disrupted? Does the HB network percolate also at supercritical conditions?

This issue has been investigated by molecular dynamics simulations (Pártay and Jedlovszky 2005, 2007) and a percolation line, which separates two structurally different fluids, respectively below and above the so-called percolation threshold has been identified. This threshold is defined as the locus of thermodynamic states where the distribution of cluster size, $P(n)$, obeys a power law:

$$P(n) \propto n^\tau \tag{5}$$

where n is the cluster size and the exponent τ is 2.19 for three-dimensional systems (Stauffer 1985). For instance, at ambient conditions, the average number of HB per water molecule exceeds the critical value $n_c = 1.55$ and consequently, water is a fully percolating system (Blumberg1984). At higher temperatures, and particularly in the vicinity of the critical point, the application of the percolation model to water is justified by the persistence of a weak HB peak in the O_wH_w PDF (Bellissent-Funel et al. 1997, Botti et al. 1998, Tassaing et al. 1998), corresponding to a number of HB per water molecule ≥ 1.5 at all states investigated in ref. (Soper et al. 1997). This peak allows us to define a H-bonded pair, based on purely geometrical considerations. Consequently, neutron diffraction data of water at supercritical conditions have been analyzed within this scenario (Bernabei 2008).

Data have been collected on isotopic mixtures of H_2O and D_2O, at three state points in the liquid-like supercritical region and one in the gas-like region. The O_wH_w PDFs are shown in Fig. 4A, in order to highlight, through the HB peak, the persistence of H-bonding at all investigated states, and the clear gas-like shape of the PDF at T = 673 K and P = 250 bar. The other data have been collected at T = 673 K, P = 500 bar, and 1500 bar, and at T = 753 K, P = 2450 bar, giving a factor between 5 and 7 in density with respect to the gas phase, which incidentally has a density one order of magnitude lower than ambient water.

From the $P(n)$ functions reported in Fig. 4 clearly emerges that the water network percolates also at supercritical temperatures, provided that the fluid density is high enough to be considered liquid-like. These differences between liquid-like and gas-like supercritical water support the existence of a percolation line, which separates two fluids with distinct differences in their three-dimensional structure. In particular, the preferred tetrahedral arrangement of first neighboring molecules persists at liquid-like densities on one side of the percolation line,while on the other side, at gas-like densities, water molecules are coordinated in a triangular symmetry (data not shown, see Bernabei 2008). This implies a consistent decrease of the number of HBs per molecule, which goes from ≈ 2 to ≈ 0.8 upon the transition. Consequently, while liquid-like supercritical water can still be considered a three-dimensional percolating system, at gas-like densities,water molecules are isolated or organized in small sheet-like oligomers. Unfortunately, this structural change is accompanied by an increased corrosivity of water, which is the principal obstacle to its industrial applications.

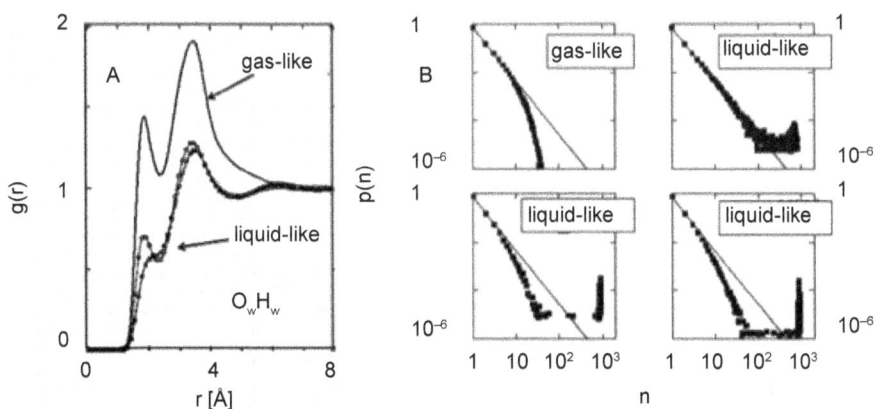

Fig. 4: A—O_wH_w PDFs at supercritical conditions. Data in the gas-like region are reported as solid line, those in the liquid-like ones as lines plus symbols. B—Distribution of the water cluster size at four supercritical states: only in the case of the gas-like fluid, the HB network does not percolate (Reprinted from Bernabei 2008).

Ionic Solution

Ion-water interactions are a central topic in chemistry and biology and the scientific community has addressed a long-lasting debate about this issue, particularly as far as the concepts of structure making/breaking ions is concerned (Jones and Dole 1929). This concept itself is somehow ambiguous, as an ion can be considered a structure maker, because it builds a structure around itself, and at the same time a structure breaker, because in doing so it breaks the structure of the solvent; nevertheless, this concept has been widely used in the past literature and is still used. A macroscopic measure of whether a solute is a structure maker or breaker is its effect on the viscosity of the solution compared to the viscosity of pure water: in this framework, solutes are said to be structure makers or structure breakers, if the viscosity of the solution increases or decreases, compared to that for pure water. The structure making/breaking character of ions has been correlated to the ability of ions to salting in/salting out proteins in solution (Hofmeister 1888, Jungwirth and Cremer 2014) and a ranking between the ions has been proposed.

The most direct test of the above hypothesis is obviously performed by neutron diffraction experiments, thus, since the pioneering work of Leberman and Soper (Leberman and Soper 1995), there is plenty of literature on this argument (Botti et al. 2004a,b, Imberti et al. 2005, Botti et al. 2006, McLainet al. 2006, Mancinelli et al. 2007a,b, Mancinelli et al. 2009, Corridoni et al. 2011, Winkel et al. 2011, Lenton et al. 2017). All these experiments confirm the strength of the HB: indeed, modifications of the HB-peaks are in all cases minimal, while the O_wO_w PSF shows severe modifications at the highest ion concentrations. Figure 5A reports, as a paradigmatic example, the case of solutions of NaCl and KCl at different concentrations (Mancinelli et al. 2007a,b). We notice that, with increasing solute concentration, the second peak of this function moves to shorter distances with respect to its position in pure water, as if an external pressure was applied to the liquid. Based on this observation, it is possible to define an effective pressure exerted by a specific ion

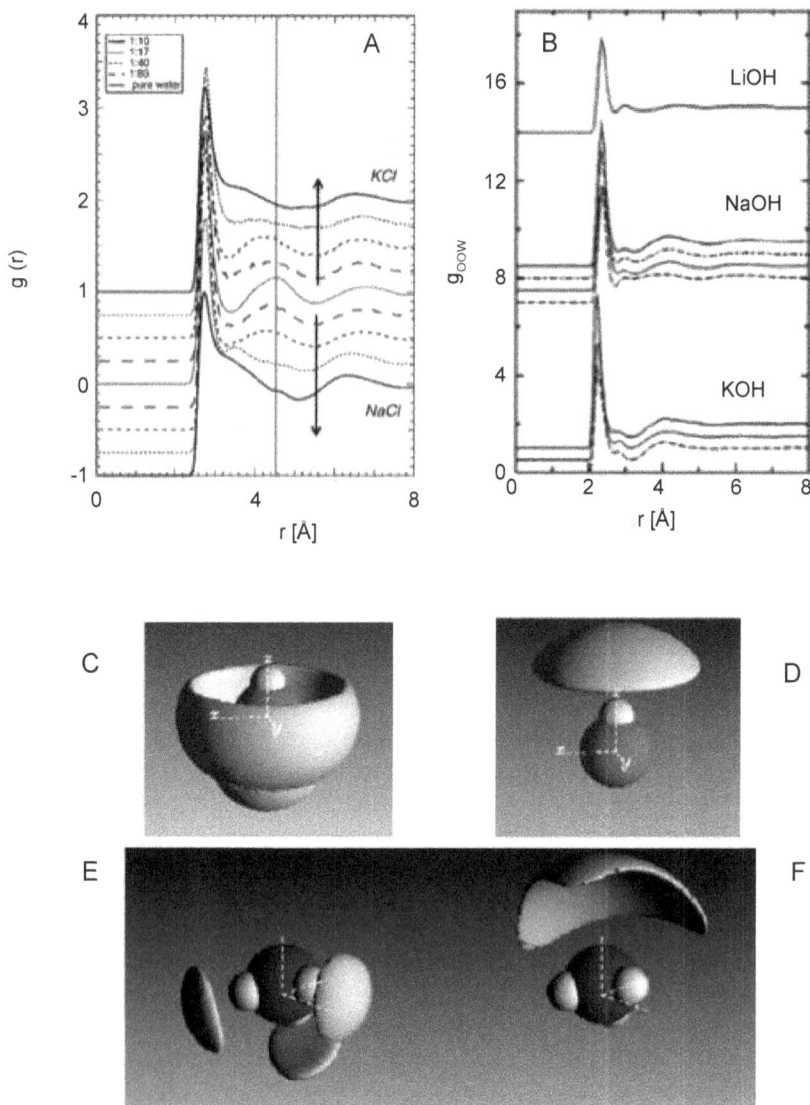

Fig. 5: A—The O_wO_w PDFs for aqueous solutions of NaCl and KCl at several concentrations (Mancinelli et al. 2007a,b). The case of pure water is represented as a solid line; the solute concentration increases along the direction of the arrows for the two salts. The vertical solid line indicates the position of the second peak of the PDF of pure water. B—The PDFs of hydroxyl oxygens (O) and water oxygens (Ow), for solutions of different hydroxides, namely LiOH, NaOH, and KOH, at different solute concentrations (for each solute the highest concentration is reported at the bottom of the group, as a long-dashed line, and the concentration decreases with increasing the offset). Notice the small peak, between the first and second hydration shell, due to the presence of the fifth neighbour molecule. C—SDF for water molecules hydrating the OH$^-$ ion at distance of $2.0 \leq r \leq 2.75$ Å from the oxygen. D—SDF for water molecules hydrating the OH$^-$ ion in the distance range $2.75 \leq r \leq 3.2$ Å from the oxygen, corresponding to 1 molecule facing the H site of the ion. E—SDFs for the hydronium ion in solution, showing the three regions of space with the highest probability of finding the H-bonded water molecules in the range of distances $2.0 \leq r \leq 2.8$ Å. F—SDFs for the hydronium ion in solution in the distance range $2.8 \leq r \leq 3.7$ Å, where the fourth neighbor molecule is sitting.

on water, depending on its concentration, and eventually define a ranking of ions as "structure breakers". As a matter of fact, if we look at the effects of ions on the water structure, all ions, independently of their charge (positive or negative), are structure breakers, although the average number of HB per molecule does not sensibly change, while the HB network is progressively distorted. On the other hand, the same ion has a distinct breaking efficacy, depending on its counterion.

Above studies have also evidenced differences within the hydration shell of individual ions. Apart from differences in the position of the first ion-water peak, that is a trivial effect of the different ionic radii, the distribution function of the angles formed by the O_w-ion director and the water dipole moment depends on the ion (data not shown). In the case of K^+ and Na^+ in particular, this changes with the ion size, and is broader for K^+. Thus, water molecules in the hydration shell of the K^+ ions, at variance with those hydrating a Na^+ ion, are orientationally more disordered and tend to bring their dipole moments more tangential to the hydration shell.

The results shown here along with those of similar studies performed on the other electrolyte solutions, quoted above, enveil the weakness and contradictions of the classical concepts of "structure maker/breaker". In particular, if we look at the hydration shell of cations, as for instance K^+ and Na^+, the orientational distribution of water molecules is broader around a K^+ ion compared with a Na^+ one. On the basis of this observation, one could infer a "structure breaker" character of K as opposed to the "structure maker" character of Na, although both ions cause a strong distortion of the HB network. Incidentally, the observed differences of their hydration shells are in agreement with faster exchange of water molecules between the first and the second shell of K^+ (Impey et al. 1983, Ramaniah et al. 1999), at difference with the case of Na^+. It is worth to notice also that these differences are relevant to understanding the diffusion of such ions through the cell membranes, explaining why the apparently smaller Na^+ ions cannot enter the K-channels.

Hydration of Water Ions

The study of aqueous solution of hydroxides and acids, by neutron diffraction augmented by computer simulations, gives access to the investigation of the hydration shell of the water ions, namely H^+ and OH^-, which are present in pure water in concentrations below the threshold of instrumental sensitivity. The hydration shell of these ions plays a relevant role in the charge transport dynamics. Indeed, this process is promoted by the continuous breaking and reforming dynamics of HBs, which favors proton transfer along the bonds, in a sort of structural diffusion. Diffusion of water ions in the bulk liquid proceeds through a Grotthuss-like mechanism: this implies that it is the pattern of hydrogen bonds, characterizing each hydration complex, that migrates, in a continuous exchange between covalent bonds and HBs, and not its individual constituents. This picture, according to several *ab initio* simulations (Tuckerman et al. 1995, Marx et al. 1999, Chen et al. 2002, Tuckerman et al. 2002, Zhu and Tuckermann 2002, Marx 2006), requires a particular shape of the water ions hydration shell and an exact number of water molecules nearby the ions. Interestingly, both ions interact with a water molecule that initially is at intermediate distance between the first and second hydration shell and takes a relevant role in the charge transfer. The predictions of the above quoted simulations has been confirmed by the neutron diffraction experiments (Botti et al. 2004a,b, Imberti et al. 2005, Botti et al. 2006).

In detail, the OH$^-$ ion is H-bonded with 4 first neighbor water molecules, at an average OO$_w$ of 2.3 Å, corresponding to a H-bond length of 1.4 Å (see Fig. 5 B), while a fifth water molecule is found at ≈ 2.9 Å. The latter molecule is not H-bonded, but orientationally correlated with the OH$^-$ dipole moment. The proton transfer is a thermaly assisted phenomenon and starts when one of the 4 first neighbours shares its proton with the ion and eventually leaves it to form a neutral water molecule in place of the OH$^-$. At this point, the fifth water molecule can come closer and form an HB with this molecule, while the OH$^-$ ion has indeed moved. The SDF reported in Fig. 5C shows the spatial distribution of first neighbour water molcules around the OH$^-$ ion ($2.0 \leq r \leq 2.75$), according to the EPSR refinement of the diffraction data. The characteristic cup-shaped SDF suggests that the four HBs between water and the hydroxyl oxygen are not coplanar. Figure 5D shows the SDF in the region between 2.75 and 3.2 Å, where the fifth water molecule is found in the vicinity of the hydroxyl hydrogen, with orientation correlated with the OH$^-$ dipole.

As far as the hydration of the H$^+$ ion is concerned, on one side, it is clear that this ion can be readly captured by a water molecule to form a hydronium ion (H3O$^+$), on the other side, an intense debate on whether hydronium preferentially coordinates into Eigen (H9O4$^+$) or Zundel (H5O2$^+$) complexes is found in the literature, between the end of the '90s and the first decade of this century (Tuckermann et al. 1995, Agmon 1998, Marx et al. 1999, Asthagiri et al. 2005, Headrick et al. 2005, Botti et al. 2006, and references therein). *Ab initio* simulations (Marx et al. 1999, Marx 2006) have shown that both complexes are needed to describe the Grotthus process, driving proton transport. Indeed,this proceedes via a structural fluctuation from an Eigen like complex, through a Zundel-like one, to eventually form an Eigen-like complex with another water molecule. Our experiment on this subject (Botti et al. 2004a,b) was interpreted by data fitting with an EPSR simulation box containing hydronium ions solvated in water. This gave evidence for three strong and short HBs between the hydronium hydrogens and first neighbours water oxygens within a shell centered at ≈ 2.48 Å from the hydronium oxygen site. Conversely, we didn't find evidence for HBs between the hydronium oxygen and water hydrogens, but similarly to the case of the OH$^-$ ion, we found on average, one water molecule quite close to the oxygen and orientationally correlated with the hydronium dipole moment. This molecule determines a small structure in the OO$_w$ PDF (data not shown), at intermediate distances between the first and second peaks, similarly to the case of the OH$^-$-water PDF of Fig. 5B. The three-dimensional shape of the hydronium hydration shell is shown in Fig. 5C and D. These results fit the structure of Eigen complexes, while we did not find at that time evidence for the presence of Zundel complexes. The same diffraction data have been reanalized later, by using an EPSR box containing only H$^+$ ions and water, in order to verify that previous results were not biased by the presence of H3O$^+$ ions (Botti et al. 2006). The simulation box with bare protons allows more flexibility to the system, so that both Eigen and Zundel complexes can be formed. As a matter of fact, within the simulation box and within the same cluster of H-bonded water molecules, both ionic species can be identified, although somehow distorted. Conversely, symmetric Zundel ions are not likely to occur in this situation and Eigen complexes are favored. In conclusion, at the present state, according to neutron diffraction data, the distinction between the two complexes, Eigen and Zundel, seems to depend sensitively on how the two ions are defined.

The picture that emerges from these studies agrees with *ab initio* simulations of Marx and coworkers (see for instance Marx et al. 1999), showing that the proton transfer starts as the hydronium ion shares one of its hydrogens with one of the three H-bonded molecules. As soon as the proton is transferred to the second molecule, forming a new hydronium ion, the fourth weakly interacting molecule completes the first neighbouring shell of the water molecule left by the proton.

Hydration of Monosaccharides

The last issue that I want to address in this review is the hydration of monosaccharides, as an example of the role played by water in biological function, available from neutron diffraction experiments.

The case of monosaccharides is particularly interesting and intriguing, as these molecules are all quite similar with respect to their molecular structure, nevertheless their taste may dramatically differ. This observation implies that they interact with the proteins responsible for the elicitation of the sweet taste in quite distinct manner. This interaction takes place through H bonds (Shallenberger and Acree 1967, Kier 1972, Nofre and Tinti 1996, Eggers et al. 2000), thus may be direct or mediated by water, as it always takes place in an aqueous environment. In particular, in order to elicit the sweet taste, a molecule must expose to the taste receptor a proton donor site, a proton acceptor site, and a third hydrophoic site: these may be easily identified, by determining the hydration shell of the saccharide in solution. In the case of monosaccharides, namely fructose, glucose and mannose, neutron diffraction has shown that all these molecules have a hydrophobic side, close to the oxygen in the saccharide ring, while all hydroxyl sites form HBs, as donors or acceptors (Bruni et al. 2018). Figure 6 shows the PDFs relative to H-bonded water hydrogens and hydroxyl oxygens on the sugar. While the number of water-sugar HBs does not dramatically change in these solutions, there is a clear difference among the three monosaccharides as far as the bond length, and thus, the strength of the bonds, is concerned.

Interestingly, the length of the HB is correlated with the sugar sweetness. Indeed, at odds with glucose and fructose, mannose exhibits the weakest bonds and is tasteless, with a bitter aftertaste. Differences in the strength of interaction of these monosaccharides and water have been confmed by Raman spectroscopy (Ruggiero et al. 2018).

In conclusion, sweeter-tasting sugars form tighter and stronger bonds. Thus, it seems that the strength of the HB may be the primary factor in determining a sugar's level of sweetness. Further studies on disaccharides and artificial sweeteners are in progress in order to get better insight into this issue, which is of great interest for food and pharmaceutical industries.

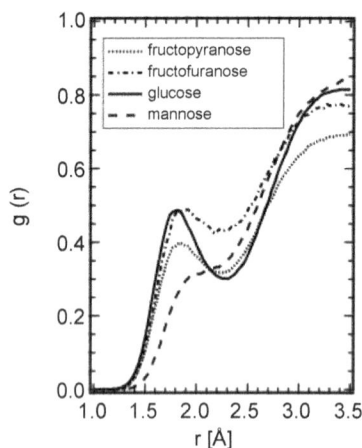

Fig. 6: PDFs of water hydrogens and hydroxyl oxygens of three monosaccharides, showing the dependence of the HB length.

Conclusions

The studies reported in this chapter have hopefully demonstrated the degree of information that can be extracted from neutron diffraction experiments on water and aqueous solutions, with the instruments and computer faciliticies currently available.

We have learned that the HBs in pure water break and reform with a fast dynamic, so that their average number per molecule does not dramatically change in the entire range of existence of the liquid Additionally, these bonds form a continuous percolating network through the sample at all temperatures and pressures below the liquid-gas critical point, and also above this temperature at sufficiently high pressures. We have understood at the atomic level why water is denser than ice: it is because the network of HBs is distorted compared to the perfect tetrahedral symmetry of the crystal, thus accommodating on average more than four molecules around each water molecule. As a matter of fact, the tetrahedral symmetry of the first shell of water molecules is broken only at supercritical temperatures in the gas-like density region, where water adopts a triangular symmetry. We have learned that pressure is more effective than temperature in disturbing the three-dimensional water network, which means that HBs can be more easily distorted than broken. We have seen that the solvation of ions in water is quite effective in disturbing the tetrahedral arrangement of molecules, and this method may be sometimes more convenient than applying pressure.

The most promising and stimulating results are those relative to the hydration shells of solutes. Indeed, their shape, the strength of the HB, and the orientational correlations between water and solute help understanding at the atomistic scale phenomena that are known at a larger scale. In this respect, we have understood why Na^+ cannot enter the membrane K-channels, although it's a smaller size compared with K^+. This is because it binds its neighbouring water molecules much strongly than K^+ does. We have understood how is it possible that protons move in water orders of magnitude faster than other ions: indeed, the hydration shells of both H^+ and

OH⁻ ions are designed in the proper manner to favor Grotthuss like charge transfer. Finally, we have seen that the hydration shell of sugars can tell us something about their taste and degree of sweetness. ... All these findings suggest that water molecules in the hydration shell of a solute play a role in the way this solute interacts with the environment: this is perhaps the atomistic definition of water as the solvent of life.

Acknowledgments

All the work reported above has been performed in collaboration with students and collegues, who are warmly acknowledged: in particular, I want to mention the longstanding collaboration with Prof. F. Bruni and Prof. A. K. Soper.

References

Agmon, N. 1998. Structure of Concentrated HCl Solutions. J. Phys. Chem. A 102: 192–199.

Asthagiri, D., L. R. Pratt and J. D. Kress. 2005. *Ab initio* molecular dynamics and quasichemical study of H⁺(aq). Proc. Natl. Acad. Sci. 102: 6704–6708.

Ball, P. and E. Ben-Jacob. 2014. Water as the fabric of life. Eur. Phys. J.: Spec. Top. 223: 849–852.

Bellissent-Funel, M. C., T. Tassaing, H. Zhao, D. Beysens, B. Guillot and Y. Guissani. 1997. The structure of supercritical heavy water as studied by neutron diffraction. J. Chem. Phys. 107: 2942–2949.

Bernabei, M., A. Botti, F. Bruni, M. A. Ricci and A. K. Soper. 2008a. Percolation and three-dimensional structure of supercritical water. Phys. Rev. E 78: 021505-9.

Bernabei, M. and M. A. Ricci. 2008b. Percolation and clustering in supercritical aqueous fluids. J. Phys.: Condens. Matter 20: 494208–4942216.

Blumberg, R. L. and H. E. Stanley. 1984. Connectivity of hydrogen bonds in liquid water. J. Chem. Phys. 80: 5230–5241.

Botti, A., F. Bruni, M. A. Ricci and A. K. Soper. 1998. Neutron diffraction study of high density supercritical water. J. Chem. Phys. 109: 3180–3184.

Botti, A., F. Bruni, A. Isopo, M. A. Ricci and A. K. Soper. 2002. Experimental determination of the site–site radial distribution functions of supercooled ultrapure bulk water. J. Chem. Phys. 117: 6196–6199.

Botti, A., F. Bruni, S. Imberti, M. A. Ricci and A. K. Soper. 2004a. Ions in water: The microscopic structure of concentrated NaOH solutions. J. Chem. Phys. 120: 10154–10162.

Botti, A., F. Bruni, S. Imberti, M. A. Ricci and A. K. Soper. 2004b. Ions in water: The microscopic structure of a concentrated HCl solution. J. Chem. Phys. 121: 7840–7848.

Botti, A., F. Bruni, M. A. Ricci and A. K. Soper. 2006. Eigen versus Zundel complexes in HCl-water mixtures. J. Chem. Phys. 125: 014508: 9.

Bruni, F., C. Di Mino, S. Imberti, S. E. McLain, N. H. Rhys and M. A. Ricci. 2018. Hydrogen bond length as a key to understanding sweetness. J. Phys. Chem. Lett. 9: 3667–3672.

Chen, B., J. M. Park, I. Ivanov, G. Tabacchi, M. L. Klein and M. Parrinello. 2002. First-principles study of aqueous hydroxide solutions. J. Am. Chem. Soc. 124: 8534–8535.

Conde, O. and J. Teixeira. 1984. Hydrogen bond dynamics in water studied by depolarized Rayleigh scattering. Journal de Physique 44: 525–529.

Corridoni, T., R. Mancinelli, M. A. Ricci and F. Bruni. 2011. Viscosity of aqueous solutions and local microscopic structure. J. Phys. Chem. B. 115: 14008–14013.

Eggers, S. C., T. E. Acree and R. S. Shallenberger. 2000. Sweetness chemoreception theory and sweetness transduction. Food Chem. 68: 45–49.

Fernandez-Alonso, F. and D. L. Price (eds.). 2017. Neutron Scattering—Applications in Biology, Chemistry, and Materials Science. Experimental Methods in the Physical Sciences, vol. 49, Academic Press Inc./Elsevier Science, San Diego, CA.

Franck, E. U. 2000. Supercritical water and other fluids—A historical perspective. *In*: Kiran, E., P. G. Debenedetti and C. J. Peters (eds.). Supercritical Fluids. NATO Science Series (Series E: Applied Sciences), vol. 366. Springer, Dordrecht.

Geiger, A. and H. E. Stanley. 1982. Tests of universality of percolation exponents for a three-dimensional continuum system of interacting waterlike particles. Phys. Rev. Lett. 49: 1895–1898.

Hansen, J.-P. and I. R. McDonald. 2013. Theory of simple liquids. 4th Edition. Academic Press. Elsevier.

Harris, K. R. and P. J. Newitt. 1997. Self-diffusion of water at low temperatures and high pressure. J. Chem. Eng. Data 42: 346–349.

Headrick, J. M., E. G., Diken, R. S. Walters, N. I. Hammer, R. A. Christie, J. Cui et al. 2005. Spectral signatures of hydrated proton vibrations in water clusters. Science 308: 1765–1769.

Hofmeister, F. 1888. Zur Lehre von der Wirkung der Salze. Arch. Exp. Pathol. Pharmakol. 24: 247–260.

Imberti, S., A. Botti, F. Bruni, G. Cappa, M. A. Ricci and A. K. Soper. 2005. Ions in water: The microscopic structure of concentrated hydroxide solutions. J. Chem. Phys. 122: 194509: 1–9.

Impey, R. W., P. A. Madden and I. R. McDonald. 1983. Hydration and mobility of ions in solution. J. Phys. Chem. 87: 5071–5083.

Jones, G. and M. Dole. 1929. The viscosity of aqueous solutions of strong electrolytes with special reference to barium chloride. J. Am. Chem. Soc. 41: 2950–2964.

Jungwirth, P. and P. S. Cremer. 2014. Beyond hofmeister. Nature Chemistry 6: 261–263.

Kier, L. B. 1972. A molecular theory of sweet taste. J. Pharm. Sci. 61: 1394–1397.

Leberman, R. and A. K. Soper. 1995. Nature 378: 364–366.

Lenton, S., N. H. Rhys, J. J. Towey, A. K. Soper and L. Dougan. 2017. Highly compressed water structure observed in a perchlorate aqueous solution. Nature Communications 8: 919–5.

Lovesey, S. W. 1986. The Theory of Neutron Scattering from Condensed Matter: Volume II (The International Series of Monographs on Physics). Oxford Science Publications.

Mancinelli, R., A. Botti, F. Bruni, M. A. Ricci and A. K. Soper. 2007a. Perturbation of water structure due to monovalent ions in solution. Phys. Chem. Chem. Phys. 9: 2959–2967.

Mancinelli, R., A. Botti, F. Bruni, M. A. Ricci and A. K. Soper. 2007b. Hydration of sodium, potassium, and chloride ions in solution and the concept of structure maker/breaker. J. Phys. Chem. B. 111: 13570–13577.

Mancinelli, R., A. Sodo, F. Bruni, M. A. Ricci and A. K. Soper. 2009. Influence of concentration and anion size on hydration of H+ ions and water structure. J. Phys. Chem. B. 113: 4075–4081.

McGreevy, R. L. and L. Pusztai. 1988. Reverse Monte Carlo simulation: a new technique for the determination of disordered structures. Mol. Simul. 1: 359–367.

McLain, S. E., S. Imberti, A. K. Soper, A. Botti, F. Bruni and M. A. Ricci. 2006. Structure of a 2 Molar NaOH solution from neutron diffraction and empirical potential structure refinement. Phys. Rev. B. 74: 094201–9.

Marx, D., E. Tuckerman, J. Hutter and M. Parrinello. 1999. The nature of the hydrated excess proton in water. Nature 397: 601–604.

Marx, D. 2006. Proton transfer 200 Years after von Grotthuss: Insights from *Ab Initio* simulations. Chem. Phys. Chem. 7: 1848–1870.

Mishima O., L. D. Calvert and E. Whalley. 1985. An apparently first-order transition between two amorphous phases of ice induced by pressure. Nature 314: 76–78.

Nofre, C. and J.-M. Tinti. 1996. Sweetness reception in man: The multipoint attachment theory. Food Chem. 56: 263–274.

Orosei, R., S. E. Lauro, E. Pettinelli, A. Cicchetti, M. Coradini, B. Cosciotti et al. 2018. Radar evidence of subglacial liquid water on Mars. Science 361: 490–493.

Pártay, L. and P. Jedlovszky. 2005. Line of percolation in supercritical water. J. Chem. Phys. 123: 024502.

Pártay, L., P. Jedlovszky, I. Brovchenko and A. Oleinikova. 2007. Formation of mesoscopic water networks in aqueous systems. 2007. Phys. Chem. Chem. Phys. 9: 1341–1346.

Poole, P. H., F. Sciortino, U. Essmann and H.E. Stanley. 1992. Phase behaviour of metastable water. Nature 360: 324–328.

Ramaniah, L. M., M. Bernasconi and M. Parrinello. 1999. *Ab initio* molecular-dynamics simulation of K+solvation in water. J. Chem. Phys. 111: 1587–1591.

Ruggiero, L., A. Sodo, F. Bruni and M. A. Ricci. 2018. Hydration of monosaccharides studied by raman scattering. J. Raman Spectrosc. 49: 1066–1075.

Sears, V. F. 1992. Neutron scattering lengths and cross sections. Neutron News 3: 26–37.

Shallenberger, R. S. and T. E. Acree. 1967. Molecular theory of sweet taste. Nature 216: 480–482.

Soper, A. K. and R. Silver. 1982. Hydrogen-hydrogen pair correlation function in liquid water. Phys. Rev. Lett. 49: 471–474.

Soper, A. K. 1996. Empirical potential Monte Carlo simulation of fluid structure. Chem. Phys. 202: 295–306.

Soper, A. K., F. Bruni and M. A. Ricci. 1997. Site–site pair correlation functions of water from 25 to 400°C: Revised analysis of new and old diffraction data. J. Chem. Phys. 106: 247–254.

Soper, A. K. and M. A. Ricci. 2000. Structures of high-density and low-density water. Phys. Rev. Lett. 84: 2881–2884.

Soper, A. K. 2013. The radial distribution functions of water as derived from radiation total scattering experiments: Is there anything we can say for sure? ISRN Phys. Chem. 2013: 1–67.

Stanley, H. E. 1979. A polychromatic correlated-site percolation problem with possible relevance to the unusual behaviour of supercooled H_2O and D_2O. J. Phys. A 12: L329–L337.

Stanley, H. E. and J. Teixeira. 1980. Interpretation of the unusual behavior of H_2O and D_2O at low temperatures: Tests of a percolation model. J. Chem. Phys. 73: 3404–3422.

Stauffer, D. 1985. Introduction to Percolation Theory (Taylor & Francis, London).

Svishchev, I. M. and P. G. Kusalik. 1993. Structure in liquid water: A study of spatial distribution functions. J. Chem. Phys. 99: 3049–3058.

Tassaing, T., M. C. Bellissent-Funel, B. Guillot and Y. Guissani. 1998. The partial pair correlation functions of dense supercritical water. Europhys. Lett. 42: 265–270.

Tsiaras, A., I. P. Waldmann, G. Tinetti, J. Tennison and S. N. Yurchenko. 2019.Water vapour in the atmosphere of the habitable-zone eight-Earth-mass planet K2-18 b. Nature Astronomy (2019) DOI: 10.1038/s41550-019-0878-9.

Tuckermann, M. E., K. Laasonen, M. Sprik and M. Parrinello. 1995. *Ab initio* molecular dynamics simulation of the solvation and transport of hydronium and hydroxyl ions in water. J. Chem. Phys. 103: 150–161.

Tuckermann, M. E., D. Marx and M. Parrinello. 2002. The nature and transport mechanism of hydrated hydroxide ions in aqueous solutions. Nature 417: 925–929.

Zhu, Z. and M. E. Tuckermann. 2002. *Ab Initio* molecular dynamics investigation of the concentration dependence of charged defect transport in basic solutions via calculation of the infrared spectrum. J. Phys. Chem. 106: 8009–80018.

Winkel, K., M. Seidl, T. Loerting, L. E. Bove, V. Molinero et al. 2011. Structural study of low concentration LiCl aqueous solutions in the liquid, supercooled, and hyperquenched glassy states. J. Chem. Phys. 134: 024515–8.

Index

For Product Safety Concerns and Information please contact our EU
representative GPSR@taylorandfrancis.com
Taylor & Francis Verlag GmbH, Kaufingerstraße 24, 80331 München, Germany

www.ingramcontent.com/pod-product-compliance
Lightning Source LLC
Chambersburg PA
CBHW070714220326
41598CB00024BA/3150